Springer

Milan
Berlin
Heidelberg
New York
Barcelona
Hong Kong
London
Paris
Singapore
Tokyo

Giancarlo Spezie
Giuseppe M.R. Manzella
(Eds)

Oceanography of the Ross Sea *Antarctica*

Springer

GIANCARLO SPEZIE
IUN
Institute of Meteorology and Oceanography
Via Amm. Acton, 38
80133 Napoli
Italy

GIUSEPPE M.R. MANZELLA
ENEA
Marine Environment Research Centre
Loc. Pozzuolo, Forte S. Teresa
19036 S. Terenzo (La Spezia)
Italy

Softcover reprint of the hardcover 1st edition 1999

ISBN 978-88-470-2252-2 ISBN 978-88-470-2250-8 (eBook)
DOI 10.1007/978-88-470-2250-8

Library of Congress Cataloging-in-Publication Data: applied for

Typesetting: Photo Life (Milan)

Cover design: Simona Colombo

SPIN: 10694908

To Viviana and Lodovico

Preface

The era of the exploration of the World Ocean is not yet over: some areas still lack an adequate number of observations. The relationships between the physical, chemical and biological processes, which sustain the life on this planet, are not yet fully understood. In short, knowledge of the oceans is still far from satisfactory. This book covers an important period in the study of one of the last investigated areas of the World Ocean: the Ross Sea, Antarctica. During the 1990s, long-term experiments were conducted in this area as part of the Climatic Long-term Interaction for the Mass balance in Antarctica (CLIMA) project of the Italian National Programme for Antarctic Research (PNRA), providing a wealth of oceanographic information. This book is an outcome of the CLIMA field observations and international collaborations with the most important programmes in the Ross Sea–Southern Ocean area. It contributes to the studies of the spatial and temporal variability of Ross Sea water masses and circulation and their relationship with the Southern Ocean circulation. A comprehensive review of historical data is offered, and new data sets are analysed.

The studies presented in this book show that much progress has been achieved during the last decade, but large gaps in our understanding of the physical processes in the Ross Sea are still to be filled. However, these studies contribute significantly to the investigation of some specific aspects regarding the circulation of the main water masses.

One of the main aspects of the dynamics of the Ross Sea is the considerable variability over small distance. This poses serious constraints, not easily satisfied, on sampling criteria for hydrological and current measurements. Furthermore, the available CTD (Conductivity Temperature Depth) data are mainly collected during the austral summer, owing to practical reasons. This, along with the uneven sampling distribution, obviously makes the use of in situ data for the initialisation and validation of circulation models aimed at simulating and understanding dynamic processes in the area a particularly delicate issue.

The existing data sets also display a marked seasonal and interannual variability, which is typical of limited areas where distinctive water formation and transformation are occurring. The studies reported in this book show that the Ross Sea has a complex circulation at all depths driven by a combination of factors operating on different, but critically related, time scales. Thus, climatological averages may not be a significant means of characterising the circulation when the investigation is focused on particular aspects of it. Also considered in this book are the relationships between the circulation around Antarctica and in the

Ross Sea, which develop by dynamic coupling between the Antarctic Circumpolar Current and the Ross Sea Gyre as well as by fluid exchange between offshore larger-scale circulation and local ones. Research on the horizontal and vertical particle flux has increased our knowledge of the intensity of the horizontal and vertical transport of biogenic elements from suspended particulate matter. The data will surely prove very valuable also in the design and implementation of ecological models.

The International Conference on the Oceanography of the Ross Sea held in Lerici, Italy, in March 1997 has constituted the first occasion on which data, model results, and ideas sprung from the activities carried out in the framework of the CLIMA project were presented and discussed in an organic manner. The state of the knowledge achieved within the project is demonstrated by the contributions in this volume, along with the extent of the collaborations which flourished between the project itself, or individual activities conducted in its framework, and major international scientific initiatives. Just to mention a few examples, repeated expendable bathythermograph (XBT) deployments along the World Ocean Circulation Experiment (WOCE) P14 transect between New Zealand and Antarctica have been carried out in the framework of the WOCE; the CLIMA project is also an official contributor to WOCE activities, with the surface drifter data gathered in the Pacific sector of the Antarctic Circumpolar Current, and these data are available to the scientific community through the WOCE–TOGA Global Drifter Data Center located in and hosted by the National Oceanic and Atmospheric Administration (NOAA) in Miami; they also constitute part of the Italian contribution to the WCRP International Programme for Antarctic Buoys (IPAB). Particular attention, in this respect, has been devoted to making data collected within CLIMA activities available to the community at large, again through bilateral collaborations or international data centers.

As mentioned above, this has been the first opportunity to summarise the achieved results, and the stage of the synthesis has not been reached yet, i.e., the point when scientists from the different disciplines can consider the entire set of data and plan its use for the implementation of predictive models of polynya formation, interannual variability of the circulation, and links between physics, chemistry and biology within the Ross Sea ecosystem.

It is also necessary to underline a very crucial point which clearly results from the data analysis: the Ross Sea appears to be a quite complex system, whose continuous and exhaustive monitoring is not feasible, obviously. In the best case, we can expect to be able to carry out measurement campaigns for 2 or maybe 3 months a year. We therefore have to optimise men and equipment resources in order to take advantage of the short "fair weather" (so to speak) time we have for our observations. At the same time, we also need to make the best out of the long months we cannot be in the field. Remotely sensed data collection will fill the gap between campaigns, and therefore we will have to give impulse to the connected analysis and research. Also, we need to keep developing strong numerical modelling activities, at different levels of complexity, with the final goal of covering as much of the complexity of the Ross Sea system as possible.

It has clearly been shown that the Ross Sea can be a good reference site to monitor possible climatic trends, looking at the variability of water mass formation as an indicator of global warming. To this end there is the need to pool together the different data sets available and at the same time it is crucial to acquire time series of the Ross Sea characteristics long enough to be able to estimate its contribution to the Southern Ocean system as a whole.

The International Conference on the Oceanography of the Ross Sea developed a number of topics laid out in seven work sessions. We started with a general description and discussion of the physical oceanography of the Southern Ocean and of its most important dynamic feature, the Antarctic Circumpolar Current. Then, we zoomed in and looked at the Ross Sea, the focus of the interest of most of us, in terms of its hydrological as well as its biogeochemical characteristics, and of its circulation. We finished with a session on remote sensing, which is a way to reconnect local analysis to the big picture of the circulation, and, more generally of the overall climatic conditions of the Southern Ocean, of its Pacific sector, and, in particular, once again, of our area of interest: the Ross Sea.

The work discussed over 5 days has involved, and is still involving, more than 120 researchers, and has provided a quite exhaustive picture of the oceanography of the Ross Sea. At the same time, presentations, discussions and round tables have made us aware of a number of still open questions, whose answers will provide a more complete understanding of how the Ross Sea works as a dynamic and ecological system. Also, in this sense, we believe we can view the conference as a success, since it has provided us with some issues which will be our research challenges for the future.

December 1998

G. Spezie
G.M.R. Manzella

Contents

General Aspects

Thermohaline Data and Ocean Circulation on the Ross Sea Continental Shelf

S.S. Jacobs and C.F. Giulivi

Lamont-Doherty Earth Observatory of Columbia University, Palisades, New York 10964, USA

Abstract

The ocean station data base on the Ross Sea continental shelf includes measurements from more than fifty cruises, and a few observations through ice holes. Using representative salinity and temperature transects from the summers of 1962-63, 1976-77 and 1983-84, we briefly review the ocean stratification and circulation, identifying five primary water masses. Year-round salinity profiles at a single location in 1960-61 and early winter temperature profiles along the dateline in 1995 are discussed in relation to anticipated seasonal changes. A strong zonal salinity gradient and substantial differences between two representations of the shelf data base reveal the need for caution in modelling and bottom water investigations. Long term observations of shelf water salinity and studies of the inflow of modified deep water would help in the evaluation of recent evidence for marked interannual and decadal variability on the continental shelf.

1 Introduction

Analyses of historical ocean temperature and salinity measurements on the Antarctic continental shelf require a careful consideration of original sources, data processing, and variability in space and time. The effects of stronger winter forcing must be inferred from predominantly late summer observations. Considerable variability over small distances can influence the choice of areas over which data can be meaningfully averaged. Lengthening time series display substantial variability on interannual and longer temporal scales. Here we revisit these caveats in the context of the increasing utilization of gridded, archived data for modelling and climate change work. Available ocean station data on the Ross Sea continental shelf are summarized, and some differences that can arise from alternate treatments of a similar database are illustrated. We also evaluate a venerable set of year-long salinity profiles, and a transect of recent May temperature measurements.

The region north of the Ross Ice Shelf is better sampled than most of the remaining circumpolar continental shelf, thanks to a minimal sea ice cover during the austral summer and a location adjacent to McMurdo Station and other Antarctic bases. Aspects of its ocean circulation have been described in numerous papers (e.g., [1-14]) and the references therein. Here we present a comparable set of representative zonal and meridional transects, along with a brief description of the primary water masses and their probable sources and sinks. Temporal variability in the ocean thermohaline data over recent decades in this sector is discussed in more detail by Jacobs and Giulivi [15].

2 The Data Base and Seasonal Variability

Temperature and salinity data south of the 600-m isobath near the continental shelf break were winnowed from National Oceanographic Data Center (NODC) archives and supplemented by additional observations (Fig. 1). Since the Ross Ice Shelf ice front advances for decades, followed by rapid retreat [16], some stations appear south of the apparent coastline. The seasons (years) and months that ocean station data were obtained are listed in Table 1, along with ship name or other source. The vast majority of the ~1200 profiles utilized were occupied from December through February, and have an areal distribution that is biased toward the southern and western sectors. We did not use data from far beneath the

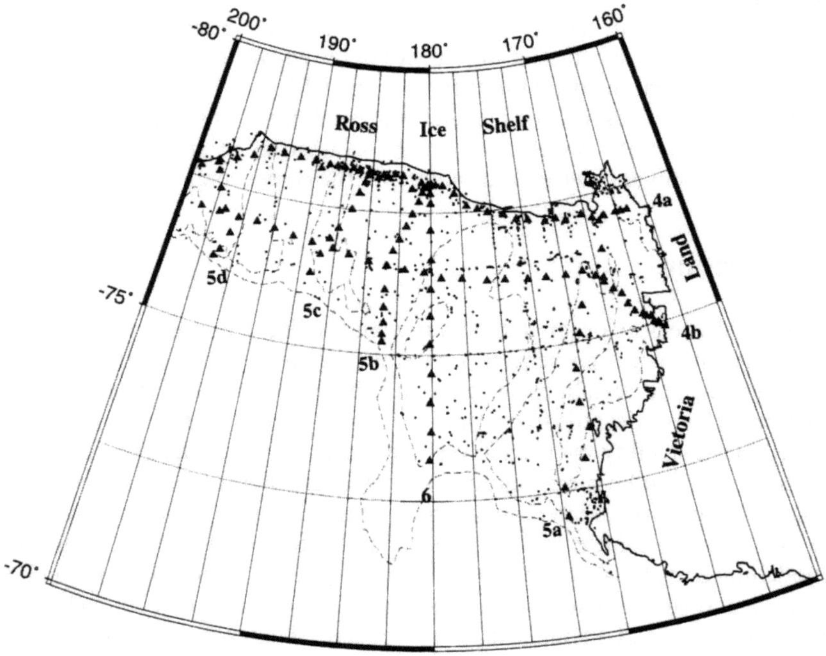

Fig. 1. Distribution of ocean stations (*small dots*) summarised in Table 1, superimposed on 500- and 1000-m bathymetric contours. More detailed bathymetric data may be found in Davey [43]. Coordinates are negative south and positive east, with 190 and 200 equivalent to 170°W and 160°W. McMurdo Station, on the west side of McMurdo Sound and Southeast end of Ross Island, is located near 77°51'S, 166°40'E. *Larger symbols* at northern and western ends of the transects show the stations in Figs. 4-6

Season	Months	Ship/Source	Station	Season	Months	Ship/Source	Station
1928-1929	12,1	Larson	8	1967-1968	1,2	Eltanin	32
1935-1936	1	Discovery	14		2	Burton Island	28
1954-1955	1	Atka	7	1968-1969	2	Burton Island	21
1955-1956	1,2	Edisto	7		2	Pelamida	12
	1,3	Glacier	3	1969-1970	2	Edisto	23
	3	*(Unknown)*	1	1970-1971	1	[47]	6
1956-1957	11,12	Glacier	7	1971-1972	2,3	Eltanin	11
	3	Northwind	2	1974-1975	9,10	Ice Island	8
1957-1958	12,2	Glacier	13	1976-1977	12	Northwind	42
	1	Burton Island	3	1976-1977		[45]	
	12,1	Atka	23	1977-1978	1,2	Burton Island	68
1958-1959	11,12	Glacier	18	1978-1979	2	Glacier	12
	12,1	Northwind	9	1981-1982	1	Glacier	36
	12,1,3	Staten Island	4	1982-1983	2,3	Glacier	17
1959-1960	12	Glacier	6	1982-1983		[7]	
	5	[20]	3	1983-1984	1,2	Polar Sea	107
1960-1961	6-2	[20]	24	1984-1985	1,2	Polar Star	35
	12	Staten Island	8	1985-1986	2	Vieze	2
1960-1961		[48]		1987-1988	2	Fedorov	1
1961-1962	1,2	Burton Island	28	1989-1990	12,1	Cariboo	27
	2	Eastwind	23		1,2	Polar Duke	26
	1,2	Glacier	9	1990-1991	12,1	Cariboo	17
1961-1962		[48]		1992		Polar Duke	
1962-1963	2,3	Edisto	101	1993-1994	2	Polar Sea	42
1963-1964	2,3	Atka	36		2	Palmer	2
1964-1965	2	Staten Island	24	1994-1995	11,12	Palmer	151
1965-1966	2	Atka	24	1995-1996	5	Palmer 9503	11
1966-1967	1	Eltanin	22			Palmer 9508	
	1,2	Glacier	13	1996-1997		Palmer (JGOFS)	
	2	Staten Island	26			Palmer (ROAVERRS)	

Table 1. A chronological summary of ocean stations with temperature and salinity data on the Ross Sea continental shelf, from 1928-1929 to 1996-1997. In this compilation each season (year) begins on 1 June (month 1). Identified data sets for which the stations have yet to be merged with this file appear as *blanks* under *Stations*. Most ship data are available from the National Oceanographic Data Center, and some also in data reports (e.g., [17, 44, 46])

ice shelf [17], at instrument moorings [18, e.g.] or from very recent cruises (http://wwwl.whoi.edu/jgofs.html).

Data quality and seasonal cycles are important issues in studies of coastal regions. We have long been aware of apparently large (> 0.15) salinity changes from one summer to the next in the deep shelf water of the southwest Ross Sea [19]. While these changes exceeded the measurement accuracy by an order of magnitude, they seemed confounded by a year-round series of station measurements taken through the sea ice in McMurdo Sound (Fig. 2). Extending to depths exceeding 500 m, the temperatures showed relatively little variability, but the salinities displayed a greater range than the interannual summer changes noted above. As in the summer profiles [19], temporal shifts were simultaneous over

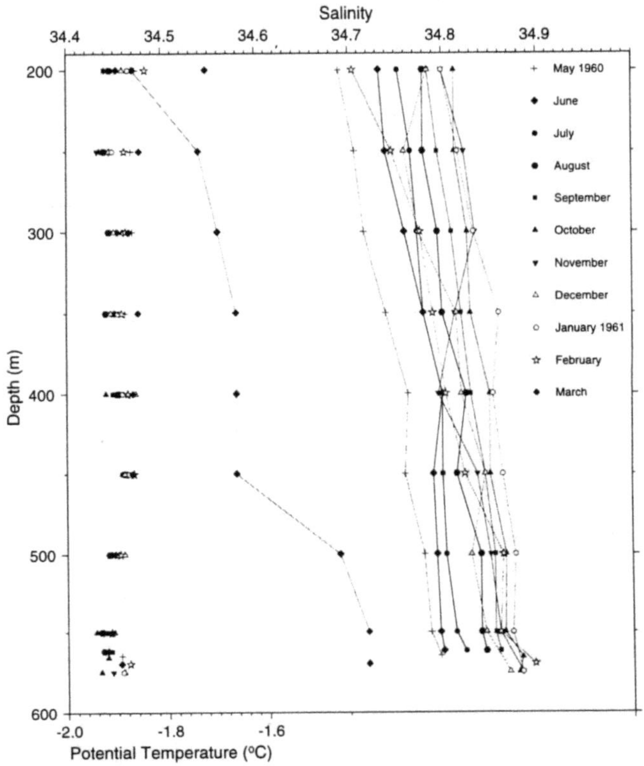

Fig. 2. Temperature (*left, symbols* only) and salinity (right, *connected symbols*) measurements through an ice hole at ~77°53'S, 166°53'E, in water ~575 m deep in McMurdo Sound, from May 1960 to March 1961 [20]. For most months, the standard-depth observations have been averaged from two or three separate casts each a week or two apart

most of the water column, with salinities increasing from May through October, and decreasing in December and February. An unexpected January reversal resulted in the highest annual values from 350 to 550 m, and between 21 February and 7 March most of the water column plunged by more than 0.2 in salinity.

A closer look at the Tressler and Ommundsen report [20] indicates that the authors made their last measurements on 21 February and then left McMurdo, presumably with the salinity samples taken up to that time. A follow-on program by another group was unceremoniously terminated after one cast (March in Fig. 2) by a storm that swept away the ice floe, hole and hut. The anomalously low salinity of the March profile may thus have resulted from the program change, perhaps related to different sampling and analysis protocols. This shows the value of thorough data reports (metadata) detailing the methods and problems, which are unlikely to appear in large ocean data archives. Nevertheless, this year-round series of ocean station measurements in McMurdo Sound remains a valuable and unique data set, suggesting a strong seasonal variability in response to surface forcing and circulation within the shelf water.

3 Data Products and Areal Variability

Large portions of the Antarctic continental shelf serve as production sites and temporary storage reservoirs for dense shelf waters. Formed mainly by winter cooling and freezing at the sea surface, shelf waters supply much of the salt that drives thermohaline circulation beneath the large ice shelves, and the properties by which new bottom water is identified [6, 21]. At the cold temperatures that characterize the Ross Sea continental shelf, salinity variations exert a greater control over the density field. Ocean general circulation models often "restore" salinity near Antarctica to observed conditions, e.g., as represented by the NODC

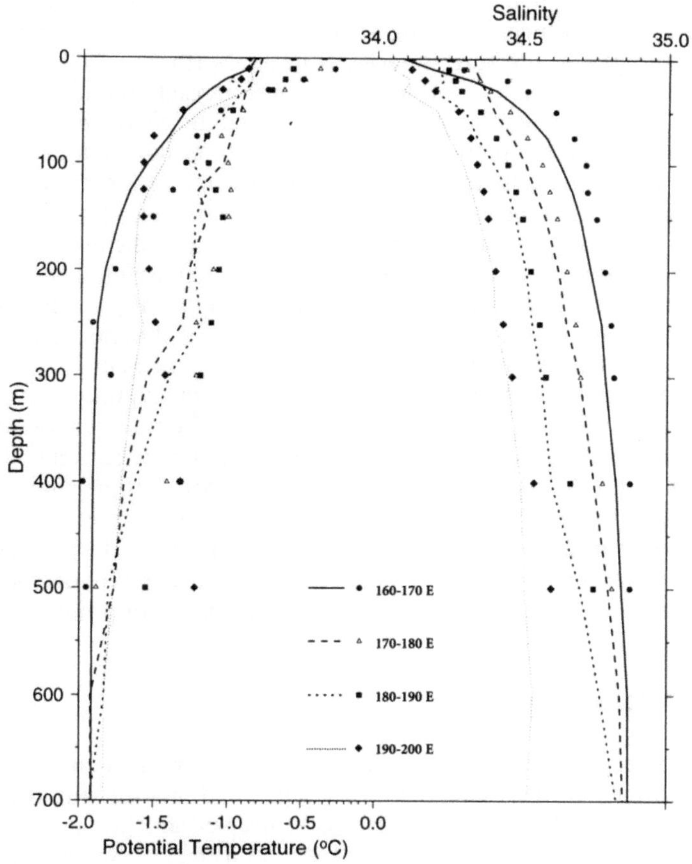

Fig. 3. Depth-averaged temperature (*left*) and salinity (*right*) from summer (DJF) stations in Table 1 (*continuous lines*) and from data objectively interpolated into 1° grids at standard levels in the NODC *World Ocean Atlas* [22] (*discrete symbols*). Both data sets are divided into 10° meridional bands and restricted to summer stations/grids south of the continental shelf break. This figure has been truncated at 700 m, below which measurements in the two western bands extend ~monotonically to 900 and 1200 m

World Ocean Atlas [22]. Model results can vary markedly and have global impacts [23, 24] depending upon whether salinities are restored to values as low as 34.0, characteristic of zonal average summer surface waters, or as high as 35.0, which exceed anything reported on the shelf for the past three decades.

Substantial differences are also evident between our summer observations and NODC Atlas objectively averaged 1-degree grid data over the same months in each meridional band (Fig. 3). The Atlas data are consistently saltier, perhaps because more recent, lower-salinity measurements [15] were not in the NODC archive. Atlas temperatures also tend to be higher and more variable and, like salinities, do not extend below 500 m. These anomalies might result from mislocated stations, from the incorporation of bathythermograph data into the Atlas, and from irregular shelf depths over which isolated deep values are removed [25]. Atlas data on the shelf may also be biased by different water types from north of the shelf break, near which the strong Antarctic Slope Front [26] could also influence data averaged on pressure surfaces, as in the North Atlantic [27]. Gridded and averaged data are valuable for many purposes, but caution must be exercised in their use for modelling or other work on the Antarctic continental shelf.

Annual averages are compromised by the scarcity of non-summer measurements, and spatial averages can show a high variance due to the strong east-west gradient resulting from the variable salt fluxes and wind fields [2, 28]. The zonal gradients in Fig. 3 are largest below the upper 100 m, with the highest salinities and coldest temperatures in the western sector. The salinity gradient may have implications for classical bottom water formation arguments, which frequently hinge on much lower changes in density (salinity). Indeed, the entire region below 350 m is saltier than the Pacific continental shelf average of 34.46 [29]. If bottom water formation simply required a shelf component with a salinity above ~34.51 [30], then most water on the Ross Sea continental shelf would qualify. However, the Ross Sea is widely believed to produce much less bottom water than the Weddell Sea (e.g., [29]), in spite of similar environments. This may in part be an artifact of water mass definitions, or be a consequence of the greater warmth of the Ross Gyre, elements of which intrude onto the continental shelf. An apparent drift toward lower shelf water salinities in the Ross Sea over recent decades might also have altered the regional bottom water production rate or properties [15].

4 Water masses

Circumpolar Deep Water evolves into several new water masses on and near the Antarctic continental shelves. As these waters mix and interact with the atmosphere, sea ice and shelf ice, some ventilating the deep ocean, their properties and variability are of more than taxonomic interest in this region. Names and definitions have evolved over time, but common use has tended to follow Carmack [29]. The north-south and east-west temperature and salinity sections across the shelf in Figs. 4 and 5 are plotted at the same vertical scale, with the following water mass abbreviations and characteristics:

HSSW (High Salinity Shelf Water), identified by salinities of > 34.6 and temperatures at the sea surface freezing point, dominates the subsurface western sector in summer. Previously referred to as Ross Sea Shelf Water (and Western Shelf Water in the Weddell Sea), HSSW appears on all sections but the one in Fig. 5d.

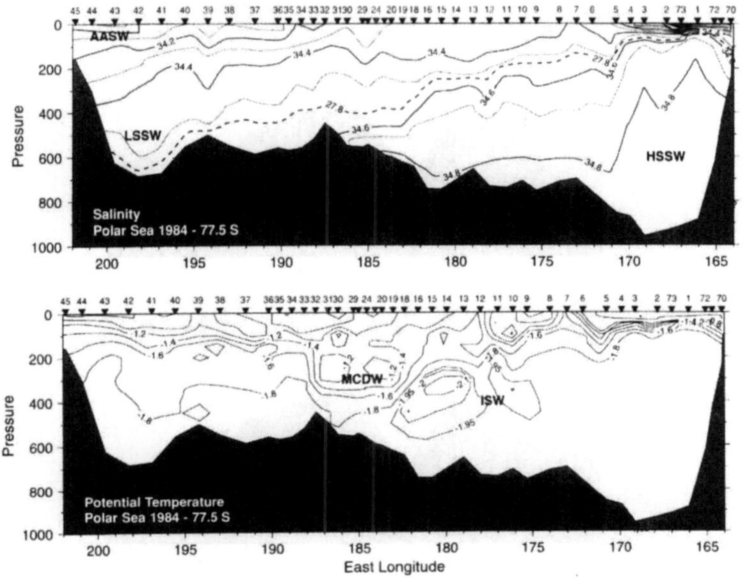

Fig. 4a. Salinity and potential temperature along the Ross Ice Shelf front and westward across McMurdo Sound (*right*) from ocean stations (*numbers* at *top*; locations as in Fig. 1) in early February 1984. From continuous vertical measurements at each CTD station, projected onto 77.5°S, a 50-dbar by 1° longitude grid was derived and machine-contoured, removing some of the spatial variability. Pressure (in decibars) is roughly equal to depth in meters; potential temperature (°C) is within 0.03 of in-situ temperature at these depths; salinity is in practical salinity units. The 27.8 isopycnal illustrates the similarity between definition of density and salinity fields (see text for water mass abbreviations)

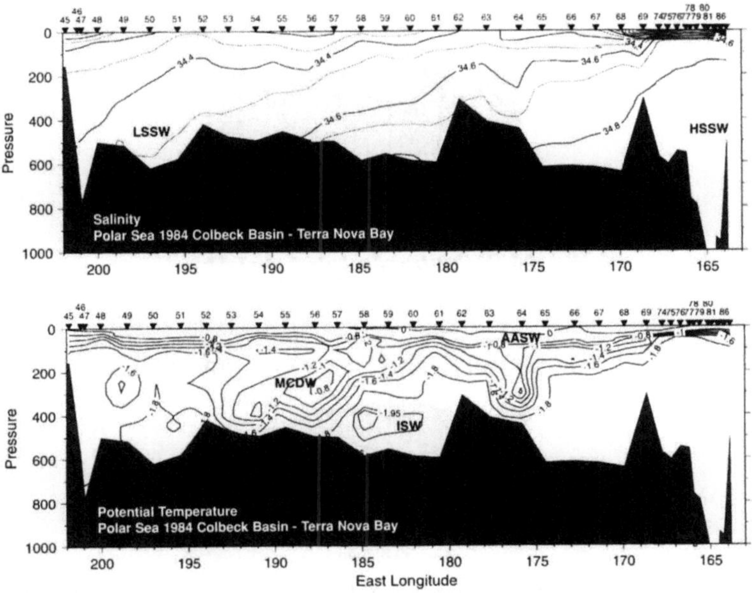

Fig. 4b. East-west section as in Fig. 4a, but further north (Fig. 1) extending northwest at the right end into Terra Nova Bay (74°50'S, 164°30'E)

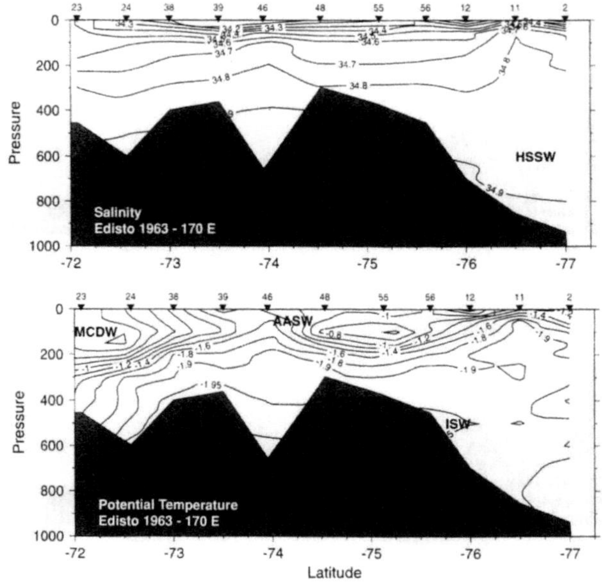

Fig. 5a. As in Fig. 4a, but constructed from bottle measurements made in February 1963, and projected along a constant longitude. This western Ross Sea section extends from just south of the continental shelf break (*left*) to the vicinity of Ross Island (*right*; Fig. 1).The horizontal contouring grids were 0.25° and 0.5° in latitude for temperature and salinity

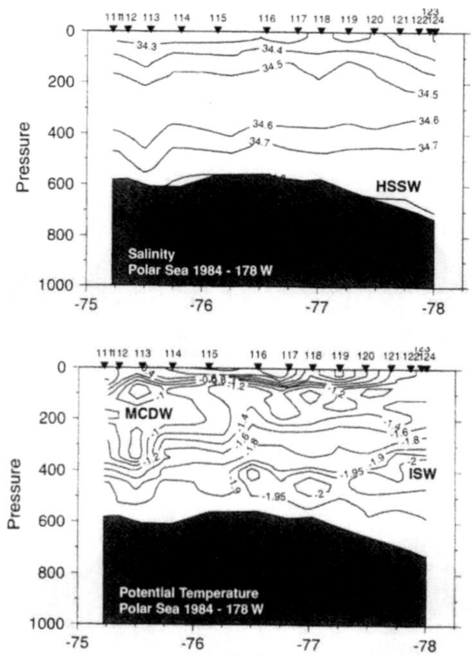

Fig. 5b. As in Fig. 4a, from CTD data taken in February 1984 between the continental shelf break (*left*) and the Ross Ice Shelf (*right*; Fig. 1). Projected on constant longitude with 25-dbar and 0.25° latitude gridding

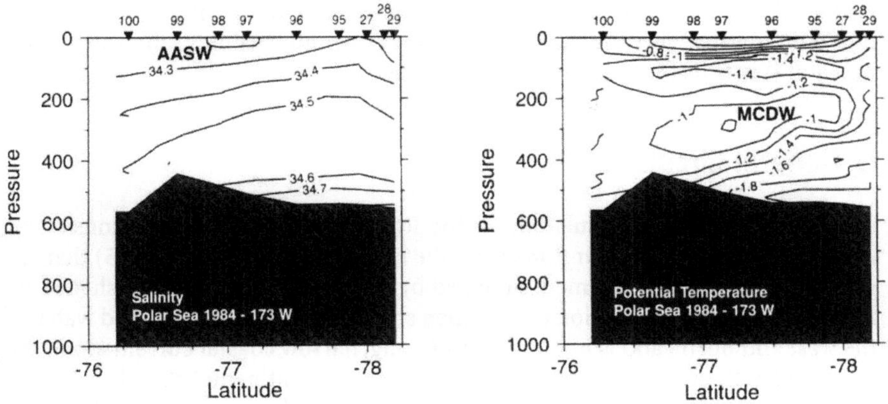

Fig. 5c. As in Fig. 5b, from CTD data taken in February 1984 between the continental shelf break and the Ross Ice Shelf (Fig. 1)

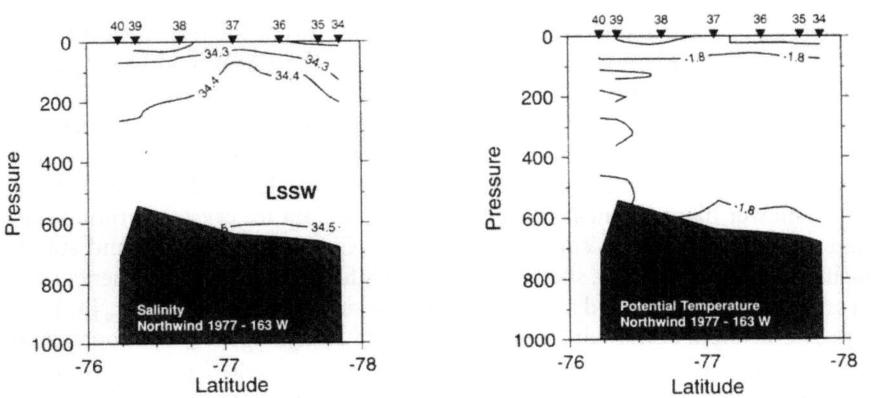

Fig. 5d. As in Fig. 5b, from CTD data taken in late December 1976 between the continental shelf break and the Ross Ice Shelf (Fig. 1)

LSSW (Low Salinity Shelf Water), at a slightly warmer sea surface freezing temperature corresponding to its lower (~34.4–34.6) salinities, is best developed on Figs. 5c and 5d. Similar to Eastern Shelf Water in the Weddell Sea, LSSW is hard to differentiate in T/S space from the "winter water" temperature minimum at the base of the surface water.

ISW (Ice Shelf Water), with subsurface temperatures below the sea surface freezing point, is rarely supercooled in situ. ISW is concentrated in the west-central sector (Fig. 5b), where it emerges from beneath the Ross Ice Shelf.

MCDW (Modified Circumpolar Deep Water), derived from Circumpolar Deep Water near the continental shelf break and typified by temperatures between +1.0 and −1.5 °C, intrudes at least to the ice shelf front in Fig. 5c. MCDW has been referred to as a Warm Core in the Ross Sea (and as Modified Warm or Weddell Deep Water in the Weddell Sea).

AASW (Antarctic Surface Water), typically warmer (to ~+2 °C) and fresher (< 34.5) in summer than the deeper shelf waters from which some of it evolves, is comprised of a mixed layer of variable thickness, with or without an underlying "winter water" layer. As AASW cools, thickens, and becomes saltier in winter, much of it is transformed back into shelf waters.

5 General Circulation

The large-scale ocean circulation on the Ross Sea continental shelf consists of AASW, LSSW and MCDW inflows from the north and east (Figs. 4 and 5) that are seasonally cooled and warmed, salinized by sea ice formation and freshened by meltwater. Surface circulation on the open shelf generally moves ice and water to the west and north, and is bounded by a strong, narrow coastal current along the Ross Ice Shelf and another westward flow near the continental shelf break [26]. Intense winter sea ice formation and its removal from the coastlines produces the nearly isothermal LSSW and HSSW. Coastal upwelling in these same regions is consistent with the large-scale density field [2] and with predominant offshore winds.

Shelf waters contribute directly to bottom water formation near the continental shelf break [1, 6]. In addition, some portion of the HSSW and LSSW drains into the deeper cavity beneath the Ross Ice Shelf, where glacial ice is melted, probably most strongly near the deep grounding lines. This occurs because seawater freezing temperature decreases with increasing pressure, and results in a deep ISW plume that emerges in the central Ross Sea (Fig. 4). Both this outflow and shallower filaments near the ice front may contain ice crystals, produced as the rising water reacts to supercooling [7, 31]. Circulation, melting and subsequent freezing beneath the shelf ice have been inferred from measurements near, on and beneath the ice, and from models of the sub-ice circulation (e.g., [3, 6, 32-38]). The strength of the sub-ice circulation and its exchanges with the open Ross Sea are likely to vary considerably over interannual and longer periods in response to changes in HSSW salinity and volume [15].

The Ross Sea continental shelf is covered by sea ice for most of the year, with decay and growth both starting near the ice shelf front, typically in late October and late February. Northward transport of sea ice is strongest in the western sector, where barrier and katabatic winds maintain the large Ross Sea Polynya adjacent to the ice shelf front [39] and a small, persistent polynya near 75°S on the Victoria Land coast [40]. The volume and salinity of HSSW will depend in part on the volume of sea ice produced on the continental shelf and exported from it each year. Detailed records of ice transport are not yet available, however, and sea ice thickness data are scarce [41]. The length of the sea ice season on the shelf varied by about a month from 1979 to 1987, with little apparent interannual change in winter ice concentration [9]. A longer Comiso sea ice record from the entire Ross Sea displays strong interannual changes in the sea ice extent [15] at a period comparable to that of the "Antarctic circumpolar wave" [42].

MCDW intrudes year-round at intermediate depths onto the continental shelf [10, 18], and overrides the HSSW at some locations in the western sector. Since the HSSW shoals rapidly westward, the "warm" MCDW that penetrates this sector will lie directly beneath the AASW, from which it cannot easily be discriminated from summer temperature and salinity measurements. MCDW may extend for some distance beneath the ice shelf, as current measurements accompanying the stations in Fig. 4a revealed inflow and outflow, respectively, in the eastern and western "cores" defined by the −1.2 °C contours. This recirculation appears to be persistent, and may account for a slight thinning of the ice shelf in this region [10, 16].

Temperatures from a May 1995 section across the shelf (Fig. 6) were obtained about 2 months after sea ice had covered this region. By this time any traces of summer AASW and shallow ISW had been eliminated, and the mixed layer was deepest near the Ross Ice Shelf where surface forcing is strongest. The main ISW outflow remained around 450 m, and the MCDW inflow with temperatures warmer than −1 °C appeared on several stations. In some locations this MCDW reached above 100 m, probably accounting for mixed layer temperatures warmer than the surface freezing point. The corresponding salinities (not shown) were

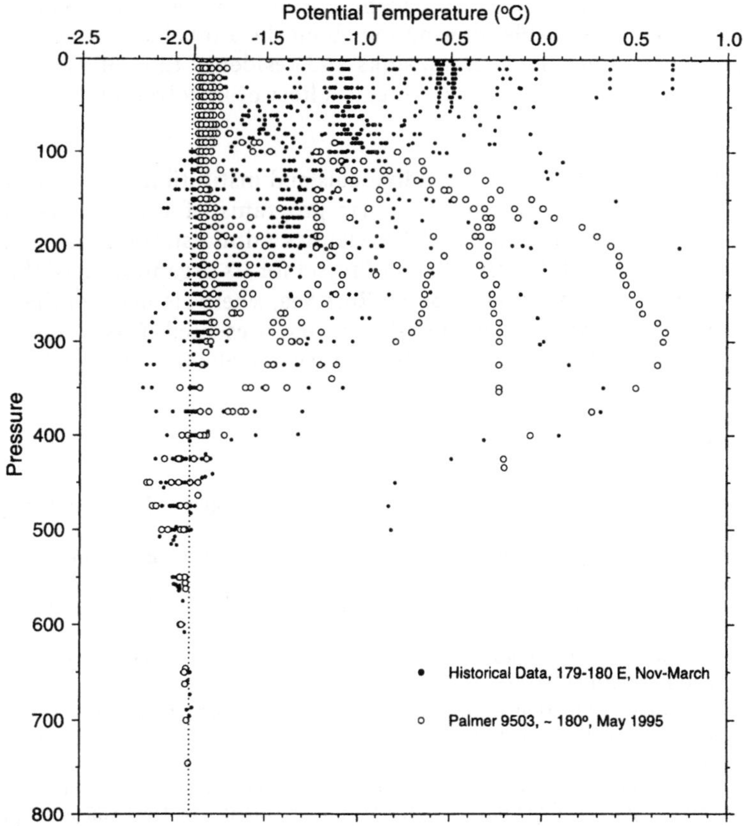

Fig. 6. Potential temperature vs. pressure (dbar) near the dateline in Fig. 1. May 1995 stations from the "NB Palmer" appear against a background of historical summer data from the same meridional band. The *vertical dotted line* is the sea surface freezing point

relatively low, either due to an apparent interdecadal decline [15] or because 3-4 months of winter brine production remained ahead.

6 Future Work

An improved understanding of the ocean circulation on this shelf could be gained by additional long-term high-quality measurements, particularly of salinity in the HSSW. A detailed study of the seasonal and interannual variability of the Antarctic Slope Front, including the cross-slope transport of water masses and sea ice, could also be extremely valuable. At present the imported MCDW may contribute more to the maintenance of the Ross Sea Polynya than to basal melting of the Ross Ice Shelf, but neither impact is well documented. Exported shelf waters contribute to bottom water formation, but neither the volume nor the spatial and temporal variability of deep ocean ventilation are known in this sector. AASW properties and biological productivity will respond to interannual changes in the sea ice formation and concentration. Satellite data continue to provide valuable information about the annual sea ice cycle, but we know much less about the sources, sinks and residence time of AASW on the continental shelf. One thing not lacking, is a host of unsolved problems for the coming millennium.

Acknowledgements. We thank the individuals and institutions that provided the data utilized in this study, and the polar programs of the National Science Foundation and National Space and Aeronautics Administration that supported our work. We are grateful to the CLIMA Project for its organization of the Lerici Conference on the Oceanography of the Ross Sea, where elements of the material herein and in Jacobs and Giulivi [15] were first presented. This is contribution 5761 of the Lamont-Doherty Earth Observatory of Columbia University.

References

1. Jacobs SS, Amos AF, Bruchhausen PM (1970) Ross Sea oceanography and Antarctic bottom water formation. Deep Sea Res 17:935-962
2. Killworth PD (1974) A baroclinic model of motions on Antarctic continental shelves. Deep Sea Res 21(10):815-838
3. Jacobs SS, Gordon AL, Ardai JL (1979) Circulation and melting beneath the Ross Ice Shelf. Science 203:439-442
4. Ainley DG, Jacobs SS (1981) Sea-bird affinities for ocean and ice boundaries in the Antarctic. Deep Sea Res 28A(10):1173-1185
5. Dunbar RB, Anderson JB, Domack EW, Jacobs SS (1985) Oceanographic influences on sedimentation along the Antarctic continental shelf. In: Jacobs SS (ed) Oceanology of the Antarctic Continental Shelf. Ant Res Ser 43. AGU, Washington, pp 291-312
6. Jacobs SS, Fairbanks RG, Horibe Y (1985) Origin and evolution of water masses near the Antarctic continental margin: evidence from $H_2{}^{18}O/H_2{}^{16}O$ ratios in seawater. In: Jacobs SS (ed) Oceanology of the Antarctic Continental Shelf. Ant Res Ser 43. AGU, Washington, pp 59-85
7. Lewis EL, Perkin RG (1985) The winter oceanography of McMurdo Sound, Antarctica. In: Jacobs SS (ed) Oceanology of the Antarctic Continental Shelf, Ant Res Ser 43. AGU, Washington, pp 145-166

8. Jacobs SS (1989) Marine controls on modern sedimentation on the Antarctic continental shelf. Mar Geol 85:121-153
9. Jacobs SS, Comiso JC (1989) Sea ice and oceanic processes on the Ross Sea continental shelf. J Geophys Res 94:18195-18211
10. Keys HJR, Jacobs SS, Barnett D (1990) The calving and drift of iceberg B-9 in the Ross Sea, Antarctica. Antarct Sci 2(3):243-257
11. Trumbore S, Jacobs SS, Smethie W (1991) Chlorofluorocarbon evidence for rapid ventilation of the Ross Sea. Deep Sea Res 38(7):845-870
12. Locarnini RA (1994) Water masses and circulation in the Ross Gyre and environs. PhD thesis, Dept. Ocean, Texas A&M University, College Station, 86 pp
13. Jaeger JM, Nittrouer CA, DeMaster DJ, Kelchner C, Dunbar RB (1996) Lateral transport of settling particles in the Ross Sea and implications for the fate of biogenic material. J Geophys Res 101(C8):18479-18488
14. Hofmann EE, Klinck JM (1998) Hydrography and circulation of the Antarctic continental shelf: 150°E eastward to the Greenwich Meridian. In: Robinson AR, Brink KH (eds) The sea II. The global coastal ocean, regional studies and synthesis. Wiley, New York, pp 997-1042
15. Jacobs SS, Giulivi CF (1998) Interannual ocean and sea ice variability in the Ross Sea. In: Jacobs SS, Weiss RF (eds). Ant Res Ser 75. AGU, Washington, pp 135-150
16. Keys HJR, Jacobs SS, Brigham LW (1997) Continued northward expansion of the Ross Ice Shelf. Ann Glaciol 27, in press
17. Jacobs SS, Haines WH (1982) Ross Ice Shelf Project, oceanographic data, 1976-1979. TR LDGO-82-1 Columbia Univ., Palisades, 505 pp
18. Pillsbury RD, Jacobs SS (1985) Preliminary observations from long-term current meter moorings near the Ross Ice Shelf, Antarctica. In: Jacobs SS (ed) Oceanology of the Antarctic Continental Shelf. Ant Res Ser 43. AGU, Washington, pp 87-107
19. Jacobs SS (1985) Oceanographic evidence for land ice/ocean interactions in the Southern Ocean. In: Glaciers, ice sheets, and sea level, effects of a CO_2 induced climatic change. Nat Acad Press, Washington, pp 116-128
20. Tressler WL, Ommundsen AM (1962) Seasonal oceanographic studies in McMurdo Sound, Antarctica. Tech Rep 125. US Navy Hydrographic Office, Washington, 144 pp
21. Foldvik A, Gammelsrod T, Torresen T (1985) Circulation and water masses on the southern Weddell Sea shelf. In: Jacobs SS (ed) Oceanology of the Antarctic Continental Shelf. Ant Res Ser 43. AGU, Washington, pp 5-20
22. NODC (1994) World Ocean atlas, CD-ROM data set documentation. Ocean Clim Lab, National Oceanographic Data Center IR13, Washington, 30 pp
23. Stocker TF, Wright DG, Broecker WS (1992) The influence of high-latitude surface forcing on the global thermohaline circulation. Paleoceanography 7(5):529-541
24. Toggweiler JR, Samuels AM (1995) Effect of sea ice on the salinity of Antarctic bottom water. J Phys Ocean 25:1980-1997
25. Levitus S (1982) Climatological atlas of the World Ocean. NOAA Prof Paper 13, Rockville, 173 pp
26. Jacobs SS (1991) On the nature and significance of the Antarctic slope front. In: Treguer P, Queguiner B (eds) Biochemistry and circulation of water masses in the Southern Ocean. Mar Chem 35:9-24
27. Lozier MS, McCartney MS, Owens WB (1994) Anomalous anomalies in averaged hydrographic data. J Phys Ocean 24:2624-2638
28. Grumbine RW (1991) Model of the formation of high-salinity shelf water on polar continental shelves. J Geophys Res 96(C12):22049-22062
29. Carmack EC (1977) Water characteristics of the Southern Ocean south of the Polar Front. In: Angel M (ed) A voyage of discovery. G. Deacon 70th Anniv Vol Suppl to Deep Sea Res. Pergamon, New York, pp 15-42
30. Fofonoff NP (1956) Some properties of seawater influencing the formation of Antarctic bottom water. Deep Sea Res 4:32-66
31. Foldvik A, Kvinge T (1974) Conditional stability of sea water at the freezing point. Deep Sea Res 21(3):169-174

32. Jacobs SS, Hellmer HH, Doake CSM, Jenkins A, Frolich R (1992) Melting of ice shelves and the mass balance of Antarctica. J Glaciol 38:75-387
33. Williams RT, Robinson ES (1980) Ocean tide in the southern Ross Sea. J Geophys Res 85(C11):6689-6696
34. MacAyeal DR (1985) Evolution of tidally triggered meltwater plumes below ice shelves. In: Jacobs SS (ed) Oceanology of the Antarctic Continental Shelf. Ant Res Ser 43. AGU, Washington, pp 133-143
35. MacAyeal DR (1985) Tidal rectification below the Ross Ice Shelf, Antarctica. In: Jacobs SS (ed) Oceanology of the Antarctic Continental Shelf. Ant Res Ser 43. AGU, Washington, pp 109-132
36. Scheduikat M, Olbers DJ (1990) A one-dimensional mixed layer model beneath the Ross Ice Shelf with tidally induced vertical mixing. Antarct Sci 2:29-42
37. Nost OA, Foldvik A (1994) A model of ice shelf-ocean interaction with application to the Filchner-Ronne and Ross Ice Shelves. J Geophys Res 99(C7):14243-14254
38. Hellmer HH, Jacobs SS (1995) Seasonal circulation under the eastern Ross Ice Shelf, Antarctica. J Geophys Res 100(C6):10873-10885
39. Zwally HJ, Comiso JC, Gordon AL (1985) Antarctic offshore leads and polynyas and oceanographic effects. In: Jacobs SS (ed) Oceanology of the Antarctic Continental Shelf, Ant Res Ser 43. AGU, Washington, pp 203-226
40. Kurtz DD, Bromwich DH (1985) A recurring, atmospherically forced polynya in Terra Nova Bay. In: Jacobs SS (ed) Oceanology of the Antarctic Continental Shelf, Ant Res Ser 43. AGU, Washington, pp 177-201
41. Jeffries MO, Adolphs U (1997) Early winter ice and snow thickness distribution, ice structure and development of the western Ross Sea pack ice between the ice edge and the Ross Ice Shelf. Antarct Sci 9(2):188-200
42. White WB, Peterson RG (1996) An Antarctic circumpolar wave in surface pressure, wind, temperature and sea ice extent. Nature 380:699-702
43. Davey FJ (1995) Bathymetry, seismic stratigraphic atlas of the Ross Sea, Antarctica. In: Cooper A, Barker P, Brancolini G (eds) Geology and seismic stratigraphy of the Antarctic margin. Ant Res Ser 68. AGU, Washington, Plate 1, Map 1b
44. Biggs DC, Amos AF (1983) Oceanographic data from the southwestern Ross Sea, January 1982, STD and nutrient chemistry programs. Ref 83-3-T Dept Ocean. Texas A&M University, College Station, 167 pp
45. El-Sayed SZ, Biggs DC, Holm-Hansen O (1983) Phytoplankton standing crop, primary productivity, and near surface nitrogenous nutrient fields in the Ross Sea, Antarctica. Deep Sea Res 30(8A):871-886
46. Giulivi CF, Jacobs SS, O'Hara SH, Ardai JL (1995) Ross Sea polynya project 1994, oceanographic data, NB Palmer Cruise 9406. TR LDEO-95-1 Lamont Earth Observatory, Palisades, 315 pp
47. Heath RA (1971) Circulation and hydrology under the seasonal ice of McMurdo Sound, Antarctica. Nz J Mar Freshw Res 5(3/4):497-515
48. Littlepage JL (1965) Oceanographic investigations in McMurdo Sound, Antarctica. In: Pawson DL (ed) Biology of the Antarctica Seas II. Ant Res Ser 5. Agu, Washington, pp 1-37

Mapping the Thickness of Pancake Ice Using Ocean Wave Dispersion in SAR Imagery

P. Wadhams[1], F. Parmiggiani[2], G. de Carolis[3] and M. Tadross[1]

[1]Scott Polar Research Institute, University of Cambridge, Cambridge CB2 1ER, UK
[2]IMGA-CNR, Via Gobetti 101, 40129 Bologna, Italy
[3]ITIS-CNR, P.O. Box 11, 75100 Matera, Italy

Abstract

During the early to midwinter period pancake ice is a major component of the Antarctic sea ice cover, occupying a belt extending 200-300 km from the outer ice edge with an area of about 6 million km^2. Experience in the Arctic suggests that the sea ice thickness in this region can be mapped by monitoring the penetration of ocean waves into it. Spectral analysis of subscenes from ERS-2 synthetic aperture radar (SAR) images yields the wavelength and direction of the principal spectral component both outside and inside the ice cover. There is a change of wavelength at the ice edge associated with a change in the wave dispersion, which can be quantitatively related to the thickness of the ice. The analysis is complex because the true wave spectrum must be retrieved from the SAR spectrum, which involves an inversion technique requiring a "first-guess" spectrum. The analysis technique is described, and the wave theory which predicts the change in wavelength. Experience in the Chukchi Sea and especially in the Odden ice tongue in the Greenland Sea is reviewed. Preliminary data from the Ross Sea have been analysed, yielding principal wave components, but the ambient wavelength was too long for the effect of ice to be detectable. The next application will be to the outer ice edge, when midwinter data from this region are available at the end of the current season.

1 Introduction: Pancake Ice and Waves

1.1 Antarctic Pancake Ice

In order to study and model the properties of the Antarctic sea ice cover and its role in climate change, an essential parameter to measure is its thickness distribution. Sea ice thickness is a product of thermodynamic and dynamic processes, and is itself a determinant of ocean-atmosphere heat flux. In the Arctic most information on sea ice thickness has come from upward-looking sonar profiling carried out by military nuclear submarines. The terms of the Antarctic Treaty do not allow the use of such systems iin the Antarctic. Therefore, in attempting to measure the thickness distribution of different ice types at different seasons, there has been considerable dependence on drilling lines of closely spaced holes [1], although the use of moored upward sonar systems and sledge- or helicopter-mounted electromagnetic eddy current systems is increasing.

One particular type of Antarctic ice shows promise of being mapped by a remote sensing technique. During the Winter Weddell Sea Project cruise of FS "Polarstern" in 1986 it was discovered [1, 2] that the outermost zone of the advancing circumpolar sea ice cover in early winter is composed of pancake ice. This is a type of young ice which forms in turbulent waters. The first stage in its production is freezing at the sea surface to produce a suspension of small ice crystals called frazil ice. These crystals cannot freeze together to form a continuous sheet because of the high energy and turbulence in the Southern Ocean wave field, so instead an increasingly dense suspension of frazil ice forms out of which pancake ice congeals by wave-induced compression of the suspension, a mechanism described by Martin and Kauffman [3]. At the ice edge the pancakes are only a few centimetres in diameter, but they gradually grow in diameter and thickness with increasing distance from the ice edge, until they reach 3 to 5 m diameter and 50-cm thickness. The growth occurs by accretion from the frazil, which continues to form because the open water surface permits maximum ocean-atmosphere heat flux. The pancakes begin to freeze together in groups, but the wave field is strong enough to prevent overall freezing until a penetration of some 270 km (observed in 1986) is reached. Here the pancakes coalesce to form a continuous sheet of first-year ice. At this point, with the open water surface cut off, the growth rate drops to a very low level (estimated at 0.4 cm per day by Wadhams et al. [1]) and the ultimate thickness reached by first-year ice is only a few centimetres more than the thickness attained at the time of consolidation of the pancakes. If the observed width of 270 km is a circumpolar phenomenon in early to mid-winter, this would correspond to an area of 6 000 000 km^2 occupied by pancake ice, making this a significant component of the planet's surface.

Ice formed in this way has a different bottom morphology from Arctic ice. The pancakes at the time of consolidation are jumbled together and rafted over one another, and freeze together in this way with the frazil acting as "glue". The result is a very rough, jagged bottom, with rafted cakes doubling or tripling the normal ice thickness, and with the edges of pancakes protruding upwards to give a surface topography resembling a "stony field" [1]. The rafted bottom provides a large surface area per unit area of sea surface, providing an excellent substrate for algal growth and a refuge for krill. The thin ice permits much light to penetrate, and the result is a winter ice ecosystem which is more fertile than that seen in the Arctic. Because of the large amount of free water on and around the pancakes, the zone containing this type of ice had formerly been characterised from passive microwave data as consisting of first-year ice at low concentration [4].

It has been found that in pancake icefields ocean waves can be clearly imaged by satellite-borne synthetic aperture radar (SAR), and these observations show that the dominant waves often appear to change in wavelength (and direction) as they enter the ice. The change in dispersion is related to ice thickness, and can be used as a means of estimating the thickness of frazil and pancake icefields. Existing experience has been in the Arctic, but is now being applied to Antarctic ice.

1.2 Wave Propagation in Ice

If pancake ice is not present, then when waves enter the ice edge from the open sea they first encounter an outermost fringe of wave-broken floes of finite diameter, the marginal ice zone (MIZ), before reaching the interior of the polar ice cover, which can be thought of in a simplified way as a continuous elastic sheet. Extensive field observations of wave decay in MIZ regions have been carried out during the past three decades, with some of the more recent measurements being described by Wadhams et al. [5, 6]. The main conclusions from the observations are that the attenuation of waves with distance into the pack takes a negative exponential form, with an attenuation coefficient which decreases with increasing wave period over most of the spectral range. In heavy compact ice (e.g. East Greenland) the energy attenuation coefficient typically varies from $2 \times 10^{-4}\ m^{-1}$ for the longest swell to $8 \times 10^{-4}\ m^{-1}$ for 8 to 9 s waves, corresponding to e-folding distances of 5-1.2 km.

A simple scattering model (described in full in [7]) was developed to explain these findings. It treats each floe as an elastic floating raft, solving for velocity potentials at the leading and trailing edges of the raft. Within the raft itself energy propagates as a flexural-gravity wave with an altered dispersion relation. The treatment yields an energy reflection coefficient from which an attenuation rate can be derived. Floe diameter is a more critical parameter than ice thickness. The model agrees well with experimental results at normal and long wave periods, although it does not predict an observed increase in directional spread as waves pass into the ice. The basis for the model is that the scattering is due to a mismatch between the mode of propagation under an elastic raft and under an open water surface.

In open water a wave of angular frequency ω (= $2\pi / T$, where T is wave period) follows the familiar dispersion relation

$$k = \omega^2 / g \qquad (1)$$

if water depth is great compared to the wavelength λ (= $2\pi / k$). Under ice, however, with p(x,t) being the pressure just below the water-ice interface, we have

$$p(x,t) - L\partial^4\eta_2 / \partial x^4 = \rho_i\, h\, \partial^2\eta_2 / \partial t^2 \qquad \text{(equation of motion)}, \qquad (2)$$

$$p(x,t) = -\rho_w [g\, \eta_2 + \partial\Phi_2 / \partial t\, |_{y=0}] \qquad \text{(Bernoulli equation)}, \qquad (3)$$

$$- \partial\eta_2 / \partial t = \partial\Phi_2 / \partial y\, |_{y=0} \qquad \text{(boundary condition)}. \qquad (4)$$

Here L is the flexural rigidity of the ice, given by

$$L = E\, h^3 / 12\, (1 - \upsilon^2) \qquad (5)$$

where E is Young's modulus, h the ice thickness and υ Poisson's ratio. Φ_2 is a velocity potential for a wave of amplitude η_2 propagating under the ice in the positive x-direction with the y-axis extending vertically down from the surface; ρ_i and ρ_w

are ice and water densities. A solution for Φ_2 is a sum of potentials of form

$$\Phi^2 = \sum_n [A_n \exp(ik_nx) + B_n \exp(-ik_nx)] \exp(-k_ny) \exp(-i\omega t). \quad (6)$$

A velocity potential of this form satisfies the boundary conditions provided

$$L k_n^5 + (\rho_w g - \rho_i h \omega^2) k_n - \rho_w \omega^2 = 0. \quad (7)$$

There are three physically feasible roots for k_n, of which the real root k_o represents a conservative wave propagating through the ice sheet with a dispersion relation which differs from that of a wave in open water. This new type of wave, called a *flexural-gravity wave*, has a dispersion relation which involves a minimum value of phase and group velocity. In continuous ice sheets it can be generated by wind blowing over ice (so long as the wind velocity is sufficient), or by moving vehicles, while it also represents the mode for long-distance penetration of wave and swell energy through ice-covered polar basins.

In the context of scattering by floes, k_n defines a set of wave numbers which give the mismatch from open water propagation which is responsible for reflections at floe edges, and which can be used to predict the energy decay rate [7].

1.3 Wave Propagation in Pancake Ice

In pancake and frazil ice, the floe size is so small in relation to the wavelength that we can consider the material to be simply a mass of particles floating on the water surface. If we consider such ice to be a continuum composed of non-interacting mass points which exerts a pressure upon the water surface but which has no coherence or rheological properties as a material, then the relevant equations for propagation of waves in the ice are those developed above for the wave scattering model but setting L = 0. This theory was developed first for waves in any kind of ice [8-11] but in fact it is inapplicable to continuous ice sheets or to marginal ice zones composed of floes whose diameter is a non-negligible fraction of a wavelength.

Following Wadhams [7], we find that the wave number k_i for propagation in ice is given, setting L to zero in Eq. (7), by

$$k_i = \rho_w \omega^2 / [\rho_w g - \rho_i c h \omega^2], \quad (8)$$

where h is the thickness of the frazil slurry or pancakes and c is the average ice concentration. This implies that a wave propagating into the ice from open water will acquire a reduced wavelength $\lambda_i = 2\pi/k_i$. This has the following implications: (1) waves incident obliquely on the ice edge are refracted towards the normal, according to Snell's Law; (2) since the group velocity inside the ice is lower than in the open water, a wave entering the ice acquires a greater amplitude. This, combined with the reduced wavelength, implies a greater wave steepness; (3) there is a frequency limit implied in Eq. (8), given by

$$\omega_c^2 = \rho_w g / \rho_i c h. \quad (9)$$

Above this frequency propagation is not possible. A wave incident normally on the ice at a frequency higher than ω_c will suffer total reflection at the ice edge. Wadhams and Holt [12] pointed out an implication for frazil ice imaging on SAR. Waves at the Bragg wavelength (30 cm for Seasat SAR considered in their paper) have periods only just above T_c, so that they suffer extreme modification on passing into the ice, with a high reflection coefficient at the ice edge. There is therefore little energy present in a frazil ice slurry at these wavelengths and so frazil ice appears dark on SAR imagery. The same argument applies to pancake ice, but in this case the physical roughness of the upturned edges of the pancakes will make this type of ice appear bright to SAR.

Reflection and transmission coefficients across the ice edge were derived by Keller and Weitz [10] as

$$(1 - |R|^2) = |T|^2 \, k^2/k_i^2, \tag{10}$$

giving an energy transmission coefficient t for normal transmission of

$$t = |T|^2 \, U_i/Uw = 4 \, k \, k_i \, / \, (k + k_i)^2. \tag{11}$$

This model does not yield progressive wave decay inside the ice cover, although it does imply wave refraction at the ice edge and the total exclusion of the shortest waves from the icefield. Clearly energy decay takes place in frazil slurries, and Martin and Kauffman [3] observed an internal circulation within the slick in tank experiments, which would lead to viscous energy losses. Recent laboratory experiments by Newyear and Martin [13] on very short waves in frazil ice also suggest that the mass loading theory breaks down for wave periods of 1 s or less, when a wavelength increase is observed relative to the open water relation. To resolve this question, further laboratory experiments on the wave dispersion relation in frazil ice were carried out during February 1997 in the European ice tank at Hamburg Schiffbau-Versuchsanstalt by R. Hall, a student of P. Wadhams, with wave periods of up to 2 s available, but with results which show no measurable change in wavelength.

2 Imaging of Waves by SAR

The potential of SAR to image ocean surfaces has been recognised over the past two decades [14] and the SAR imaging theory of a moving surface is now quite well understood [15]. The ocean wave spectral information is not entirely mapped in the SAR image plane. Since SAR images provide a snapshot of the sea surface, the directional properties of the wave spectrum are lost in the SAR image plane. Moreover, SAR imaging is frequently a strongly non-linear process induced by motion effects of the ocean surface. The velocity bunching mechanism associated with the orbital motion of the long waves is the most relevant source of non-linearity which produces distortions of the SAR spectrum through an azimuthal high wave number cut-off.

Considering the velocity bunching mechanism as fully non-linear and the other modulation processes of the real cross section as linear through their modulation transfer functions (tilt and hydrodynamic), Hasselmann and Hasselmann [16] derived a closed, non-linear integral mapping transformation relating the SAR spectrum $P(\vec{k})$ to the the two-dimensional directional wave spectrum $F(\vec{k})$which can be computed by fast fourier transform using the following expression:

$$P(\vec{k}) = \frac{1}{(2\pi)^2} \exp(-k_x^2 \xi'^2) \cdot \sum_{n=1}^{\infty} \sum_{m=2n-2}^{2n} (k_x \beta)^m P_{n,m}(\vec{k}), \tag{12}$$

where $\beta = R/V$ is the slant range and platform velocity ratio, k_x is the azimuthal component of the wavenumber $\vec{k}$ in the SAR reference frame and $P_{n,m}$ are higher products of auto and covariance functions of the slant range components of the orbital velocity and real cross section modulation. The non-linear exponential factor depends on the mean square azimuthal displacement of a scattering element given by $\xi'^2 = \beta^2 \int F(\vec{k}) |T^v(\vec{k})|^2 d\vec{k}$, where $T^v(\vec{k})$ is the range velocity transfer function. The first term of Eq. (12) yields the quasi-linear approximation. Based on the forward mapping model expressed by Eq. (12), an inversion scheme from SAR spectra with respect to the wave spectrum has been proposed by Hasselmann et al. [17].

The inversion procedure involves the minimization of a cost function which penalizes the deviations between the measured and simulated SAR spectra. In this case, however, regularization terms should be also used to resolve the non-linearities affecting SAR data. In order to resolve the intrinsic 180° ambiguity of SAR imagery and to provide the missing information at high azimuthal wave number, a first regularization term including a first guess wave spectrum $\hat{F}(\vec{k})$ should be used [16, 17]. A second term, with the double function of penalizing the deviations between the observed and the simulated cut-off scales and of adjusting the energy level of the simulated spectrum, should be also introduced [17].

The cost function to be minimized can be defined as follows:

$$J = \int d\vec{k}\ [P(\vec{k}) - \hat{P}(\vec{k})]^2 \hat{P}(\vec{k}) + \mu \int d\vec{k}\ \frac{[F(\vec{k}) - \hat{F}(\vec{k})]^2}{\{B + \min[F(\vec{k}), \hat{F}(\vec{k})]\}^2} + \eta \frac{(\alpha\lambda_{cl}^2 - \hat{\lambda}_{cl}^2)}{\max\{\lambda_{cl}^4 - \hat{\lambda}_{cl}^4\}}, \tag{13}$$

where $\hat{F}(\vec{k})$ is the first guess wave spectrum; $\hat{P}(\vec{k})$ is the observed SAR spectrum; $P(\vec{k})$ is the SAR spectrum computed from the best fit wave spectrum $F(\vec{k})$; λ_{cl} and $\hat{\lambda}_{cl}$ are the clutter cut-off length of the simulated and observed SAR images, respectively; and α is an energy scaling parameter to be applied to the entire wave spectrum in order to minimize the error between the simulated and observed cut-off scales. The parameter μ should have a small value in order to reduce impact of the guess wave spectrum $\hat{F}(\vec{k})$ on the final solution, but should be large enough in order to resolve the 180° ambiguity and to provide wave information beyond the cut-off scale; the weighting factor η is introduced to make the first and last term of Eq. (13) the same order of magnitude. Finally, the constant B prevents the denominator of the second term from vanishing. Starting from the first guess wave spectrum, $F^1(\vec{k}) = \hat{F}(\vec{k})$, the minimization of the cost function J is carried out by means of an iteration scheme in which the solution $F^n(\vec{k})$ obtained at step n is updated using the following relation:

$$F^{n+1}(\vec{k}) = \alpha^n F^n(\vec{k}) + \Delta F^n(\vec{k}), \tag{14}$$

where α^n is computed after $\Delta F^n(\vec{k})$ has been determined. Assuming the increment

$$\Delta P^n(\vec{k}) = P^{n+1}(\vec{k}) - P^n(\vec{k})$$

can be computed from $\Delta F^n(\vec{k})$ using the quasi-linear SAR mapping relation, the following expressions for $\Delta F^n(\vec{k})$ and α^n hold [16, 17]:

$$\Delta F^n(\vec{k}) = \frac{A(-\vec{k})[W(\vec{k})\Delta\hat{P}^n(\vec{k}) + \mu^n \Delta\hat{F}^n(\vec{k})] - B(\vec{k})[W(-\vec{k})\Delta\hat{P}^n(\vec{k}) + \mu^n \Delta\hat{F}^n(-\vec{k})]}{A(\vec{k})A(-\vec{k}) - B^2(\vec{k})} \tag{15}$$

$$\alpha^n = \frac{\eta^n \hat{\lambda}_{cl}^2 \lambda_{cl}^{n^2} - \mu^n \int F^n(\vec{k})[\Delta F^n(\vec{k}) - \hat{F}(\vec{k})]d\vec{k}}{\eta^n \lambda_{cl}^{n^4} + \mu^n \int F^{n^2}(\vec{k})d\vec{k}}, \tag{16}$$

where

$$\Delta\hat{P}^n(\vec{k}) = \hat{P}(\vec{k}) - P^n(\vec{k}) = \hat{P}(-\vec{k}) - P^n(-\vec{k}), \tag{17}$$

$$\Delta\hat{F}^n(\vec{k}) = \hat{F}(\vec{k}) - \alpha^n F^n(\vec{k}), \tag{18}$$

$$A(\vec{k}) = W^2(\vec{k}) + 2\mu^n, \tag{19}$$

$$B(\vec{k}) = W(\vec{k})W(-\vec{k}), \tag{20}$$

$$\mu^n = \frac{\mu}{\min[F^n(\vec{k}), \hat{F}(\vec{k})]}, \tag{21}$$

$$\eta^n = \frac{\eta}{\max[\lambda_{cl}^{n^4}, \hat{\lambda}_{cl}^4]}. \tag{22}$$

Equations (15) and (16) are solved iteratively. Expression (15) is firstly computed by setting $\alpha^n = 1$ in Eq. (18); the value of α^n is then updated by using Eq. (16) and inserted in Eq. (18) to obtain a new estimation of Eq. (15). To avoid numerical divergence, the value of $\Delta F^n(\vec{k})$ is constrained to be [17]:

$$|\Delta F^n(\vec{k})| \leq \frac{1}{4} \min[\hat{F}^n(\vec{k}), \alpha F^n(\vec{k})]$$

if

$$\frac{\mu \Delta F^{n^2}(\vec{k})}{[B + \min(F(\vec{k}), \hat{F}(\vec{k}))]^2} \geq \frac{1}{4} \hat{P}(\vec{k})[P^n(\vec{k}) - \hat{P}(\vec{k})]^2.$$

The iteration is terminated when α^n changes less than 1%. Having obtained the best estimate for $\Delta F^n(\vec{k})$ and α^n, the spectrum is updated to $F^{n+1}(\vec{k})$ through Eq.

(14) and the iteration is prosecuted until convergence is reached, i.e. a minimum of the *J* function is reached.

In order to avoid the occurrence of discontinuities in the region of the inverted wave spectrum separating the neighbourhood of the azimuth cut-off length and to allow the inverted spectrum to depart, if necessary, from the first guess spectrum, at the end of the iteration step previously described, the first guess spectrum is successively modified yielding second-guess, third-guess, etc. input spectra. This is achieved using two techniques: (1) a partitioning scheme which decomposes the wave spectra into their wave systems components; and (2) an algorithm of cross assignment of the wave systems [17]. Our implementation of the algorithm was written in IDL language (Interactive Data Language) and implemented for cartesian SAR spectra like those obtained from SAR-PRI (Precision Image) images of ERS-1 (earth resources satellite). It also includes an interpolation module which works as a smoothing filter on SAR spectra sampling variability and noise. The effects of the filter are spectral increments being added to the wave spectrum at each iteration step. The updated guess spectrum, computed on a coarser resolution of the wave number domain, allows the inversion procedure to capture the features of the measured SAR spectrum in order to increase the performances of the inversion algorithm in terms of both a lower number of iteration steps and a better agreement between the simulated and measured SAR spectra.

3 Detection of Changes in Wave Dispersion

On the assumption that we have determined the shape of the correct wave number spectrum in the open sea and in ice, if necessary by the use of the inversion technique described in Section 2, the next stage is to extract the relevant data on the change in wave dispersion. Most spectra show a single strong peak. The wave number and direction of the principal wave component are extracted from the central point of this contoured peak. Equation (8) can be rearranged to yield

$$h = (k_i - k) \, / \, c \, r \, k_i \, k, \qquad (23)$$

where k is the wave number in the open sea and $r = (\rho_i \, / \, \rho_w)$. The main uncertainty in the estimation of h in Eq. (23) is thus the uncertainty in the estimation of k_i and k from imagery in which the wave number resolution is determined by the pixel size. It is clear also from Eq. (8) that the change in wave number induced by ice of a given thickness is greatest for the largest wave numbers (up to the cut-off), i.e. the shortest period waves. The technique therefore becomes less reliable if the dominant wave period is very long, as is frequently the case in the Southern Ocean. However, in a case where local wind is low, so that the spectrum has a strong swell peak at low wave number and a secondary wind-wave peak at high wave number, the secondary peak can also be used for the analysis.

A further characteristic of the wave dispersion which can be used is the refraction at the ice edge. Following Snell's Law, Eq. (8) shows that a wave should be refracted towards the normal as it crosses the ice edge. The directions of the wave peaks found in the SAR spectra will, in principle, allow the refraction to be used as another method of estimating h. However, in most cases a frazil-pancake ice-field has a very ragged or imperfectly defined edge; it is seldom straight. Therefore the angles of incidence and refraction of a wave cannot be defined accurately, so this method is usually inapplicable.

4 Results from Arctic Areas

4.1 Chukchi Sea

The first dataset to be analysed in this way was a Seasat image from the Chukchi Sea northwest of Wrangel Island, obtained in October [12]. Fringing the main Arctic Ocean ice edge was pancake ice, with frazil ice seaward of it. Six sets of sub-scenes were analysed, along lines running parallel to the main wave vector into the ice. Wavelengths involved were in the range 150-180 m. Spectra were unimodal and clearly defined. Refraction in the correct sense was observed in every case. The wavelength changes were used to compute ice thickness, which increased linearly to 2-3 m at 25 km inside the ice edge and 4-6 m at 50 km inside the ice edge. These values were felt to be high, but not necessarily impossibly high, since clearly this type of ice had amassed against the thicker polar ice due to the action of a strong on-ice wind, which might have piled it up in a wedge which could have reached such a thickness. The absence of any in-situ verification was a drawback here. Only the simple spectral analysis was carried out, without any attempt at inversion.

4.2 Greenland Sea

During the 1993-1996 ESOP (European Subpolar Ocean Programme) Project, an EU MAST-2 project to study the role of frazil-pancake ice in stimulating Greenland Sea convection, extensive direct observations were carried out in the Odden ice tongue, a protrusion of ice, usually frazil-pancake, which grows eastward from the main East Greenland ice edge in winter in the vicinity of latitudes 72-75°N [18]. These measurements included ice thickness measurements, both of the pancakes and of the frazil ice matrix. Most measurements were made on the well-developed 1993 season Odden, which provided an opportunity to make comparisons with the wave analysis of concurrent ERS-1 SAR images. The work was done in two stages. Initially, all possible ERS-1 imagery of the 1993 Odden was acquired and examined. In every case where the image showed a clear wave train entering an edge of Odden, the image was subjected to a straightforward spectral analysis. The nature of the changing SAR spectrum as the waves entered the ice was studied in order to assess the conditions under which the simple technique may be considered to have some validity. Finally, the full inversion technique was applied to selected data in order to examine the changes which this produces in the spectra.

As an example, the inversion algorithm was applied to an ERS-1 SAR PRI image of the Odden ice tongue in the Greenland Sea acquired on April 10, 1993 (orbit 9077, frame 1521). A directional wave spectrum was collected by a buoy deployed in a location inside the SAR scene but some time before the satellite pass. The scene includes a region of frazil-pancake ice which extends along the azimuth direction, mostly in the mid-range direction of the radar acquisition; regions of open ocean are present in the near range and far range region of the scene.

Following the direction of the wave number peak, a series of 15 windows each of 512 x 512 pixels have been extracted from the SAR scene to compute spectra of 256 x 256 pixels with wave number increment of 19.635 x 10^{-4} m^{-1}. The selected windows are shown in Fig. 1, the numbering order going from the near to far range so that it follows the propagation direction of the wave systems from open sea to ice-covered sea.

In general, the algorithm convergence was found to be very sensitive to the choice of parameters μ, η and B of Eq. (13). Although an optimum selection of the parameters has not been carried out, the best performance has been found with the following values:

$$\begin{cases} \mu = 0.01\,\hat{P}_{max} \\ \eta = 50\,000\,\left(\int \hat{P}(\vec{k})\,d\vec{k}\right)^3 \\ B = 10^{-3}\,\hat{F}_{max}\,. \end{cases}$$

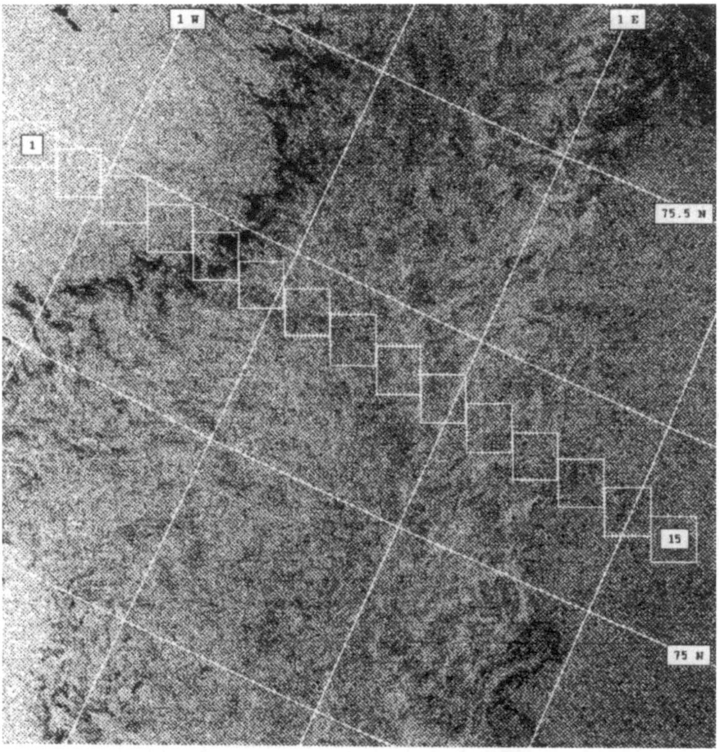

Fig. 1. ERS-1 SAR PRI scene of the Odden ice tongue, Greenland Sea, of April 10, 1993 (orbit 9077, frame 1521), with frames of the 15 analysed windows in overlay

The quality of the inversion results was evaluated considering both the squared error and the correlation index pattern between the observed SAR spectrum and the simulated SAR spectrum, as well as the cut-off wavelength measured on the simulated spectra. The correlation index and the square error are defined as

$$C = \frac{\int P(\vec{k})\hat{P}(\vec{k})d\vec{k}}{\sqrt{\int P^2(\vec{k})d\vec{k} \cdot \int \hat{P}^2(\vec{k})d\vec{k}}} \quad \text{and} \quad \varepsilon^2 = \frac{\int [P(\vec{k}) - \hat{P}(\vec{k})]^2 d\vec{k}}{\sqrt{\int P^2(\vec{k})d\vec{k} \cdot \int \hat{P}^2(\vec{k})d\vec{k}}},$$

respectively. Both parameters have been estimated in the wave number domain defined by

$$\frac{2\pi}{700} \leq |\vec{k}| \leq \frac{2\pi}{100},$$

where SAR data are expected to be more reliable.

Table 1 reports the inversion results from the 15 windows and includes the square error, the correlation index, the wavelength and propagation angle of the dominant wave computed with respect to the range direction and the ratio between the real and simulated cut-off wavelength. Results show that the dominant wavelength is affected by the presence of sea ice so that its value is lower than the one propagating in open sea, as expected. Figure 2 shows: (1) the retrieved

Table 1. Inversion results of the 15 windows

Window	λ	Φ	H_S	ε^2	$C(\%)$	ρ
1	167.49	42.88	2.08	0.093	95.38	293/265
2	173.54	40.60	2.38	0.10	94.97	292/286
3	141.42	45.00	2.52	0.097	95.41	280/283
4	145.30	50.52	2.28	0.16	95.10	271/271
5	****	****	****	****	****	****
6	128.83	49.90	1.93	0.22	92.77	257/252
7	129.24	43.36	1.78	0.27	93.47	249/203
8	140.32	52.12	1.73	0.15	94.93	253/232
9	132.87	48.36	2.15	0.095	95.80	261/268
10	132.87	48.36	2.42	0.14	95.60	267/267
11	196.57	47.49	1.71	0.17	93.54	265/262
12	153.78	54.58	2.47	0.26	90.59	283/289
13	150.51	41.18	2.62	0.12	93.72	285/285
14	187.91	40.23	2.32	0.11	94.38	316/298
15	196.57	43.12	2.37	0.17	92.77	290/275

λ, dominant wave wavelength (in metres).

Φ, propagation direction of the dominant wave relative to range direction.

H_S, estimated height (in metres) of the dominant wave.

ε^2, square error (see text).

$C(\%)$, correlation index (see text).

ρ, ratio of cut-off wavelength of SAR-simulated spectrum to cut-off wavelength of SAR-observed spectrum.

****, quality of image did not permit parameters to be extracted.

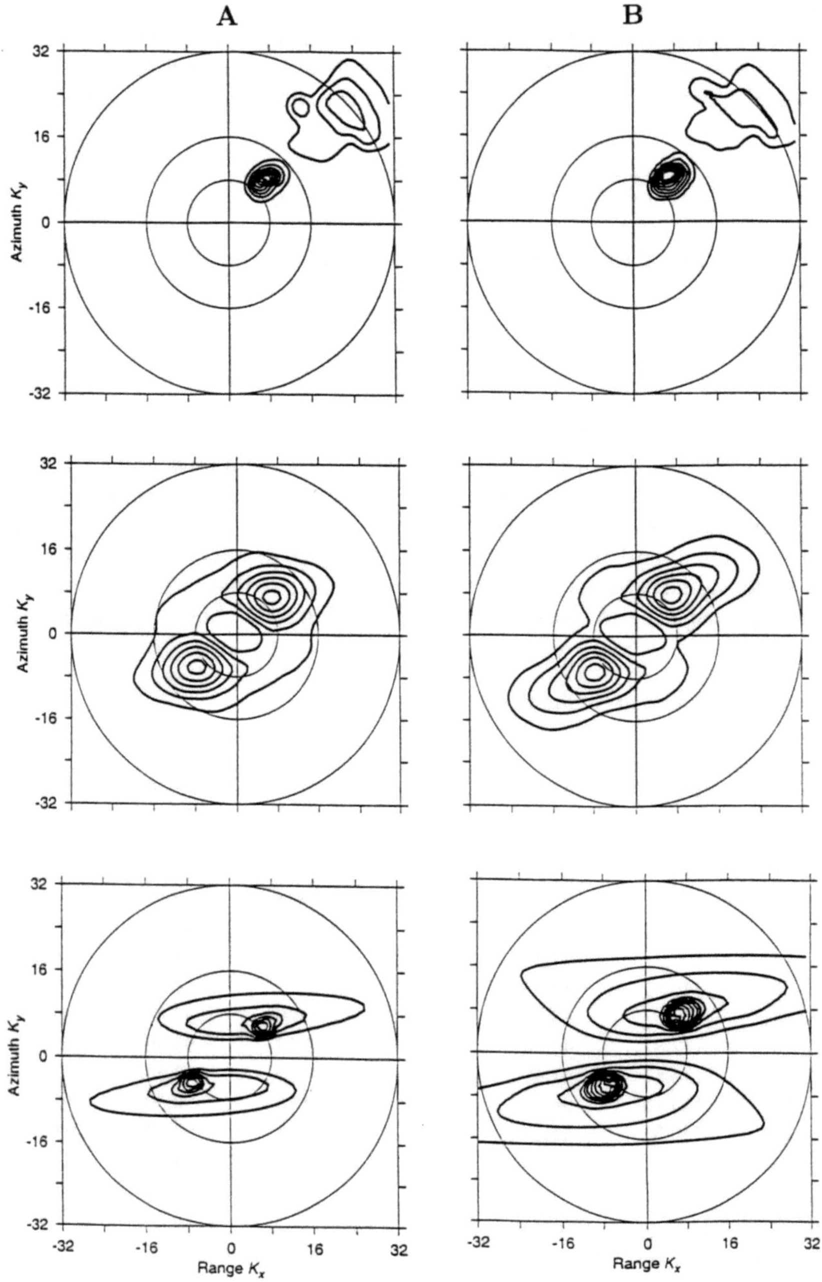

Fig. 2. Retrieved wave spectrum (*top row*), observed SAR spectrum (*middle row*) and simulated SAR spectrum (*bottom row*) for windows 3 (open sea, A), 4 (frazil ice, B), 10 (pancake ice, C) and 14 (open sea, D) as shown in Fig. 1, respectively. $\Delta k = 8/\pi\ 10^{-2}$ rad/m; thus the *three circles*, from the outer to the inner, correspond to $\lambda = 50$, 100 and 200 m

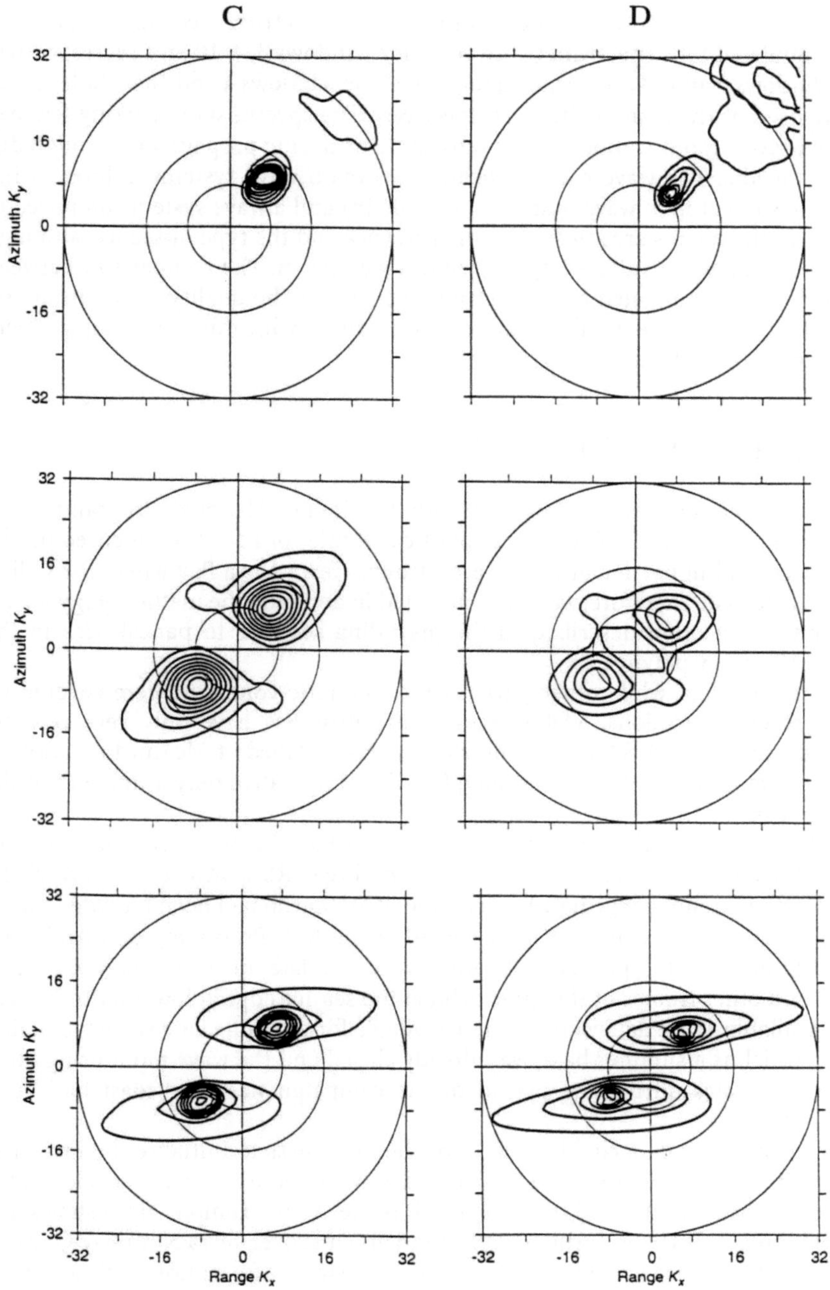

Fig. 2. Continued

wave spectrum (first row); (2) the observed SAR spectrum (second row); and (3) the simulated SAR spectrum (third row), for windows 3, 4, 10 and 14, respectively. Windows 3 and 14 are open-sea regions, while windows 4 and 10 include sea ice composed of frazil and pancake, respectively. The spectra show striking features which reveal modifications by the wave systems during propagation through different media. The wave spectra include two main wave systems: a broad wind wave system at high wave numbers ($\lambda < 100$ m) and a wave system composed of swell. Both systems are affected by the presence and the type of sea ice, although major changes are suffered by the wind wave system. The comparison between observed and simulated SAR spectra demonstrates the quality of the inversion results for each case. Preliminary results obtained with this methodology were presented in Wadhams et al. [19].

5 Application to the Antarctic

Our research groups are involved with the Italian National Programme for Antarctic Research (PNRA). The scientific activities of PNRA are focused on the Ross Sea, and in particular on the area around Terra Nova Bay where the Italian base is located. Therefore, we were interested in applying the methodology for ice thickness detection described in the preceding sections to pancake ice in the Terra Nova Bay polynya.

Unfortunately, while other parts of the Antarctic continent were covered by SAR receiving stations, SAR images of the Ross Sea have only very recently become available. A SAR receiving station was installed at McMurdo Station in the first months of 1995 and became (partially) operative only at the end of the same year.

The first acquisitions relative to Terra Nova Bay that we were able to obtain from the European Space Agency (ESA) were three ERS-2 SAR scenes of April 24, 1996. Other scenes of spring 1996 acquired at McMurdo had been recorded incorrectly on tape and could not be decoded at the ESA Processing and Archiving Facility in Frascati. April 24 is already a very late date for having pancake ice at 75°S and other frames of the same orbit at the sea-ice edge at lower latitude were not available. Wave data were also not available for the same date since the Italian base, which is a summer base, was already closed and the wave monitoring buoy which operates during the period of the campaign near the coast had been removed.

Thus no wave data could be used for the first guess to initialise the inversion algorithm described in Section 2. We could only compute the observed spectra for a set of windows extracted from each SAR scene. As an example, Fig. 3 shows one of the three SAR scenes (orbit 5289, frame 5607) just over Terra Nova Bay. One can observe that the ice was already in an advanced state of formation while a zone of open water of the polynya still remains (dark area in the centre of the image).

Five windows, 512 x 512 pixels (see Fig. 3) were extracted and analysed according to the simple procedure described in Wadhams et al. [20]. The observed SAR spectra for windows 2 and 5 are shown in Fig. 4 and reveal the presence of a small wave system with $\lambda > 250$ m. Very similar results were obtained when the same

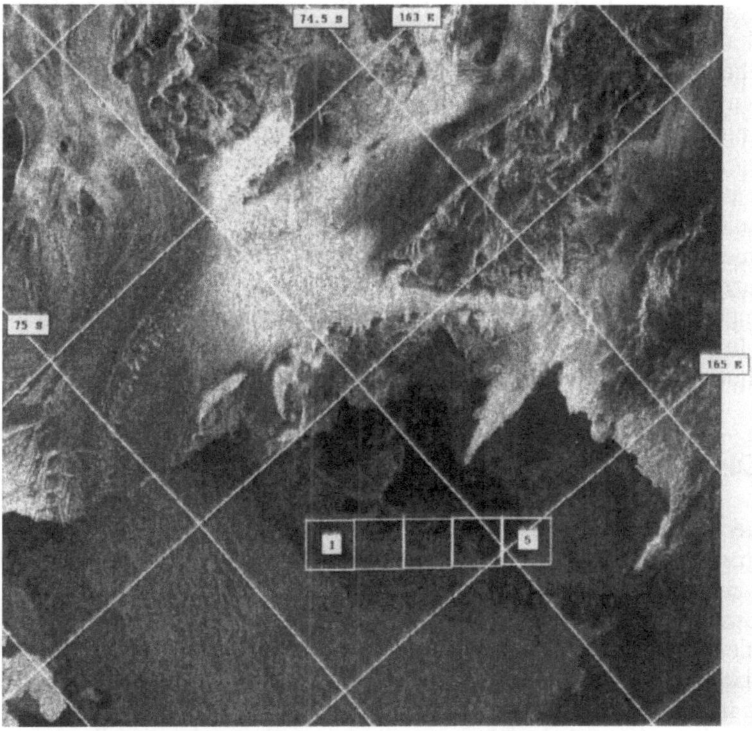

Fig. 3. ERS-2 SAR PRI scene of Terra Nova Bay of April 24, 1996 (orbit 5289, frame 5607), with frames of the five analysed windows in overlay

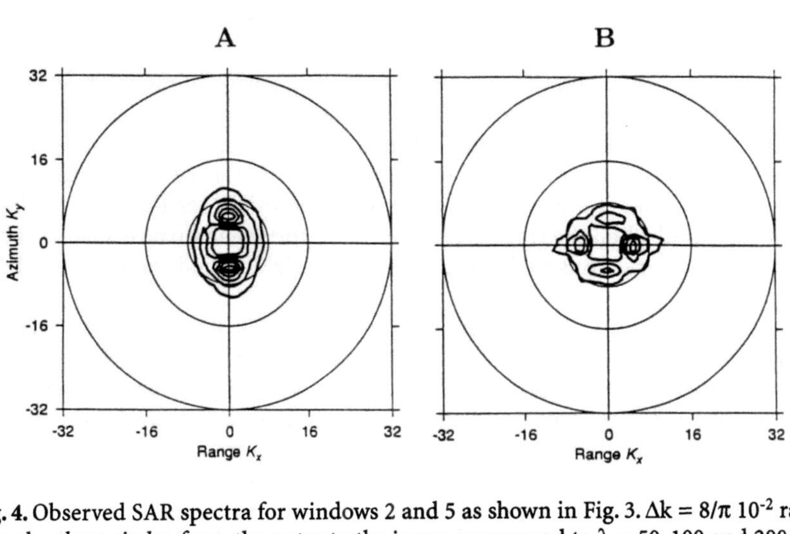

Fig. 4. Observed SAR spectra for windows 2 and 5 as shown in Fig. 3. $\Delta k = 8/\pi \; 10^{-2}$ rad/m; thus the *three circles*, from the outer to the inner, correspond to λ = 50, 100 and 200 m

analysis was applied to the other two SAR scenes of April 24 (orbit 5289, frames 5571 and 5589).

The very long wavelength observed in the spectrum is typical of a swell wave system for which the methodology to derive ice thickness developed in the preceding sections cannot be applied. The high wind speeds and effectively infinite fetches typical of the Southern Ocean generate much wave energy at long periods; at the same time, as we showed in Section 1.2, pack ice acts as a low pass filter, allowing only the longest waves to pass through it. Thus an open water environment separated from the main Southern Ocean by a wide belt of pack ice, as occurred at the time of these images, is a geometry which guarantees that only long period waves will be present, for which the dispersion relation is not sensitive to ice thickness. This technique is not applicable in such circumstances. We do, however, consider that it may be applicable to the outer ice edge.

6 Discussion

Inspection of cases where "impossibly large" changes in wave number occur on a simple analysis show that what is actually happening is that an azimuthal component is being suppressed in the open sea and is then appearing within the ice, a consequence of the azimuthal cut-off effect combined with lower energies inside the ice. This indicates the need to carry out the full inversion in such cases; the simple method can only be considered valid when the main wave direction both inside and outside the ice is close to the range direction.

The best controlled and clearest cases of changed wave dispersion in the Greenland Sea still give excessively high values for ice thickness. This agrees with the high results found by Wadhams and Holt [12]. It is clear from Greenland Sea results that the changed wave dispersion is indeed in the right direction (towards larger wave numbers), but the simple mass loading theory predicts ice thicknesses which are greater than those really occurring. Two possibilities present themselves. One is that the problem is just due to the failure to carry out a full inversion, i.e. that nonlinear effects may be moving the position of the main wave peak such as to exaggerate the real wave number change that is occurring. The other is that the theory is inadequate and requires modification. It is possible, for instance, that the sedimentation equilibrium of the frazil ice suspension in the water column gives it a greater effective dynamic thickness than the thickness that is measured when the frazil is scooped from the water surface. In situ underwater video observations of frazil ice in a wave field would be a way of determining this.

Winter conditions in the Greenland Sea or Southern Ocean may be unfavourable much of the time for the optimal use of this method. The reason is that wave energies are high (which is good for imaging the waves), but at the same time fetches are large and wind speeds are high and thus dominant wave periods are high. At high wave periods the change in wave number on entering an icefield is low, according to the mass loading theory. Ideally we need shorter waves - although not too short, because of the minimal resolution due to pixel size. For instance, in all 33 cases analysed in the Greenland Sea, the dominant wave number in the open sea lay within the range 0.02–0.06 m^{-1}. As Table 2 shows, for a realistic frazil-pancake ice thickness (e.g. 0.3 m chosen here) the relative change in wave number on entering the ice is only 0.5% for a 0.02 m^{-1} wave number and 1.6% at 0.06 m^{-1}. Such small changes are difficult to detect against natural and statistical variability.

Table 2. Wave number changes for a 0.3-m ice cover

k_w (m^{-1})	k_i (m^{-1})	T (s)
0.02	0.02011	14.2
0.03	0.03024	11.6
0.04	0.04044	10.0
0.05	0.05068	9.0
0.06	0.06099	8.2

Nevertheless, in the Antarctic Ocean there will be calm periods, even in winter, when the very wide swathe of frazil-pancake ice created by the "frazil-pancake cycle", during the expansion northward of the ice edge in turbulent wavy conditions, is temporarily subjected only to the short waves generated by a low wind speed. At these times, if they can be sampled by SAR, spectral analysis of the images should yield usable values for ice thickness. The global importance of the very large areas (possibly 6 000 000 km^2) occupied by this ice type in winter is such that an extensive examination of SAR imagery from the Antarctic marginal ice zone in winter is a worthwhile exercise.

Acknowledgements. The work described here was supported by the Commission of the European Communities under contract no. EV5V-CT94-0440 of the Environment Programme and by the Italian National Programme for Antarctic Research (PNRA).

References

1. Wadhams P, LangeMA, Ackley SF (1987) The ice thickness distribution across the Atlantic sector of the Antarctic Ocean in midwinter. J Geophys Res 92(C13):14535-14552
2. Lange MA, Ackley SF, Dieckmann GS, Eicken H, Wadhams P (1989) Development of sea ice in the Weddell Sea. Ann Glaciol 12:92-96
3. Martin S, Kauffman P (1981) A field and laboratory study of wave damping by grease ice. J Glaciol 27(96):283-313
4. Zwally HJ, Comiso JC, Parkinson CL, Campbell WJ, Carsey FD, Gloersen P (1983) Antarctic Sea Ice 1973-1976: satellite passive-microwave observations. NASA, Washington, Report SP-459
5. Wadhams P, Squire VA, Ewing JA, Pascal RW (1986) The effect of the marginal ice zone on the directional wave spectrum of the ocean. J Phys Oceanogr 6(2):358-376
6. Wadhams P, Squire VA, Goodman DJ, Cowan AM, Moore SC (1988) The attenuation rates of ocean waves in the marginal ice zone. J Geophys Res 93(C6):6799-6818
7. Wadhams P (1986) The seasonal ice zone. In: Untersteiner N (ed.) The geophysics of sea ice. Plenum Press, New York, 825-991

8. Peters AS (1950) The effect of a floating mat on water waves. Communs. Pure Appl. Math 3:319-354
9. Weitz M, Keller JB (1950) Reflection of water waves from floating ice in water of finite depth. Communs Pure Appl Math 3(3):305-318
10. Keller JB, Weitz M (1953) Reflection and transmission coefficients for waves entering or leaving an icefield. Communs Pure Appl Math 6(3):415-417
11. Shapiro A, Simpson LS (1953) The effect of a broken icefield on water waves. Trans Am Geophys U 34(1):36-42
12. Wadhams P, Holt B (1991) Waves in frazil and pancake ice and their detection in Seasat synthetic aperture radar imagery. J Geophys Res 96(C5):8835-8852
13. Newyear K, Martin S (1997) A comparison of theory and laboratory measurements of wave propagation and attenuation in grease ice. J Geophys Res 102(C11):25091-25099
14. Alpers WR, Ross DB, Rufenach CL (1981) On the detectability of ocean surface waves by real and synthetic aperture radar. J Geophys Res 86:6481-6498
15. Hasselmann K, Raney RK, Plant WJ, Alpers WR, Shuchman RA, Lyzenga DR, Rufenach CL, Tucker MJ (1985) Theory of synthetic aperture radar ocean imaging: a MARSEN view. J Geophys Res 90:4659-4686
16. Hasselmann K, Hasselmann S (1991) On the nonlinear mapping of an ocean wave spectrum into a synthetic aperture radar image spectrum and its inversion. J Geophys Res 96(C6):10,713-10,729
17. Hasselmann S, Brüning C, Hasselmann K, Heimbach P (1996) An improved algorithm for the retrieval of ocean wave spectra from synthetic aperture radar image spectra. J Geophys Res 101:16615-16629
18. Wadhams P, Comiso JC, Prussen E, Wells S, Brandon M, Aldworth E, Viehoff T, Allegrino R, Crane DR (1996) The development of the Odden ice tongue in the Greenland Sea during winter 1993 from remote sensing and field observations. J Geophys Res 101(C8):18,213-18,235
19. Wadhams P, De Carolis G, Parmiggiani F, Tadross M (1997) Wave dispersion by frazil-pancake ice from SAR imagery. Proc. IGARSS'97, Singapore, Aug. 1997, 41-43
20. Wadhams P, Parmiggiani F, Tadross M (1995) Wave spectra of SAR imagery of the Odden Ice Tongue. Proc. IGARSS'95, Florence, July 1995, I, 630-633

The Southern Ocean and the Antarctic Circumpolar Current

Observations of the Dynamics of the Antarctic Circumpolar Current in the Pacific Sector of the Southern Ocean

E. Zambianchi, G. Budillon, P. Falco and G. Spezie

Istituto di Meteorologia e Oceanografia, Istituto Universitario Navale, Via Acton 38, 80133 Napoli, Italy

Abstract

This paper describes long term measurements of the Antarctic Circumpolar Current carried out by means of surface and sub-surface drifters deployed as of the austral summer 1993-94, associated with XBT sections. At the surface, the velocity presents a gross zonal flow pattern, directed eastwards. The field is characterized by strong meridional shear and by a complex multiple jet structure, suggested in particular by the observations in the vicinity of the Antarctic-Pacific Ridge. The interaction with the bottom topography is seen to play a major role in shaping both the mean and the eddy fields, with a characteristic space lag which implies the presence of eddy-generating mechanisms in correspondence of major topographic features. Our subsurface float observations denote a strong vertical coupling in the velocity field. Velocities around 1000 m depth resemble those at the surface, with a scale factor of about 1/4 in magnitude. This vertical coherence is consistent with previous hydrographic analyses and with preliminary potential vorticity estimates drawn from our surface drifters data.

1 Introduction

The Southern Ocean, and its largest circulation feature, the Antarctic Circumpolar Current (ACC), play a fundamental role in the mass and heat budget of the world ocean. For this reason, in the framework of the Italian National Programme for Antarctic Research (PNRA), a specific project has been conceived, entitled CLIMA (Climatic Long-term Interaction for the Mass balance in Antarctica), whose focus is the definition of the main aspects of the circulation of the Ross Sea and, at a larger scale, of the ACC in the western portion of the Pacific sector of the Southern Ocean (PSSO hereafter) of which the Ross Sea constitutes a marginal basin.

In particular, research activities in the project have focused, as of 1993, on the following topics: hydrology and dynamics of the Southern Ocean (investigated both by observations and by numerical simulations); air-sea interactions and role of the meteoclimatic forcing; and fluxes of suspended matter [1].

In the framework of the CLIMA project, a Lagrangian investigation of the surface and subsurface dynamics of the ACC has been undertaken, along with an attempt to periodically characterize the hydrological structure of the current by the use of expendable Eulerian instruments. Drifter and XBT deployments have been regularly carried out on the ship route from New Zealand to Terra Nova Bay, or on the way back, one (drifters) or more times (XBTs) a year. In addition, subsurface floats were released during the expedition of the austral summer 1993-1994.

The Southern Ocean is the only region in the world ocean where the driving action of the wind on the surface waters is essentially unconstrained by the presence of continental margins [2, 3]. The direct action of the westerlies, in combination with thermohaline effects not completely understood as yet [4, 5], gives origin to the ACC, an extremely massive eastward zonal current limited in the north by the Subtropical Convergence [2, 3], or Front [6], and whose southern limit has been recently identified and denominated as the ACC Front [7]. The classical description of the surface circulation around Antarctica is that given by geostrophic analyses carried out on historical hydrographic data [8, 9]. More recently this picture has been updated taking into consideration additional hydrographic stations available up to 1990 [7].

The eastward flow associated with the ACC is characterized by relatively slow speeds at the surface (of the order of 15 cm/s, see e.g. [10]) but very strong transport (a typical value of 100 Sv above 3000 m is estimated in [7], in agreement with previous calculations; for a review of zonal transport estimates proposed in the past see [4, 11] and for a discussion of their variability range [10]).

Gordon et al. [8] show that the choice of a reference level at 1000 db yields a meaningful pattern for the surface geostrophic flow, but the eastward flow is seen to reach depths of the order of 3000 m: apart from the above hydrographic data analyses, at intermediate and deep levels the velocity field in the PSSO has been extensively investigated essentially just in the area of the Drake Passage, e.g. in the framework of the ISOS (International Southern Ocean Studies) experiment (see the reviews in [4, 11]), with particular emphasis on transport estimates. The jet structure of the ACC (discussed in Sec. 2 below) is seen to extend almost to the bottom [11]; a strong coupling in the mid-depth range has been observed in current meter data which show correlation from the surface down to depths of 2000 m and more [12].

The first aspect of the dynamics of the PSSO studied in this work is the surface circulation in correspondence of the ACC, investigated by means of surface drifter data. This is done in Section 2. In Section 3 the flow at a depth of around 1000 m is evaluated using subsurface float data. In Section 4, XBT data collected along a section approximately coinciding with the P14S WOCE section are presented; an estimate of the velocity field transverse to the section based on the geostrophic method is first given using the maximum measurement depth (750 m) as the reference level. This is later tentatively refined by forcing the flow to comply with the velocity estimates derived from the surface Lagrangian data. Finally, summary and conclusions can be found in Section 5.

2 The Lagrangian Picture of the Surface Flow

The surface drifters deployed over the years in the ACC by the Istituto Universitario Navale in the framework of the PNRA are WOCE-TOGA holey-sock standard instruments [13]. Theirs is now a standard design; their water-following characteristics minimizing any direct wind drag effects have been assessed [14, 15]. Besides the above reason, they were selected because we wanted to be able to make these data available to the broadest potentially interested scientific community, and they are in fact pooled in the WOCE-TOGA Global Drifter Center at the NOAA Atlantic Oceanographic and Meteorological Laboratory in Miami. The instruments have been put to sea on the ship route from New Zealand to the Italian base on Terra Nova Bay or vice versa.

We have been trying to keep a relatively constant deployment schedule over the years; variations of the number of deployed instruments depend on year-to-year funding and also on the fact that we have experienced a few instrument failures just after launch. The data utilized in this work are relative to 11 successfully deployed surface drifters, all manufactured by Technocean; six were deployed in March 1994, three in January 1995 and two in January 1996.

WOCE-TOGA standard drifters are localized by the Argos satellite system with a precision of about 300 m, at uneven time intervals. Typically, we collected a number of five or six fixes a day. The first operation to be done, after editing positions as well as surface temperatures in order to get rid of spurious data (this was carried out with a routine based on a simplification of the technique proposed in [16]), was a linear interpolation in time of the location data so as to have regular time series at 3-hour intervals. Thereafter, locations and velocities have been resampled every 24 hours by means of the application of a triangular weight filtering technique (M.S. Swenson, pers. comm.). Data presented in this work are relative to the period March 1994 through December 1996, even if the oceanography program in the framework of PNRA, including the Lagrangian investigation, is still ongoing and deployments have been carried out in January 1997 and 1998 and are planned for the austral summers of the next years up to 2001.

2.1 Trajectory Analysis

A visual analysis of the trajectories is sufficient to illustrate the main features of the surface dynamics of the ACC in the PSSO (for a general view of the physical oceanography of the region, see in addition to the above references [11]). A map of all the trajectories of the surface drifters examined in this work is shown in Fig. 1 (a sketch of the 3000- and 4000-m isobaths is shown in Fig. 2 along with the Lagrangian velocities, which will be discussed below). The observed overall Lagrangian flow pattern is in good agreement with the Eulerian one estimated from hydrographic data [7-9].

Our drifters sample extensively the variability band of the ACC, whose meridional limits, as redefined in [7] on the basis of the water mass characteristic zonation, approximately correspond to the southernmost and northernmost drifter trajectories over the whole PSSO. As expected, the overall flow pattern in that portion of the ocean is zonal, directed eastwards. As shown also by the results of the analysis of hydrographic data [7, 8], the band of the ACC in the PSSO is obviously affected by the presence of the Pacific-Antarctic Ridge (besides being possibly constrained by the presence of sea ice); its influence on the first portion of the tra-

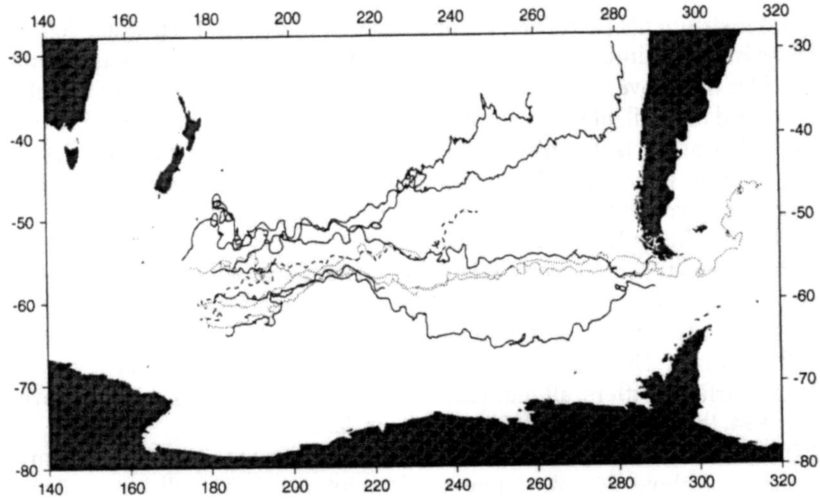

Fig. 1. Surface drifter trajectories in the PSSO, March 1994 through December 1996. *Solid lines* refer to drifters deployed in March 1994 and December 1996. *Dotted lines* to drifters deployed in January 1995; and *dashed lines* to drifters deployed in January 1996

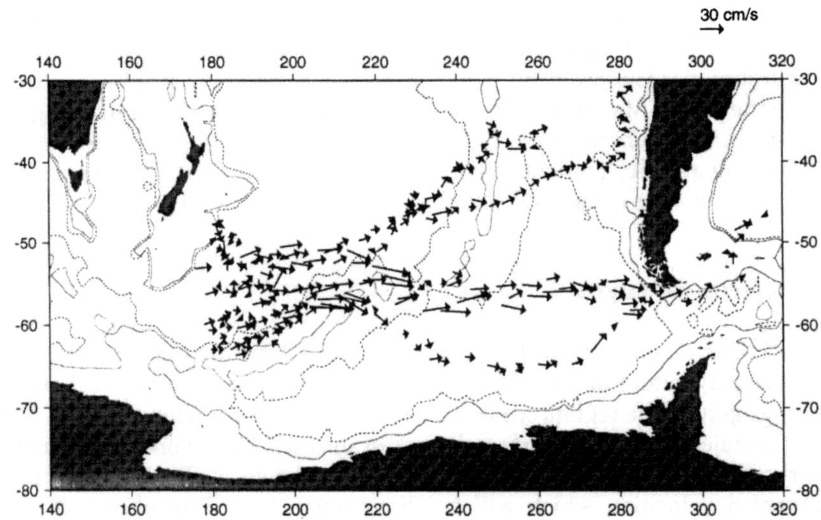

Fig. 2. Lagrangian velocities estimated from surface drifter observations and averaged over 20-day intervals. A very schematic bathymetry of the PSSO is superimposed: the *dashed line* represents the 4000-m isobath, and the *dotted line* the 3000-m isobath

jectories is seen as the surface drifters keep floating consistently north-west of the 3000-m isobath in the western half of the basin. Moreover, the trajectories show in several regions a weak meandering pattern with wavelenghts of the order of 50-100 km, as observed by several authors in various areas of the PSSO (see the summary on mesoscale variability in the area in [11]).

The eastward general flow also presents some smaller-scale recirculation zones, whose connection with the bottom topography will be explored in more detail later. The first encountered and most energetic of these areas is the one downstream of the Campbell Plateau, sampled by two surface drifters as well as by one subsurface float (see also Sec. 3). In this area the flow sampled by the drifters bends equatorward, weakens and recirculates downstream of the continental shelf. This complexity has been extensively discussed in [17], as being due to the possibility of ring formation at the Macquarie Ridge. In our case, however, the recirculations displayed in the trajectories show up well downstream with respect to the ridge. In the region west of 140°W, sea ice and bottom topography strongly confine the area spanned by the drifters. This displays a shrinking around 150°W, followed by a wide spreading of the Lagrangian paths thereafter.

The crossing of the Pacific-Antarctic Ridge takes place essentially in correspondence of the two major gaps of that topographic feature, the Eltanin and the Udintsev Fracture Zones, as happened with some of the FGGE drifters (First GARP (Global Atmospheric Research Pogram) Global Experiment) [11], with just two exceptions. In fact, east of 150°W our surface drifter set can be divided into two subsets, representative of very different flow regimes. While typically the drifters proceed following the main path of the ACC, two of them are caught in the southernmost branch of the Pacific subtropical gyre, the South Pacific Current [18], eventually entering the Peru-Chile current. The trajectories of these two instruments cross the relief farther north than the others, in the region of the East Pacific Rise, and their trajectories do not appear strongly affected by the Rise. This suggests a less strong barotropic character of the current in correspondence of the subtropical gyre [18]. A similar behaviour just occasionally occurred with FGGE drifters [6, 19], whereas hydrographic analyses do not seem to be able to resolve it [8]. It is worth noticing, however, that this is consistent with the localization of the Subtropical Front determined in [20] and refined in [7]: the front, typically located south of 40°S, is displaced equatorward beyond 30°S just in the eastern half of the South Pacific Ocean.

Finally, as to the flow in the eastern portion of the PSSO, the four drifters left in the basin show a spreading of the trajectories which is, again, consistent with the analysis in [7, 8]. The ACC is most likely constituted by individual jets coinciding with frontal positions [6], but in the eastern PSSO, i.e. in correspondence of and downstream of the Pacific-Antarctic Ridge, the continuity of the jet patterns is not demonstrated [11]. As will be seen in the discussion of the Lagrangian and pseudo-Eulerian estimated velocities, the four trajectories in that region indeed show a sensible meridional shear suggesting the presence of separated jets, but the data density is too little to draw any conclusion as to that.

Out of the four drifters approaching Drake Passage, possibly the most studied zone in the Southern Ocean (for a review see [4]), just one, in the period of observation, reaches and crosses it, and is eventually caught in the northward branch of the ACC entering the southern Atlantic and following the Malvinas Current; its path is characterized by the presence of recirculations and meanders, in particular in the Brazil/Malvinas confluence region (a general description of the flow in this area is given, e.g., in [21]).

2.2 The Velocity Field

The Lagrangian velocities obtained from interpolated drifter positions are shown in Fig. 2, averaged over intervals of 20 days, i.e. approximately 5 times the Lagrangian decorrelation time scale [22, 23]. The average surface Lagrangian velocities over the whole PSSO and the correspondent standard deviations derived by our data result (u,v)=(14±18, 1±17 cm/s). The limited number of available drifters does not allow for the identification of preferential pathways, as was possible with the FGGE data [6]; yet, the velocity field obviously shows that the ACC in the PSSO is characterized by strong meridional shear patterns and by the possible presence of multiple jets. This complexity of the flow field shows up particularly clearly in the western and central portion of the PSSO, whereas after crossing the Ridge a well defined jet is left, directed straight eastwards, in agreement with the geostrophic picture drawn in [8]. The question arises as to whether this can be an artifact due to the non-simultaneity of the Lagrangian observations, and if some degree of interannual variability is present. However, an analysis of the velocities measured by two selected pairs of drifters deployed respectively in March 1994 and in January 1995, namely the four buoys which have extensively sampled the band of the ACC up to the Drake Passage, does not show significant interannual variability in the overall field: namely, the average Lagrangian velocities computed over the trajectory pairs spanning the whole zonal extension of the PSSO are for the first drifter pair (u,v)=(14±17, 1±18 cm/s), and for the second (u,v)=(17±19, 1±17 cm/s).

Another way to look at the overall velocity field is shown in Fig. 3, where we present partial drifter trajectories corresponding to speeds, respectively, lower than 10 cm/s (Fig. 3a), between 10 and 30 cm/s (Fig. 3b), and higher than 30 cm/s (Fig. 3c). A comparison between the intermediate and larger speed pattern shows again both characteristics, namely the complexity in the west and how the ACC is clearly stronger than the South Pacific Current and the Peru-Chile Current in its southernmost branch.

The surface drifter data have been processed in order to obtain pseudo-Eulerian statistics over the whole region. The PSSO has been divided into bins measuring 5 degrees of longitude by 2.5 degrees of latitude, i.e. approximately square at these latitudes (for a critical analysis of this technique see [24] and references therein). Figure 4 shows the pseudo-Eulerian statistics as to the kinetic energy: the kinetic energy of the mean flow (MKE) is presented in Fig. 4a, the eddy kinetic energy (EKE) in Fig. 4b.

As suggested by the results of the analysis of FGGE data [25] and more recently of altimeter data [26], the PSSO is characterized by a relatively high ratio of MKE/EKE. The reason for this is probably partly to be found in the absence of a proper western boundary current in the basin, as is the case for the two other basins of the Southern Ocean: the Brazil/Malvinas Current in the Atlantic and the Agulhas Current in the Indian Ocean [18].

Our data confirm the steady character of the large-scale flow in the region sampled by our surface drifters: the MKE is typically larger than the EKE, with the exception of the areas in the lee of strong topographic features. This is particularly well verified at the crossing of the Pacific-Antarctic Ridge. There, the effect

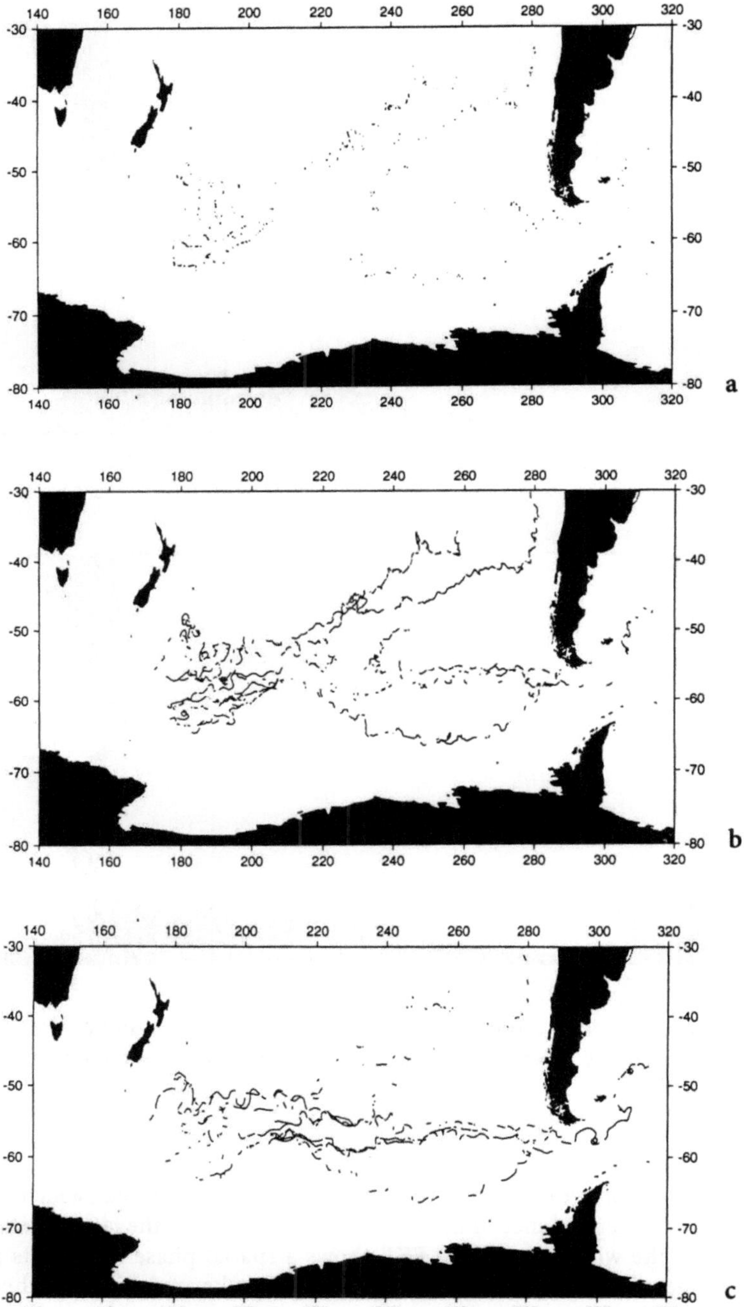

Fig. 3. Partial surface drifter trajectories characterized by speed (**a**) lower than 10 cm/s; (**b**) between 10 and 30 cm/s; (**c**) higher than 30 cm/s

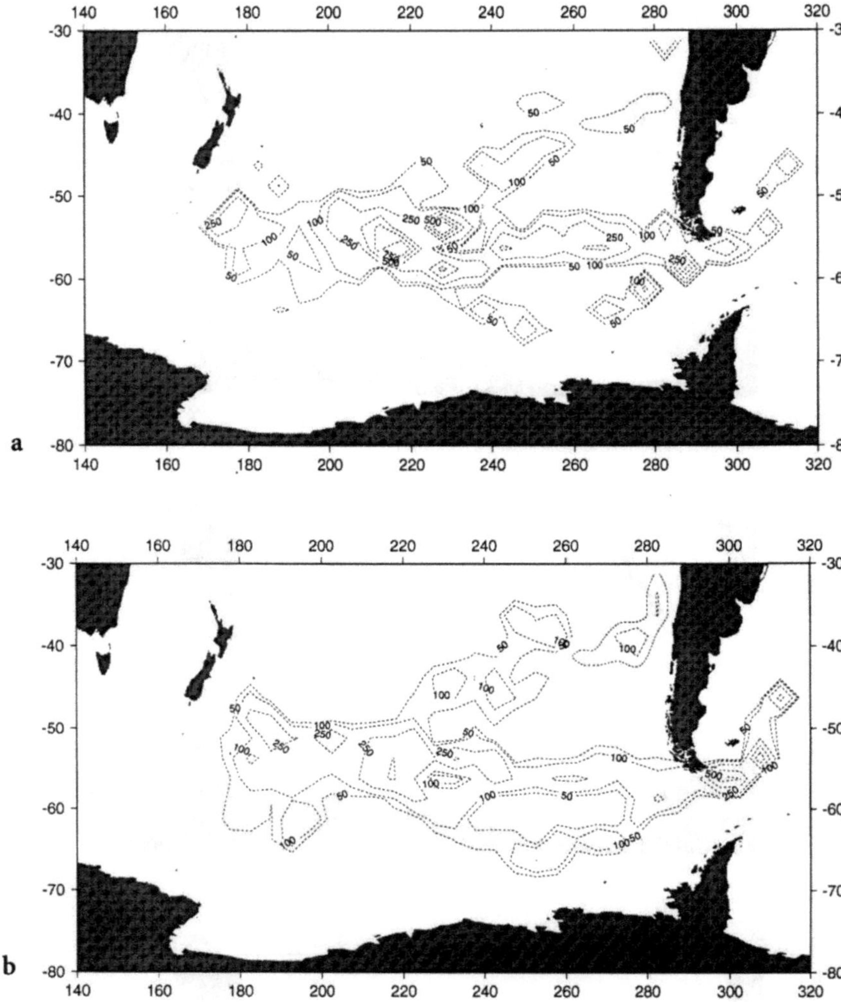

Fig. 4. Pseudo-Eulerian estimates of **a** kinetic energy of the mean flow (MKE); and **b** eddy kinetic energy (EKE), computed from surface drifter observations interpolated every 24 h

of the bottom topography is felt very evidently in the MKE field, which increases directly in correspondence of the reliefs, and there attains the largest values measured over the whole basin. The EKE shows a spatial phase shift of its maxima with respect to the maxima in the MKE, and its peaks are located further downstream. This is also true for the smaller recirculation regions downstream of the Campbell Plateau and of Drake Passage. On the contrary, as already mentioned, at the crossing of the East Pacific Rise at lower latitudes the field results much less affected by the presence of the ridge.

A thorough analysis of the importance of bottom topography in the dynamics of the ACC was carried out in [27]. From this investigation bathymetric features result particularly felt where they lie perpendicular to the flow, as in our case of the Ridge. In particular, we expect a Lagrangian analysis like ours to show stronger evidence of areas characterized by eddy downstream propagation, i.e. in the lee of topographic reliefs. The strong influence of a bottom topography whose main features in the PSSO are deeper than 2500 m on the surface velocity structures obviously suggests a strong degree of vertical coherence in the large-scale flow. This is also suggested by the results of the dynamic analyses carried out over the years on conventional and XBT hydrological data over the whole area and more specifically in the Drake Passage (see Sect. 3). Confirmation of this character of the flow regime is being sought by looking at the potential vorticity balance along the drifter trajectories, and preliminary results [28] suggest the tendency of the water column sampled by our drifters to change its relative vorticity in order to keep the ratio ζ/H constant in correspondence with the crossing of the Pacific-Antarctic Ridge, where the flow is essentially zonal and therefore changes in the planetary vorticity are negligible.

3 Lagrangian Observations at Intermediate Depth

In order to explore the vertical structure of the velocity field, five Autonomous LAgrangian Circulation Explorer (ALACE) floats were also deployed along with the first surface drifters release undertaken in 1994. Out of the five instruments released in March 1994, three of them started transmitting regularly after launch, whereas another one, for no understood reason, started working only 10 months after release. ALACEs [29] are Lagrangian subsurface instruments which float at a predetermined depth, where they are neutrally buoyant in virtue of the ballasting provided by the manufacturer. Periodically, they come to the surface where they are located by, and transmit data to, the Argos satellite system.

In our experiment the nominal depth was set at 900 m and the surfacing cycle at 30 days, in order to approximately comply with the standards used in other Lagrangian studies of the subsurface circulation in the Southern Ocean ([30] R.E. Davis and D.C. Webb, pers. comm.). The actual floating depths were within the range 950-1200 m. Figure 5a shows the trajectories followed by our ALACEs in the PSSO as long as all four floats were transmitting, i.e. until August 1996. There is an evident resemblance between the surface and the subsurface field, even though speeds are obviously quite different: the overall flow is zonal, even if a recirculation zone can be seen downstream of the Campbell Plateau, approximately in correspondence of the recirculation displayed by the surface drifters and discussed in Section 2. Recirculations are also evident in correspondence of the crossing of the ridge in the trajectories of the two ALACEs which get there.

In Fig. 5b we present the Lagrangian velocities computed from the float observations. Obviously, the calculated velocities only refer to the submerged portion of the trajectories. The resulting speeds are of the order of 1/4 of those at the surface: (u,v)=(4±5, 0.6±3 cm/s). Given the sparsity of our data, pseudo-Eulerian statistics are not meaningful in this case. A compact way to compare surface and

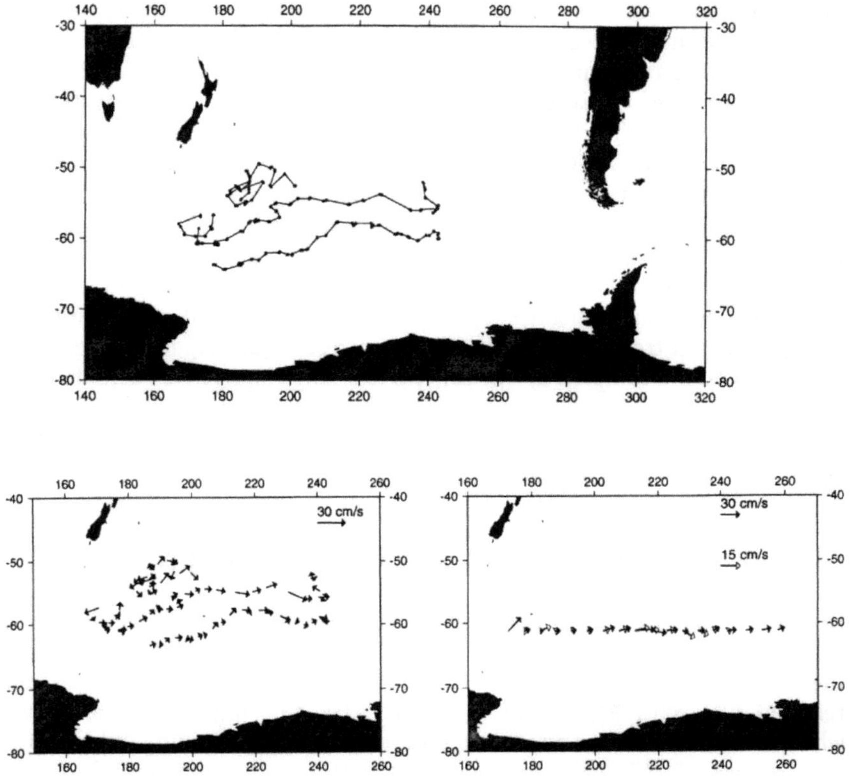

Fig. 5. a Subsurface float trajectories in the PSSO March 1994 through August 1996. *Dots* represent surfacing locations (every 30 days). **b** Lagrangian velocities estimated from subsurface float observations. **c** Surface and subsurface Lagrangian velocities averaged over meridional bins of 5 degrees

subsurface data can be that of meridional averages as shown in Fig. 5c, which again confirms the strong coupling between surface and subsurface fields.

To compare our findings with published data, it has to be kept in mind that the area in the Southern Ocean where the vertical structure of the velocity in correspondence of the ACC has been most extensively investigated is the Drake Passage ([12, 31, 32]; for a review, see [4]). In current meter data a significant correlation has been found down to depths of 2000 m and more [12]. This is also mirrored in the global analysis carried out in [8] on historical hydrographic data, as well as in [7]. In particular, Fig. 5a of [8] shows normalized geostrophic velocities relative to 3000 dbar for different station pairs in the southeastern PSSO, from which geostrophic velocities around 1000 m result of the order of a 0.2-0.4 fraction of those at the surface; this ratio is of the same order as our comparison between surface and subsurface Lagrangian measurements of the flow field.

4 An XBT Section Across the ACC

The above discussed strong vertical coupling between the surface and the subsurface flow suggests the importance of looking at the vertical structure of the mass and velocity fields across the ACC. As already mentioned, since the typical route followed by the Italian research vessel from New Zealand to Terra Nova Bay, where the Italian base is located, approximately coincides with the P14S WOCE hydrological section, an effort has been made over the years to sample the temperature field across that section as often as possible. In particular, during the cruises undertaken in the austral summers 1994-1995, 1995-1996 and 1996-1997,

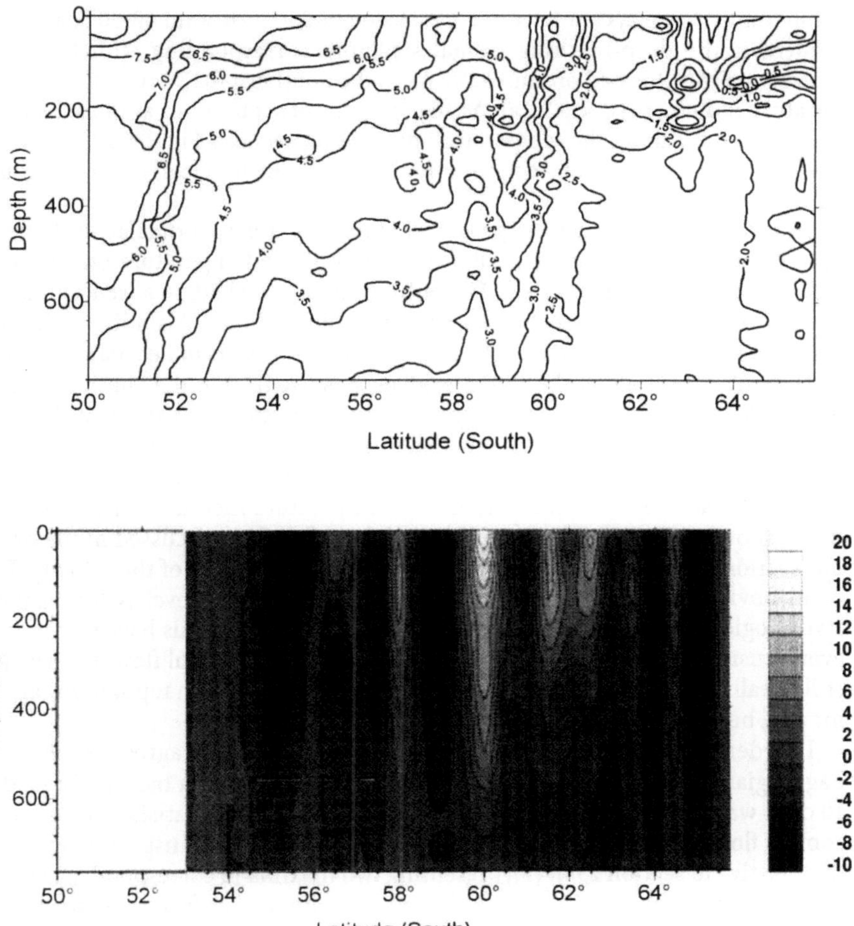

a b

Fig. 6. a Temperature section obtained from XBTs deployed along a portion of the P14S WOCE section between December 28, 1994, and January 1, 1995. **b** Eastward geostrophic flow estimated from temperature XBT data around 800 XBTs have been launched along the transect connecting Cape Adare and New Zealand, covered eight times. The most complete temperature section, gathered in February 1995 along the section, can be found in [1].

In order to get a picture as consistent as possible with our Lagrangian data, we examine here an earlier and, unfortunately, less complete section which was performed between December 28, 1994, and January 1, 1995, i.e. in coincidence with the January 1995 surface drifter deployment. The data collected by our Sippican T7 and T5 XBTs have been first of all edited in order to get rid of spikes and obviously spurious values, and then low-pass filtered. The resulting temperature section is shown in Fig. 6a, and presents very satisfactory evidence of the three fronts we expect to see across the ACC, which can be reconstructed on the basis of the definitions summarized in Table 3 of [7]: in particular, the most obvious feature in our section is the Polar Front, which can be traced in the subsurface minimum temperature tongue, characterized by a value of 2.0 °C around 60°30'S; also very evident is the deep-reaching frontal structure corresponding with the South Antarctic Front, located around 52°S. As to the Southern ACC Front, its two temperature indicators ($\theta < 0$ °C along the temperature minimum at the subsurface and $\theta > 1.8$ °C along the temperature maximum; see also Fig. 10 in [7]) do not perfectly coincide in our section, but the front can be clearly traced around 64°30'S.

In order to reconstruct the geostrophic flow field perpendicular to the transect, we needed to associate a salinity section to the temperature one, so as to infer a picture of the mass field. The only possibility we had, in absence of direct salinity measurements, was to resort to climatological data. A similar hydrological section, and precisely data collected by different vessels in February 1971 and in December 1971 along the 180° meridian, was taken from the Southern Ocean Atlas [9], checked so as to make sure that the correspondent thermal field was similar to that of 1994-1995, and digitized at 50-db depth intervals. The geostrophic field resulting from the mass field obtained by this procedure using as the reference level the maximum XBT temperature measurement depth (750 db) is shown in Fig. 6b. The complexity of the zonal flow discussed in the previous sections is shown quite evidently: we even have reversals of the velocity. This is an obvious artifact due to the fact that our reference level is too shallow. Hydrological investigations of the ACC by geostrophic analysis have shown that, even if using 1000 db as a reference level provides meaningful flow patterns [8], still a realistic reference level is more likely to be, when bottom topography allows for this, below 2500 m (see, again, [7, 8]).

In order to avoid this effect, and to get a closer quantitative agreement with the Lagrangian surface velocities, a straightforward addition of a barotropic field of 10 cm/s was tried. As to the surface flow, the results are quite satisfactory: the outcoming flow is much closer to the Lagrangian one, and the transport is similar to that relative to section 23 in [7], presenting two maxima in correspondence of the Polar (stronger) and ACC (weaker) Front and quite weaker flow in between – no salinity data were available for the latitudes of the South Antarctic Front. However, this is obviously a very crude approximation, since it implies a constant 10 cm/s velocity below 750 m; as was seen in Section 3, at 1000 m we still have eastward zonal speeds, but of the order of 4 cm/s.

5 Summary and Conclusions

The circulation in the PSSO, and in particular in correspondence of the Antarctic Circumpolar Current, has been investigated by means of surface drifters and subsurface floats. At the surface, the velocity presents a gross zonal flow pattern, directed eastwards. The field is characterized by strong meridional shear and by a complex multiple jet structure, suggested in particular by the observations in the vicinity of the Pacific-Antarctic Ridge. The interaction with the bottom topography is seen to play a major role in shaping both the mean and the eddy fields, with a characteristic space lag which implies the presence of eddy-generating mechanisms in correspondence of major topographic features. This is the case for the Pacific-Antarctic Ridge as well as for the Macquarie Ridge and for the area of the Drake Passage. Two surface drifters deployed in the ACC have been caught in the South Pacific Current, sampling a field which appears less influenced by the above effects.

Our subsurface float observations denote a strong vertical coupling in the velocity field. Velocities around 1000-m depth resemble those at the surface, with a scale factor of about 1/4 in magnitude. This vertical coherence is consistent with previous hydrographic analyses and with preliminary potential vorticity estimates drawn from our surface drifters.

An attempt has been made to connect surface and subsurface Lagrangian observations by means of evaluation of the vertical and horizontal distribution of the velocity from an XBT section along 175°E. However, the deepest reference level utilizable coincides with the maximum instrument measurement depth, i.e. only 750 m. The resulting flow picture is obviously affected by this: it is qualitatively analogous to the corresponding section presented in the literature, but the eastward velocities are systematically too low, being computed on the basis of a too shallow level of no motion. In the future it is planned to utilize deeper-reaching XBTs, so as to be able to correctly evaluate the total transport along the section, estimating the depth of the reference level from the indications coming from the Lagrangian measurements.

Acknowledgements. This work was funded by the Italian National Programme for Antarctic Research (PNRA: Programma Nazionale di Ricerche in Antartide). The authors wish to thank Elio Paschini for useful discussions and Giannetta Fusco for her valuable help in the processing and in the preliminary analysis of XBT data. Figures 1-5 were produced utilizing the GMT System [33] .

References

1. CLIMA Project Group (1997) Climatic Long-term Interaction for the Mass balance in Antarctica. Int WOCE News 27:34-35
2. Deacon GER (1937) A general account of the hydrology of the Southern Ocean. Discovery Reps 15:1-124
3. Deacon GER (1963) The Southern Ocean. In: Hill MN (ed) The sea. Interscience Publishers, New York, vol 2, pp 281-296
4. Nowlin WD Jr, Klinck JM (1986) The physics of the Antarctic Circumpolar Current. Rev Geophys 24:469-491

5. Gordon AL (1988) Spatial and temporal variability within the Southern Ocean. In: Sahrhage D (ed) Antarctic Ocean and resources variability. Springer-Verlag, Berlin, pp 43-56
6. Hofmann EE (1985) The large-scale horizontal structure of the Antarctic Circumpolar Current from FGGE drifters. J Geophys Res 90:7087-7097
7. Orsi AH, Whitworth T III, Nowlin WD Jr (1995) On the meridional extent and fronts of the Antarctic Circumpolar Current. Deep-Sea Res 42:641-673
8. Gordon AL, Molinelli E, Baker T (1978) Large-scale relative dynamic topography of the Southern Ocean. J Geophys Res 83:3023-3032
9. Gordon AL, Molinelli E (1982) Southern Ocean Atlas. Columbia University Press, New York
10. Tomczak M, Godfrey JS (1994) Regional Oceanography. Pergamon Press, Oxford
11. Patterson SL, Whitworth T (1990) Physical oceanography. In: Glasby GP (ed) Antarctic sector of the Pacific. Elsevier, Amsterdam, pp 55-93
12. Sciremammano F Jr, Pillsbury RT, Nowlin WD Jr, Whitworth T III (1980) Spatial scales of temperature and flow in Drake Passage. J Geophys Res 85:4015-4028
13. Sybrandy AL, Niiler PP (1991) WOCE/TOGA Lagrangian drifter construction manual. WOCE Rep 63, Scripps Institution of Oceanography, La Jolla, California
14. Niiler PP, Davis RE, White HJ (1987) Water-following characteristics of a mixed layer drifter. Deep Sea Res 34:1867-1881
15. Niiler PP, Sybrandy AL, Bi K, Poulain P-M, Bitterman D (1995) Measurements of the water following capability of holey sock and TRISTAR drifters. Deep Sea Res 42:1951-1964
16. Hansen DV, Poulain P-M (1996) Processing of WOCE/TOGA drifter data. J Atmos Oceanic Technol 13:900-909
17. Gordon AL (1972) On the interaction of the Antarctic Circumpolar Current and the Macquarie Ridge. Antarctic Res Ser 19:71-78
18. Stramma L, Peterson RG, Tomczak M (1995) The South Pacific Current. J Phys Oceanogr 25:77-91
19. Piola AR, Figueroa HA, Bianchi AA (1987) Some aspects of the surface circulation south of 20°S revealed by FGGE drifters. J Geophys Res 92:5101-5115
20. Deacon GER (1982) Physical and biological zonation in the Southern Ocean. Deep Sea Res 29:1-15
21. Olson DB, Podesta G, Evans R, Brown O (1988) Temporal variations in the separation of Brazil and Malvinas currents. Deep Sea Res 35:1971-1990
22. Grezio A (1995) A study of the Antarctic Circumpolar Current by Lagrangian measurements (in Italian). Degree thesis, Istituto Universitario Navale, Napoli
23. Spezie G, Zambianchi E (1996) Lagrangian measurements in the Antarctic Circumpolar Current: preliminary results (in Italian). 11 AIOL Congress Proc 701-711
24. Bauer S, Swenson MS, Griffa A, Mariano A, Owens K (1997) Eddy-mean flow decomposition and eddy-diffusivity estimates in the tropical Pacific Ocean. J Geophys Res (in press)
25. Patterson SL (1985) Surface circulation and kinetic energy distribution in the Southern hemisphere oceans from FGGE buoys. J Phys Oceanogr 15:865-884
26. Morrow R, Coleman R, Church J, Chelton D (1994) Surface eddy momentum flux and velocity variances in the Southern Ocean from GEOSAT altimetry. J Phys Oceanogr 24:2050-2061
27. Lutjeharms JRE, Baker DJ Jr (1980) A statistical analysis of the meso-scale dynamics of the Southern Ocean. Deep Sea Res 27:145-159
28. Zambianchi E, Spezie G (1997) Lagrangian observations of shear and vorticity in the Antarctic Circumpolar Current. In preparation
29. Davis RE, Webb DC, Regier LA, Dufour J (1992) The Autonomous LAgrangian Circulation Explorer (ALACE). J Atmos Oceanic Technol 9:264-285
30. Davis RE, Killworth PD, Blundell JR (1996) Comparison of Autonomous LAgrangian Circulation Explorer and Fine Resolution Antarctic Model results in the South Atlantic. J Geophys Res 101:855-884
31. Nowlin WD Jr, Whitworth T III, Pillsbury RT (1977) Structure and transport of the Antarctic Circumpolar Current at Drake Passage from short-term measurements. J Phys Oceanogr 7:788-802
32. Treshnikov AF, Pillsbury RT, Nowlin WD Jr, Sarukhanian EI, Smirnov NP (1977) A comparison of summer current measurements in the Drake Passage. J Phys Oceanogr 7:610-614
33. Wessel P, Smith WHF (1991).Free software helps map and display data. EOS Trans AGU 72:441

Altimeter Data Analysis of the Antarctic Circumpolar Current

A. Simone[1], S. Zoffoli[1], D. Iudicone[1], R. Santoleri[1] and S. Marullo[2]

[1]CNR, Istituto di Fisica dell'Atmosfera, 00144 Rome, Italy
[2]ENEA, Casaccia S. Maria di Galeria, 00060 Rome, Italy

Abstract

The first two years of TOPEX/POSEIDON altimeter measurements of the Southern Ocean, between 30° S and 65° S were analyzed in order two study the mesoscale characteristics of the basin. Objective maps of sea level anomalies were computed for each cycle, showing synoptic view of the mesoscale activity of the study area. These maps underlined a population of mesoscale eddies, with a typical diameter of 100 ÷ 200 Km in the sector 10° ÷ 50° E and 30° ÷ 60° W, possibly shed from the meanders of the Antarctic Circumpolar Current. The rings seem to move southward confined to a rather narrow corridor (about 400Km). Sea level variability maps were computed to characterize the mesoscale activity in the study area. The highest values of more 25 cm are found between 10° ÷ 60° E and near 50° W at a latitude of 40° ÷ 50° S. A high variability area extends to at least 60° ÷ 65° S. The TOPEX/POSEIDON cross-over data were used to resolve both magnitude and direction of residual geostrophic velocities. The time series are then used to determine surface eddy statistics in the Southern Ocean, hence the spatial distribution in the surface Reynolds stresses (u'^2, v'^2, $u'v'$). According with the information available so far, we found that the complex spatial distribution of surface eddy momentum flux is strongly influenced by bottom topography and the position of the mean current.

1 Introduction

Earth-observing satellite is providing a significant quantity of high quality synoptic data for the investigation of the oceanographic phenomena on spatial and temporal scales unattainable with in situ measurements. Satellite altimeter provides a direct measurement of the sea surface slope associated with the geostrophic current variability since the component of the absolute slope, due to geoid and mean geostrophic current contribution, is subtracted. Altimeter data have already proved to be able to resolve sea surface variability in regions corresponding to the major western boundary currents (the Gulf Stream, Kuroshio, Malvinas and Agulhas) and the Antarctic Circumpolar Current.

The launch of TOPEX/POSEIDON (T/P in the following) satellite ushered in a new era in satellite oceanography by greatly improving the accuracy of the altimeter measurements for each source of errors. The orbits for T/P are estimated to have a radial accuracy of 3-4 cm rms [1]. For this reason T/P is the most accurate altimeter satellite flown up to date. In this paper we analyzed the T/P dataset relative to the period October 1992-September 1994 in order to study the mesoscale variability of the Southern Ocean.

The Antarctic Circumpolar Current (ACC) is a strong eastward flow associated with a region of transition between surface water masses with antarctic and subantarctic characteristics [2]. The position of the flow is seen to vary considerably in latitude, covering a band between 40° and 60°S all around Antarctica. The ACC is the strongest current in the World Ocean. The estimated transport in the Drake Passage, where there are no barriers to the zonal flow, is about 130 Sv (1 Sv=10^6 m^3/s). In situ measurements and numerical models show that the strong persistent winds of the polar cyclone provide the main surface wind stress which drives the ACC. Nevertheless, the role of the coupling and the relative importance of the wind and thermohaline forcing has not been understood yet.

An important characteristic of the ACC is the concentration of the transport in narrow jets associated with frontal zones. This "zonation" of the ACC is clearly visible along the cores of the current, with pronounced gradients of density and other characteristics such as temperature (T), salinity (S) and nutrients [2]. Going poleward, we can identify three fronts separating three water mass zones (Fig. 1); the Subtropical Front (STF), the Subantarctic Zone (SAZ), the Subantarctic Front (SAF), the Polar Front Zone (PFZ), the Polar Front (PF), and the Antarctic Zone (AZ).

In the Drake Passage the fronts are rather narrow being about 50 km wide [3]; at the other longitudes, in open ocean, they tend to be more widely separated and about 100-150 km wide (i.e. 2-3 times the local deformation radius). The maximum geostrophic surface speeds in the ACC cores in Drake Passage are 25-45 cm/s.

Munk and Palmen [4] suggest that the main sink for momentum seems to be the topographic form stress. The current is for most of its path in near Sverdrup balance (southward drift); in some areas it is steered by topography (following contours of f/H where f is the coriolis parameter and H the depth) and in most cases (e.g. the South America shelf) the current flows northward, moreover balancing the deep ocean Sverdrup southward drift.

Dissipation, mostly lateral, seems to play a role only in Drake Passage. Following Hughes and Killworth [5], the bottom pressure torque, considered as a second order process, is concentrated at the points where the current leaves the isobaths toward the deep basins, mainly the eastern South America shelf, the Campbell plateau, and the Pacific-Antarctic Ridge. So most of the global dynamics is explained by localized processes. Almost the same conclusions have been reached by Wells and de Cuevas [6] for the ACC vertically averaged vorticity budget.

In this dynamic scenario, strong eddy field becomes necessary because it could represent the way in which the surface wind stress is transferred to the bottom, via the action of the interfacial form stress [7]. The nature of the eddy-mean flow interaction is still unclear. QG eddy-resolving experiments with zonal flows [7] or zonal-averaged balances from the Fine Resolution Antarctic Model (FRAM) [8] found that transient eddies accelerate the flow. However, again from FRAM, if the balances are carried along the mean path of the current, the net effect is a drag-

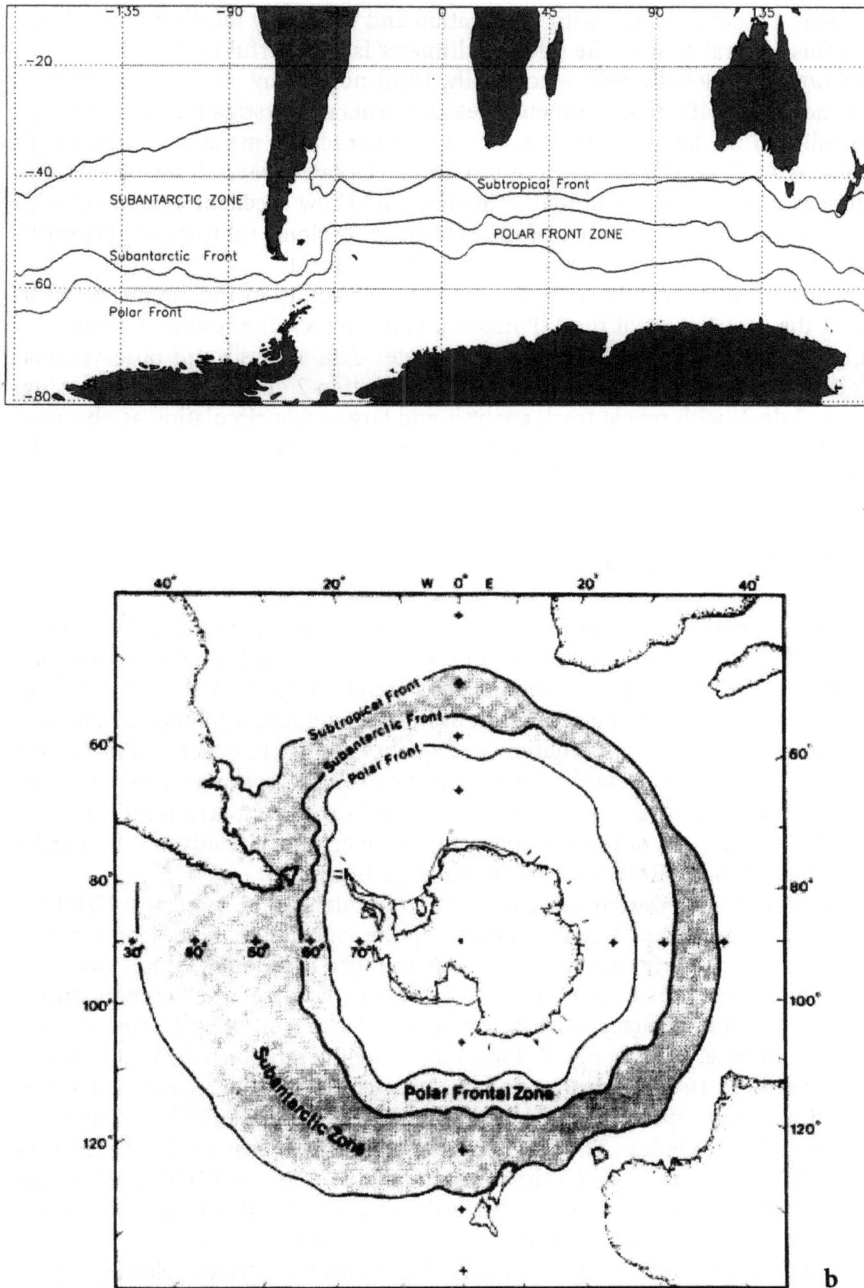

Fig. 1. Surface regimes of the Southern Ocean (adapted from [2])

ging [9]. In general, the role of the eddies, apart from the vertical transfer of momentum, seems to vary with the location and to be only locally relevant [10].

In this general context the satellite altimeter is a powerful tool to observe sea level variation globally and synoptically. Until now many studies [11-14] have taken advantage of Geosat altimeter measurements to investigate sea surface high variability in the Southern Ocean and to reconstruct the mean sea surface high across the ACC. T/P data in the Southern Indian Ocean were analyzed by Park and Gamberoni [15]; they found that altimetric data show excellent agreement with the numerical model solution and indicate an anticyclonic subtropical gyre north of the ACC and two cyclonic gyres south of this current.

In this work we will examine the sea level variability of the Southern Ocean during the first 2 years of the T/P mission and compare the resulting pictures of the surface circulation with previous altimeter data analysis and observational evidences. The paper is organized as follows. Section 2 describes the processing. Section 3 deals with repeat track analysis and large-scale circulation as observed by T/P. Cross-over analysis and Reynolds stress are discussed in Section 4. The results are summarized in the Section 5.

2 T/P Data Processing

Sea level anomaly (SLA) files, i.e., sea level measurements by the T/P altimeter, were used to study the temporal variability of the ACC. Such a data set, provided by the AVISO (Archivage Validation et Interprétation des Données des Satellites Océanographiques) [16] on CD-ROM, has been produced applying all the geophysical corrections, data validation, and quality control in order to allow direct use of the data by the scientific community. No orbit correction was applied since tests based on data fits, covariance analysis, and orbit comparison indicate that the radial component of the T/P spacecraft is determined, relative to the Earth's mass center, with an RMS error in the range of 3-4 cm [1].

The altimetric measurements made at two frequencies (5.3 and 13.6 GHz) are combined to minimize the errors caused by the ionospheric free electrons. The measurements at 10 Hz data rate have been averaged to obtain a sample each second (about 6 km in distance along a track), i.e., at 1 Hz data rate. This adjustment has improved the altimeter noise figure by 20%. Since temporal and spatial resolution are inversely proportional (for a single satellite mission), the best choice is a compromise: a 10-day repeat period has been chosen, which results in an equatorial cross-track separation of 316 km (the average is about 200 km). The exact orbit altitude is 1336 km. The RMS accuracy of a single pass sea level measurement is 4.7 cm for the TOPEX altimeter system and 5.1 for the POSEIDON system [17]. The following data processing is based on 2 years of T/P radar altimeter data, from the start of a 10-day repeat period cycle, i.e., October 1992.

We analyzed the first 74 cycles of SLA (254 tracks each cycle), selecting a geographic window of the Southern Ocean between 30° and 65°S. These data cover the period October 1992-October 1994. We studied first of all the spatial autocorrelation functions of the tracks. Thus fitting the mean autocorrelation function with a simple exponential we got an *e-folding distance* of about 100 km (Fig. 2).

This parameter is an important statistical estimate of the typical horizontal scale of the mesoscale structures under investigation. We used this parameter in applying an objective analysis scheme [18] to the dataset, being our main goal to optimally interpolate horizontally an irregular two-dimensional array of data points in order to construct synoptic maps of the mesoscale eddy field connected with the ACC. SLA maps relative to each cycle were then reconstructed by binning the data on a 0.5 x 0.5 degree grid and then applying objective analysis algorithm.

Finally, the SLA was used to compute sea level variability (SLV) along each track. SLV maps of the entire Southern Ocean were then obtained using the same procedure applied to produce SLA maps.

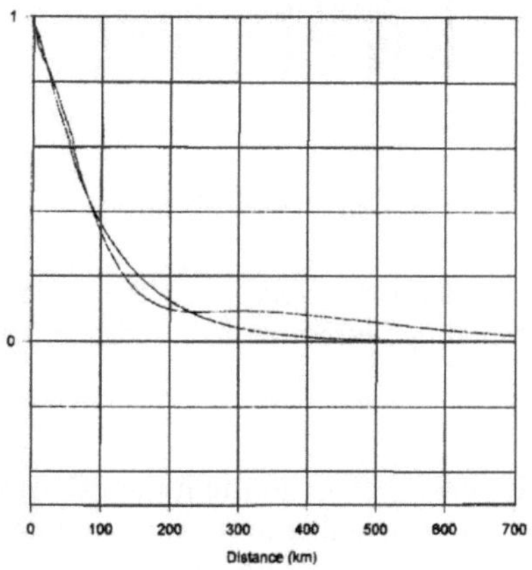

Fig. 2. Mean autocorrelation function with superimposed the exponential fit

3 Variability of Ocean Topography

The sea level variability (SLV) is a statistical measure of temporal variation of the main circulation system. It mainly depends on the mesoscale eddy activity, such as meandering of the currents, speed variations and eddy shedding from the currents cores. Figure 3 shows the SLV map computed from the cycles 3-76 of T/P. Further, we superimposed the digitalized fronts of the ACC (PF, SAF, and STF) [2] and found it impressive how, in many zones of the Southern Ocean, they bound the highest variability area.

The three western boundary currents in the subtropical gyres (the Agulhas, the Malvinas and the East Australian Currents) are the source of most of the variability in the Southern Ocean. These currents are characterized by the existence of a distinct return flow and all three systems interact with the ACC. The region with the highest variability (more than 25 cm) is confined within a latitude band of about 10-40°S in correspondence with three major western boundary currents and the ACC. Most of the high values of the SLV (as for the stresses, as will be shown below) are also in deep basins. The other parts of the Southern Ocean are dominated by relatively low values of SLV.

High values of SLV can be found in the area from 20 to 70°E, linked to the presence of the Agulhas Current and due to the wind-driven anticyclonic cell of the Southern Indian Ocean. This current, after leaving the western boundary and

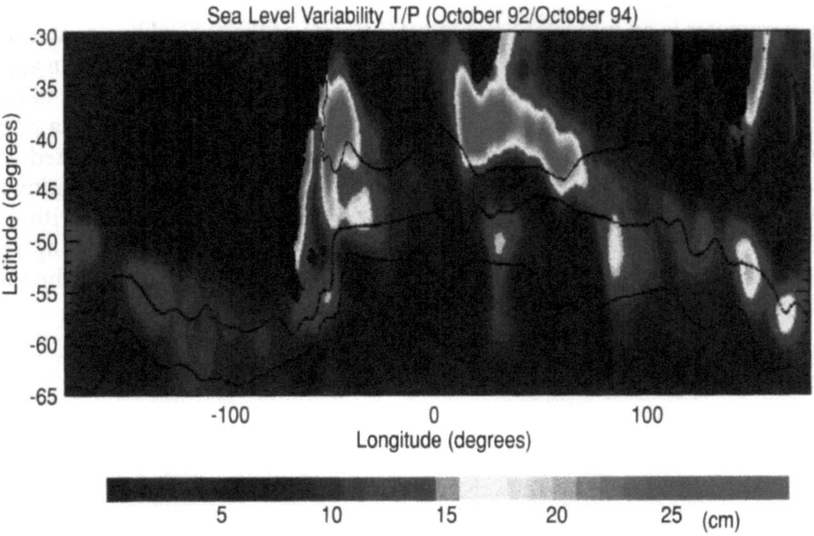

Fig. 3. Sea Level Variability (SLV) map from Topex/Poseidon relative to the period November 1992-October 1994; continuous lines superimposed to the map indicate the mean position of the main fronts

becoming a free jet, makes a turn of about 180° flowing back to the Indian Ocean (Agulhas Retroflection). The return current is more unstable and nearly regular large eddy shedding is frequently observed, mainly pinched off by interaction with the source current. Variability, on smaller scales, is also due to the baroclinic instability of the southern front of this system. The resulting zonal front is contiguous to the ACC front and eddies are observed to move along this frontal system [19]. Moreover, moving eastward around 52°S we found another high variability region (less than the former) just east of the Kerguelen Plateau. South of the Agulhas Current, a local maximum of SLV is located immediately downstream of a topographic gap in the Southwestern Ridge.

High values of SLV can be observed along the eastern coast of Australia. Here the East Australian Current flows southward along the continental slope until 33°S where it detaches from the coast, flowing toward New Zealand and forming with its southmost boundary the Tasmanian Front. This current system is often observed to produce large eddies, mainly warm, subject to subsequent interactions with the current itself and, different to other eddy sources in the area, without moving away from this region. The formation of eddies is clearly visible in the T/P SLA maps (see below).

East of 75°E and Drake Passage, at a latitude south of 45°S, the SLV higher values are along the SPF with the PF acting as a southern boundary. A clear maximum is between 150°W and 120°W, at 57°S, where the main flow crosses the Pacific-Antarctic Ridge and much of the variability is downstream of the Ridge. East of Drake Passage the variability is partly distributed along the bottom contours and is mostly in the Malvinas confluence. This could be due to the fact that the Malvinas Current and the Brazil Current converge, forming a strong thermal front, localized between 38°S and 46°S. A high variability area extends to at least 65°S, where altimeter data are no longer reliable because of the presence of sea ice.

Figure 4 shows two maps of SLV computed for the 2 years (1992-1993 and 1993-1994) separately. The two pictures are very similar; both show the same areas of high and low variability. This could agree with the fact that the bottom topography plays an important role in the variability's development.

The maps of SLV show very good similarity with the map previously presented by Chelton et al. [11], based on Geosat data. This is consistent with the idea that

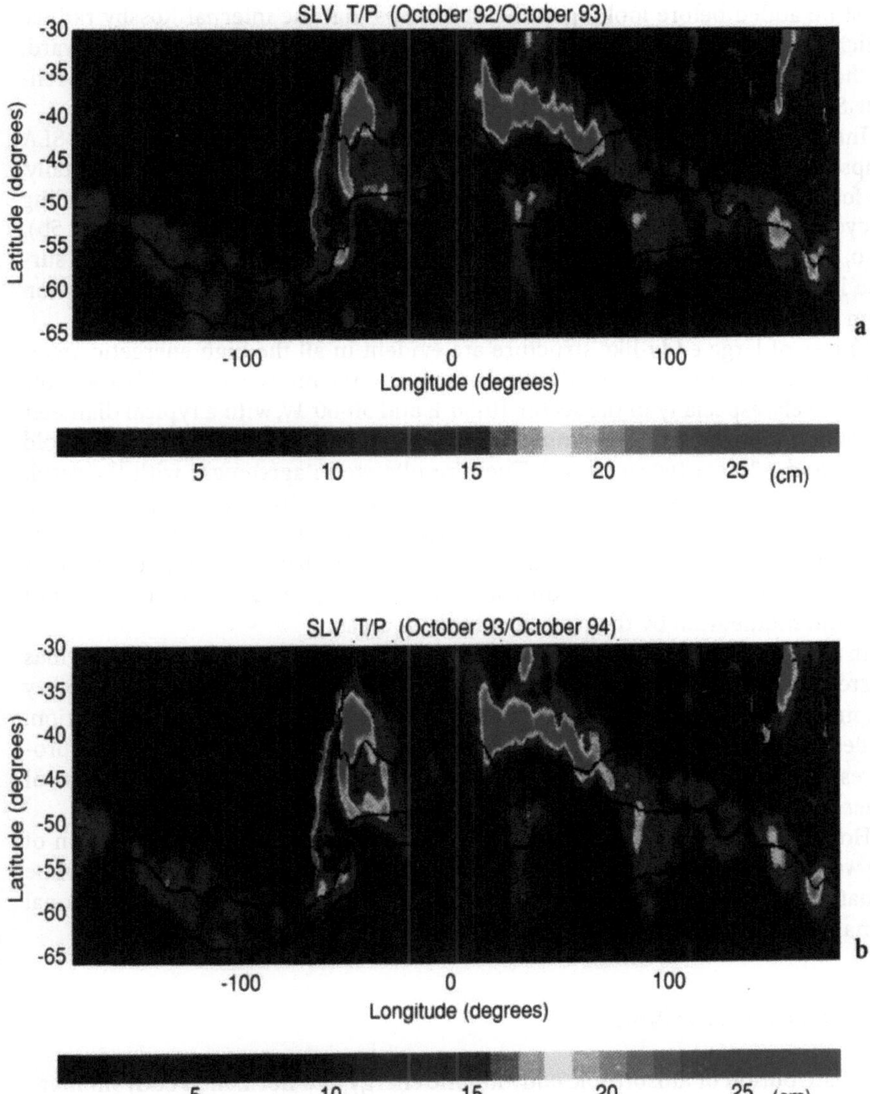

Fig. 4. As Figure 3 but **a)** SLV for the first year and **b)** for the second year

the ACC eddy variability is not affected by large seasonal and interannual variation, even if the main forcing of ACC is the wind stress. This is in agreement with the hypothesis that ACC is a partial topographical controlled. The geographic distribution of SLV, in which most of the high variability values are downstream of topographic features, confirms this hypothesis. The increased role of bottom topography would reduce also the role played by baroclinic instabilities in the formation of eddies released by the ACC [15].

Consequently, we expect to see the major eddy system in the proximity of these areas of high variability in the SLA maps. Nevertheless, a consideration must be added before looking at the SLA maps that the internal Rossby radius, which is supposed to be the preferred scale for eddies, decreases going southward. At the latitudes of Antarctica it is about 10 km, reducing the possibilities of a reliable SLA map reconstruction using actual altimeter data.

Interesting known features described above are easily recognizable in the SLA maps computed every 10 days using 74 cycles of T/P altimeter data. Incidentally, we found it useful to show in this paper only some of these maps, corresponding to cycles 12 (austral summer 1993, Fig. 5a) and 33 (austral winter 1993, Fig. 5b). Also, it is worthwhile underlining that from altimetric data, since the geoid surface is not adequately known yet, we can get only information about deviation from the mean sea surface level.

Traces of large eddy-like structure are evident in all the high energetic areas. The presence of warm and cold rings is shown on the maps white and black spots respectively, especially in the sector 10-50°E and 30-60°W, with a typical diameter of 150-200 km and sea level amplitude relative to the background <-30 cm (cold rings) and >30 cm (warm rings). These results are in agreement with the result obtained by Gouretsky and Danilov [19] using Geosat altimeter data.

T/P altimetry data indicate that in the South Indian Ocean, most of the warm-core rings remain detached from the ACC moving southward, confined to a rather narrow corridor of about 400 km wide. This fact implies poleward transport of heat and momentum by the rings just in these areas.

In the southwest Atlantic Ocean eddies can be noted in the area of the Malvinas Current and Brazil Current confluence. Here annual and semiannual variability are mostly due to the meridional variation of the Malvinas Current penetration, while, on scales of about 150 days and 300 km, the Brazil Current oscillation produces rings, mainly with warm core. Indeed, these are scales larger than those of other southern areas.

However, analysis of the SLA maps reveals a quite clear seasonal variation of the warm-rings activities; indeed, it improves during the summer and it might be a matter for future investigations to understand the coupling with the regional climatological regime.

4 Crossover Analysis

The distribution of anisotropic eddy kinetic energy and horizontal eddy momentum flux has been difficult to determine from surface altimeter data. This is due to the problem of resolving north/east directional components of the surface velocities from the along-track altimeter data. The traditional approach in altimetric studies is to resolve the across-track component of geostrophic velocity from the along-track sea surface slope, and then assume that the flow is isotropic to estimate the eddy kinetic energy. The assumption of isotropic variability may not be valid, especially in regions of strong currents or close to topography.

In this paper, following Morrow et al. [10], we adopted an alternative technique. The components of the geostrophic velocity perpendicular to each track (ascend-

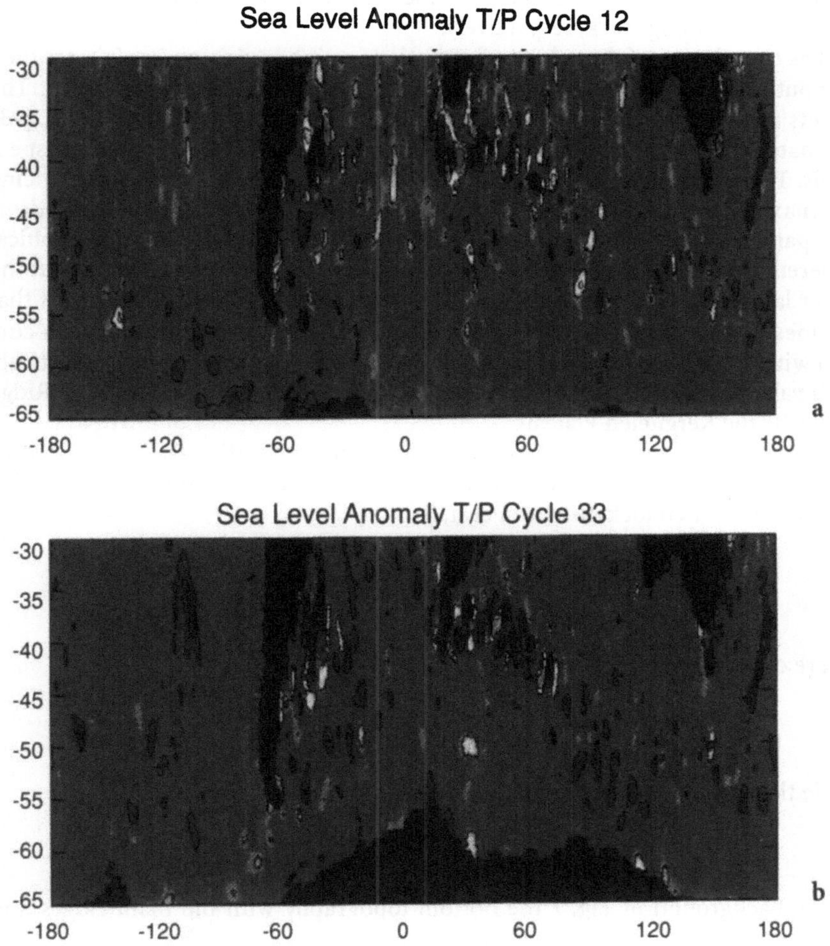

Fig. 5. Sea level anomaly maps from Topex/Poseidon. **a)** SLA for cycle 12 (January 1993); **b)** SLA for cycle 33 (August 1993)

ing and descending) were calculated at each cross-over point. Then, they were interpolated linearly to common time (every 5 days) and used to derive the zonal (u′) and meridional (u') components of the geostrophic velocity fluctuations:

$$u' = \frac{(-V_a' + V_d')}{2\cos\varphi} \qquad v' = \frac{(V_a' - V_d')}{2\sin\varphi},$$

where φ is the angle between ground track and the north meridian, and V_a and V_d are the components of the geostrophic velocity perpendicular to the ascending and descending track, respectively.

The covariances of the velocity fluctuations $<u'u'>$, $<v'v'>$ and $<u'v'>$ are then computed from the 2-year time series of u' and v' velocity components. The results are displayed in Fig. 6. It is important to underline that Johnson et al. [13] estimated that the decorrelation times from u′ and v′ are 31 and 14 days, respectively. This gives high reliability to results from analysis of T/P crossover, being the maximum delay between the two passes at the crossover less than 10 days. The patterns of $<u'u'>$, $<v'v'>$ and $<u'v'>$ (see Fig. 6) show a fair geographical coherence with scales of hundreds of kilometres. In several areas, mostly at the lower latitudes, $<v'v'>$ exceeds $<u'u'>$ as already noticed [10]. This means that the mesoscale eddies are strongly anisotropic. The interaction of the mean current with the bottom topography produces a complex pattern of $<u'v'>$. Relatively high values of $<u'v'>$ are alearly visible in the area of the Pacific-Antarctic Ridge and near the Kerguelen Plateau.

The magnitude and the direction of the eddy variability are represented in Fig. 7 by the variance ellipses. The direction of the axis of principal variability, measured counterclockwise from east, is:

$$\tan\vartheta = \frac{\sigma_{11} - <u'u'>}{<u'v'>},$$

where σ_{11} is the variance along the major axis and is given by

$$\sigma_{11} = \frac{1}{2}\ (<u'u'>+<v'v'>+\sqrt{(<u'u'>-<v'v'>)^2+4<u'v'>^2}$$

while the variance along the minor axis is

$$\sigma_{22} = (<u'u'>^2+<v'v'>^2) - \sigma_{11}.$$

In the background of Fig. 7 the bottom topography with the main ridges and shallow plateaus is shown using a gray scale.

An anisotropy flow is represented by elongated ellipses, with the principal direction of the velocity variance aligned with the direction of the major axis; ellipses with a major axis oriented toward the northeast quadrant have a positive $<u'v'>$, while ellipses with a major axis oriented toward the southeast quadrant have a negative $<u'v'>$.

In Fig. 7 distinct anisotropic variability is most evident in the high eddy regions of the Agulhas Retroflection and southwest of the Atlantic region.

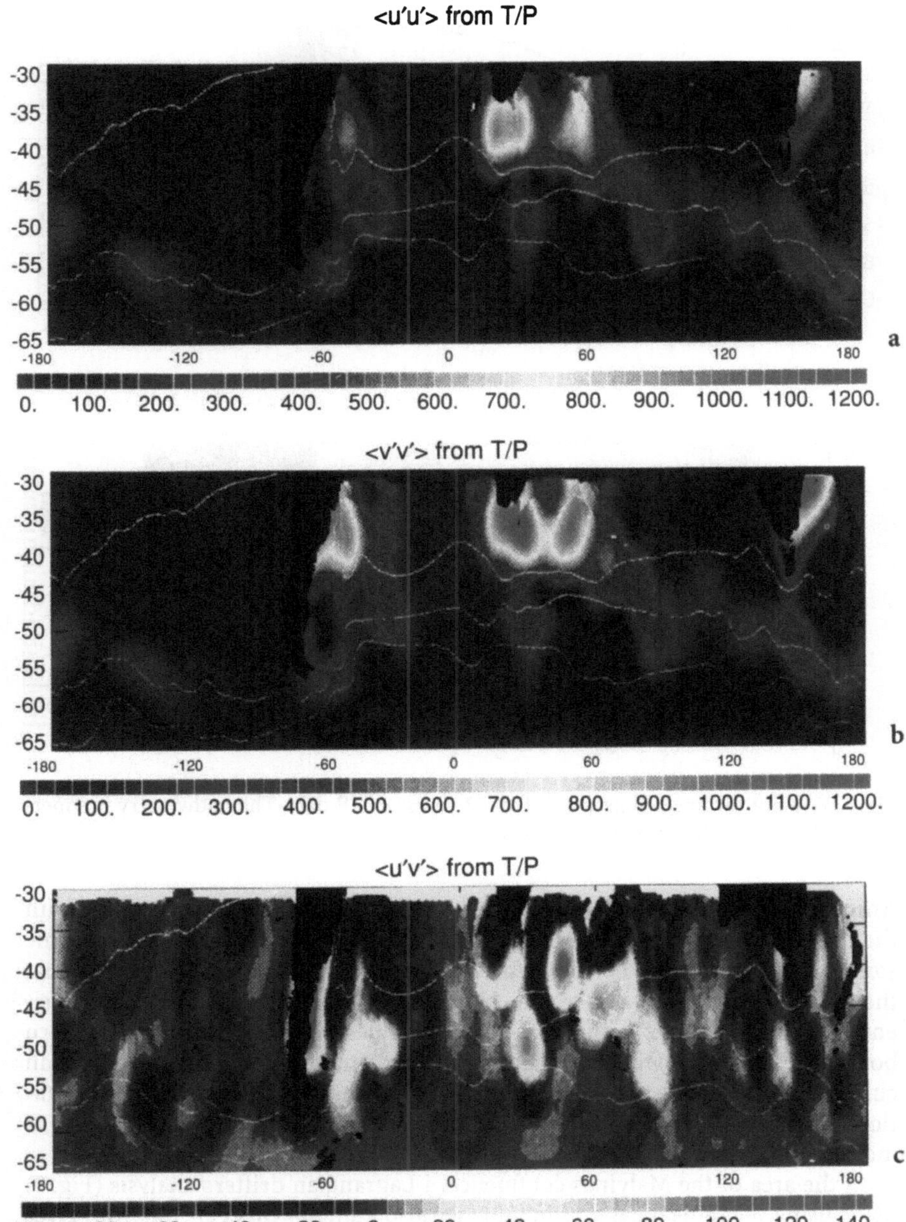

Fig. 6. Maps of Reynolds stresses **a)** <u'u'>, **b)** <v'v'> and **c)** <u'v'>

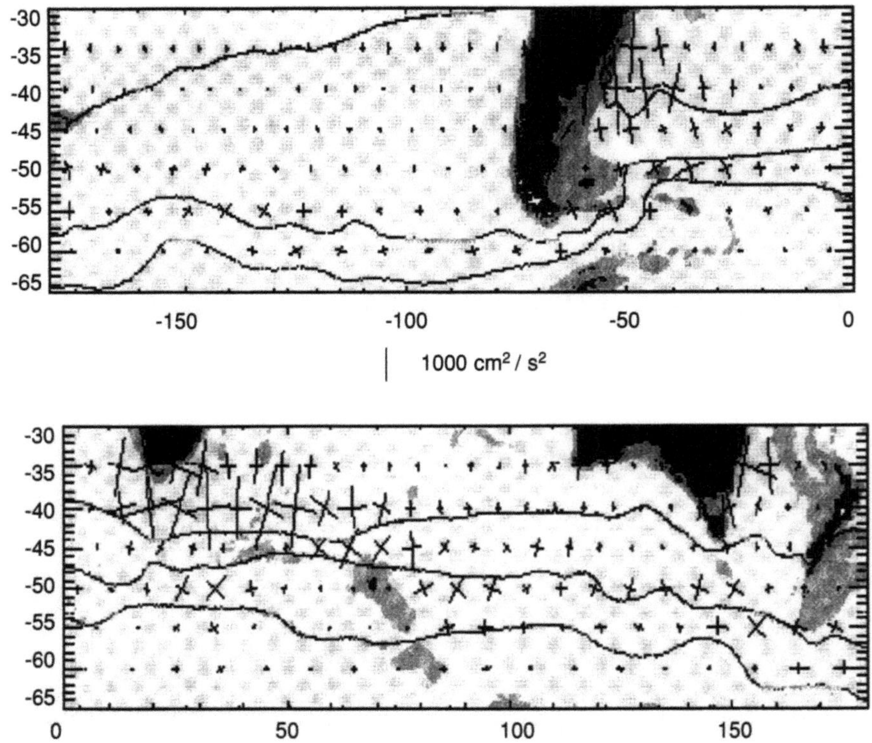

Fig. 7. Velocity variance ellipses axis from 2 years of T/P data. The bathymetry is superimposed with grey levels of 2000 m each

Anisotropic ellipses axes are also noted where the mean current interacts with the bottom topography. Close to the coast, where there is a strong topographic steering, the ellipses are orientated in the sense of the main flow indicating that the variability is mainly due to the change of current intensity. The ellipses orientation shows a predominance of negative values near the subtropical western boundaries currents, with a particular by clear situation for the East Australian current. Following the previous interpretation of the ellipses in terms of orientation, i.e. the main currents being mostly southward, this would mean a loss of momentum due to the intense eddy field.

In the area of the Malvinas confluence, a Lagrangian drifters analysis (Fig. 8) made using the same dataset as in [20] (see also Zambianchi et al., this Vol.) gives results similar to the Eulerian ones here presented, at least from a qualitative point of view. However, a careful analysis of the two maps reveals differences in orientation and intensity of some of the ellipses, like around 45°S where drifters data indicate higher values for the variance. There the ACC fronts are vary narrow and it could happen that the drifters displace themselves along fronts, producing a bias in the estimate (e.g. [12]). The different binning applied to the two datasets could also be a cause of the discrepancy.

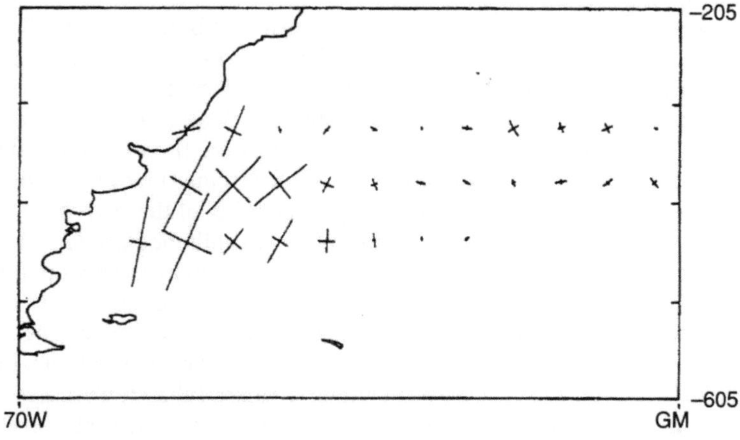

Fig. 8. Velocity variance ellipses axis derived from lagrangian buoy data (see text)

5 Conclusions

The first 2 years of T/P altimeter measurements all around the Southern Ocean, between 30°S and 65°S, were analyzed in order to study the mesoscale characteristics of the basin. Sea level anomalies, distributed by the AVISO, already processed with quality control and geophysical correction, were used. Given the high quality of the employed dataset, the data processing consisted of a resampling of the data by means of a binning algorithm. Furthermore, an objective analysis scheme was applied to compute sea level anomalies maps of the entire Southern Ocean. Thus, since T/P repeat period is 10 days, the data processing yielded 74 maps of SLA, one for each cycle, showing a synoptic view of the mesoscale activity of the study area. These maps showed evidence of a population of mesoscale eddies, with a typical diameter of 100-200 km in the sector 10-50°E and 30-60°W, possibly shed from the meanders of the Antarctic Circumpolar Current.

Both model and observation in the Southern Ocean circulation suggest an important role for mesoscale eddies in the meridional momentum transport. However, there are few observations of individual mesoscale features originating at the ACC. Analysis of T/P shows evidence that the eddies were located mainly in the Agulhas region and in the southwest Atlantic. The rings showed on these maps seem to move southward, confined to a rather narrow corridor (about 400 km). It is worthwhile noting how the SAF represents a kind of southern boundary of the high variability area. This analysis confirmed the role of the mesoscale eddies shed from the ACC in transporting either heat or momentum, though additional investigations are needed to estimate their overall importance in transporting heat and momentum south of the SAF. To characterize the mesoscale activity in the study area, maps of the standard deviation of the SLA for a 2-year period covered by T/P observations are presented.

The SLV maps show high variability values at the boundary of the three main fronts. The highest values of more 25 cm are found between 10-60°E and near 50°W at a latitude of 40-50°S. A high variability area extends to at least 60–65°S. The 2-yearly SLV maps show very similar patterns (Figs. 4 and 5). These maps are also in good agreement with previous analysis of Geosat satellite altimeter. This indicates that interannual variability does not affect the ACC eddy field. This could be due either to a lack of variability of the forcing (mainly the wind stress) or to a not negligible role of the topographic steering of the current.

Then we adopted a technique [10] for resolving both magnitude and direction of residual surface geostrophic velocities at T/P altimeter crossover points on the basis of the 2-year time series with a temporal resolution of 10 days and horizontal resolution of about 200 km (spacing between tracks). The time series are then used to determine surface eddy statistics in the Southern Ocean, hence the spatial distribution in the surface Reynolds stresses (u'^2, v'^2, $u'v'$). According to the information available so far, we found that the complex spatial distribution of surface eddy momentum flux is strongly influenced by bottom topography and the position of the mean current.

References

1. Tapley B.D, Ries JC, Davis GW, Eanes RJ, Schutz BE, Shum CK, Watkins MM, Marshall JA, Nerem RS, Putney BH, Klosko SM, Luthcke SB, Pavlis D, Williamson RG, Zelensky NP (1994) Precision orbit determination for TOPEX/POSEIDON. J Geophys Res 99(C12):24383-24404
2. Nowlin WD Jr, Klinck JM(1986) The physics of the Antarctic Circumpolar Current. Rev Geophys 24:469-491
3. Nowlin WD Jr, Clifford M (1982) The Kinematic and thermohaline zonation of the Antarctic Circumpolar Current at Drake Passage. J Mar Res 40S:481-507
4. Munk WH, Palmen E (1951) Note on the dynamics of the Antarctic Circumpolar Current. Tellus 3:53-55
5. Hughes CW, Killworth PD (1995) Effects of bottom topography in the large-scale circulation of the Southern Ocean. JPO 25:2485-2497
6. Wells NC, de Cuevas BA (1995) Depth-integrated vorticity budget of the Southern Ocean from a general circulation model. JPO 25:2569-2582
7. McWilliams JC, Holand WR, Chow JS (1978) A description of numerical Antarctic Circumpolar Current, Dyn. Atmos Oceans 2:213-291
8. Stevens DP, Ivchenko VD (1996) The zonal momentum balance in a realistic eddy resolving general circulation model of the Southern Ocean. Quart J Roy Meteor Soc, submitted
9. Ivchenko VO, Kelvin JR, Davin PS (1996) The dynamics of the ACC. JPO 26:753-774
10. Morrow R, Coleman R, Church J, Chelton D (1994) Surface eddy momentum flux and velocity variance in the southern ocean from Geosat altimetry. J Phys Oceanogr 24:2050-2071
11. Chelton DB, Schlax MG, Witter DL, Richman JC (1990) Geosat altimeter observation of the surface circulation of the Southern Ocean. J Geophys Res 95(C10):877-904
12. Morrow R, Church J, Coleman R, Chelton D, White N (1992) Eddy momentum flux and its contribution to the Southern Ocean momentum balance. Nature 357:482-484
13. Johnson TJ, Stewart RH, Shum CK, Tapley BD (1992) Distribution of Reynolds stress carried by mesoscale variability in the Antartic Circumpolar Current, Geophys Res Lett 19:1201-1204
14. Gille ST (1994) Mean surface height of the Antarctic Circumpolar Current. J Geophys Res 99 (C9):255-273
15. Park Y, Gamberoni L (1995) Large-scale circulation and its variability in the south Indian Ocean from TOPEX/POSEIDON altimetry. J Geophys Res 100(C12):911-929

16. AA.VV., AVISO user handbook: merged TOPEX/POSEIDON products (1992) AVI-NT-02-101-CN, 2nd Ed, CNES, Toulouse
17. Fu LL, Christensen EJ, Yamarone CA, Lefebvre M, Menard Y, Dorrer M, Escudier P (1994) Topex/Poseidon mission overview. J Geophys Res 99(C12):24,369-24,381
18. Bretherton FP, Davis RE, Frandy CB (1976) A technique for objective analysis and design of oceanographic experiments applied to MODE-73. Deep sea Res 23:559-582
19 Gouretsky VV, Danilov AL (1994) Characteristic of warm rings in the African sector of the Antarctic Circumpolar Current. Deep sea Res 40:1131-1157
20 Figueroa HA, Olson DB (1989) Lagrangian statistic in the South Atlantic as derived from SOS and FGGE drifters. J Marine Res 47:525-546

Upper Ocean Thermal Structure and Fronts Between New Zealand and the Ross Sea (Austral Summer 1994-1995 and 1995-1996)

A. Russo[1], A. Artegiani[1], G. Budillon[2], E. Paschini[1] and G. Spezie[2]

[1]Istituto di Ricerche sulla Pesca Marittima, CNR Ancona, Italy
[2]Istituto di Meteorologia e Oceanografia, IUN Napoli, Italy

Abstract

This paper describes the upper ocean thermal structure between New Zealand and the Ross Sea, and its evolution from the austral spring to summer 1994-95 on the basis of four high resolution XBT sections, and during the summer 1995-96 with two high resolution XBT sections. The area of investigation is almost completely (with the exception of the southernmost part) interested by the Antarctic Circumpolar Current (ACC). The main fronts (SubAntarctic Front, Polar Front and southern front) included in the ACC are individuated, and a new front, supposed to derive from a splitting of the SAF in a Northern SAF (NSAF) and a Southern SAF (SSAF), is identified. The NSAF is always located at the end of the Campbell Plateau, on the continental slope, and the SSAF is located around 58 °S, with the exception of the December 1994-January 1995 cruises. Going from spring to the end of the austral summer, the PF appears to retreat from north to south, while the southern front shows a stable position on the northern flanks of the Pacific-Antarctic and Southeast Indian Ridges. The thermal structure of the six sections reveals the presence of intense mesoscale dynamics, with cold nuclea of Antarctic Surface Water (AASW) detached from the PF and noticed up to 1000 km north of it.

1 Introduction

The Antarctic Circumpolar Current (ACC) is the main feature of the Southern Ocean and its relevance is recognised by a number of scientific works. This current flows all around Antarctica, going from west to east. From earlier observations, the ACC flow has been related to the presence of fronts, whose number, names and definition criteria change with time and investigators [1,2]. In the present work, we define these fronts using the most recent criteria available [2–4].

According to Orsi et al. [3], the ACC has as its northern limit the Subtropical Front (STF) and as southern limit the Boundary (Bdy). Within these limits, three other fronts have been individuated, each of them corresponding to a local maximum of the geostrophic transport. From north to south, they are the SubAntarctic Front (SAF), the Polar Front (PF) and the southern front.

In the area covered by our data, from 172 to 180°E, the STF is very close to the

New Zealand coast and our XBT launches started quite offshore, so STF was not detected. In the same area, the southern front and the Boundary are very close to each other and are not distinguished, so we will deal with the southern front only, intending that the Boundary is close or coincident with it.

To define these fronts, a number of tracers has been used (temperature, salinity, dissolved oxygen, nutrient salts content, etc.). We have only upper-layer temperature data available; nevertheless, they can be identified using the following definitions derived from the previously reported authors:

SAF: $T > 4$ °C at 400 m [3]
PF: northern limit of subsurface $T_{min} < 2$ °C at $z < 200$ m [2, 3]
Southern front: northern limit of subsurface $T_{min} < 0$ °C and $T > 1.8$ °C at depth > 500 m [3]

A brief description of data set and methodology is reported in Section 2. In Section 3 the fronts are defined, and a splitting into a "northern" and a "southern" part of the SAF as a permanent feature of this sector, rather than occasional, as observed in the Tasmanian sector by Belkin and Gordon [2] and Rintoul et al. [4], is sustained; consequently, names and definitions for these two fronts are given. A description of the thermal characteristics of the six sections follows, with a number of observed cold eddies and/or meanders, in particular water bodies detached from the Polar Front. In the final section, results are summarised.

2 Data

In the framework of the CLIMA (Climatic Long-term Interaction for the Mass balance in Antarctica) project, six high-resolution XBT sections were obtained by the R/V "Italica" during its routes from New Zealand to the Ross Sea and return. Four sections were collected during the austral spring/summer 1994-1995 and two sections during the austral summer 1995-1996 (see Fig. 1 for the local bathimetry, XBT launch positions and cruise dates). The used XBTs were Sippican T-7, characterised by a maximum depth of 760 m and a depth resolution of 0.6 m. The probes were not calibrated onboard; we accepted the factory declared accuracy of 0.15 °C as they were built and shipped by the factory just before the beginning of each expedition. Spacing between launches varied from 10 to 40 km, with an increased resolution in the area in which the PF was supposed to be. In a few cases the resolution was coarser due to poor weather or to marine ice cover; most of the faulty probes (less than 10% in total) were recognised during the data acquisition and these launches were immediately repeated. In the post-processing phase, other erroneous profiles were detected and partially or totally rejected.

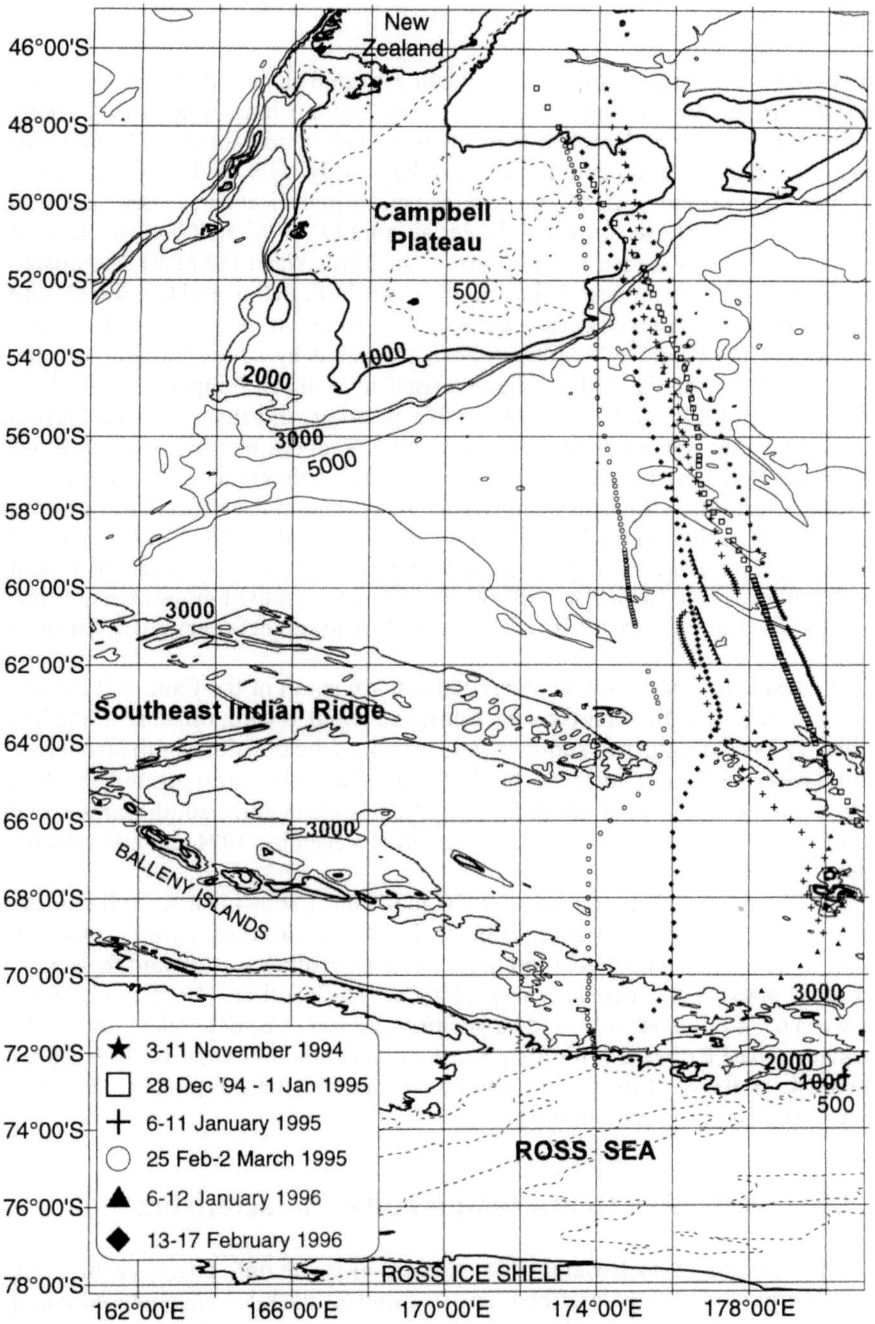

Fig. 1. Local bathymetry, XBT launch positions and cruise dates. *Simbols* identify the six sections

3 Fronts

In the New Zealand sector, SAF has been reported in two different locations: roughly around 60° S in the Orsi et al. [3] map, and much more northward, along the shelf break of the Campbell Plateau, in the Belkin and Gordon [2] map. Our data show that the two positions are simultaneously right, and the difference is probably due to some lack of data coverage of the historical dataset and above all to the definition chosen by the different authors ($T > 4$ °C at 400 m for the first location and the existence of the SubAntarctic Mode Water (SAMW) thermostad north of the SAF for the second location). In fact, both Belkin and Gordon [2] and Rintoul et al. [4] observed occasionally in the Tasmanian sector the SAF divided into a "northern" and a "southern" part, separated by about 150-200 km; the "northern" part of the SAF is characterised by a sharp temperature gradient between 6 and 8 °C and a relatively stable position, and the "southern" part is defined by a weaker gradient between 3 and 5 °C and a more variable position [4]. In the New Zealand sector, we can observe this behaviour of the SAF in each of the six sections, with lower temperature values and much wider spacing (from 400 to 800 km) than in the Tasmanian sector. The "northern" SAF (which we will call NSAF) is individuated by the transition between 4.5 and 6.5 °C at 400 m and the "southern" SAF (which we will call SSAF) by the transition between 3.5 and 4.5 °C at the same depth. Mostly, the temperature gradient is very sharp for both fronts.

The NSAF is always located about 200-300 km south of the Campbell Plateau edge, closer to it in the first three sections, and about 100 km offshore in the late February 1995 and the two summer 1996 sections (the positions of the fronts are summarised in Table 1 and reported in Fig. 2). The SSAF varies around 58–59°S, with a northernmost position in the November section and a southernmost one in the two sections collected between the end of December 1994 and mid January 1995.

The PF varied its location from 59° 50' to 63° 00' S, showing a general tendency to retreat toward south going from spring to the summer season. In fact we observe its northernmost position in the early November section and the southernmost ones in the two February sections. The southern front is observed between 63° 40' and 64° 30' S, and its position appears to be dependent not on the season as the PF does, but on the geographic location. In fact, it is observed on the northern flanks of the Pacific-Antarctic Ridge and, for the February 1995 section, of the Southeast Indian Ridge.

4 Observed Upper Ocean Temperature Characteristics

The first section (3-11 November 1994; Fig. 2.a) is the only one in spring. The southern part is poorly described, due to compact pack ice encountered going from 64° S southward; nevertheless, it is evident that the main difference between this section and the summer ones is that the subsurface temperature minimum is not confined between 50 and 200 m but covers the whole upper 200-m layer. This is due to the presence of the ice cover, which does not allow for the heating of the

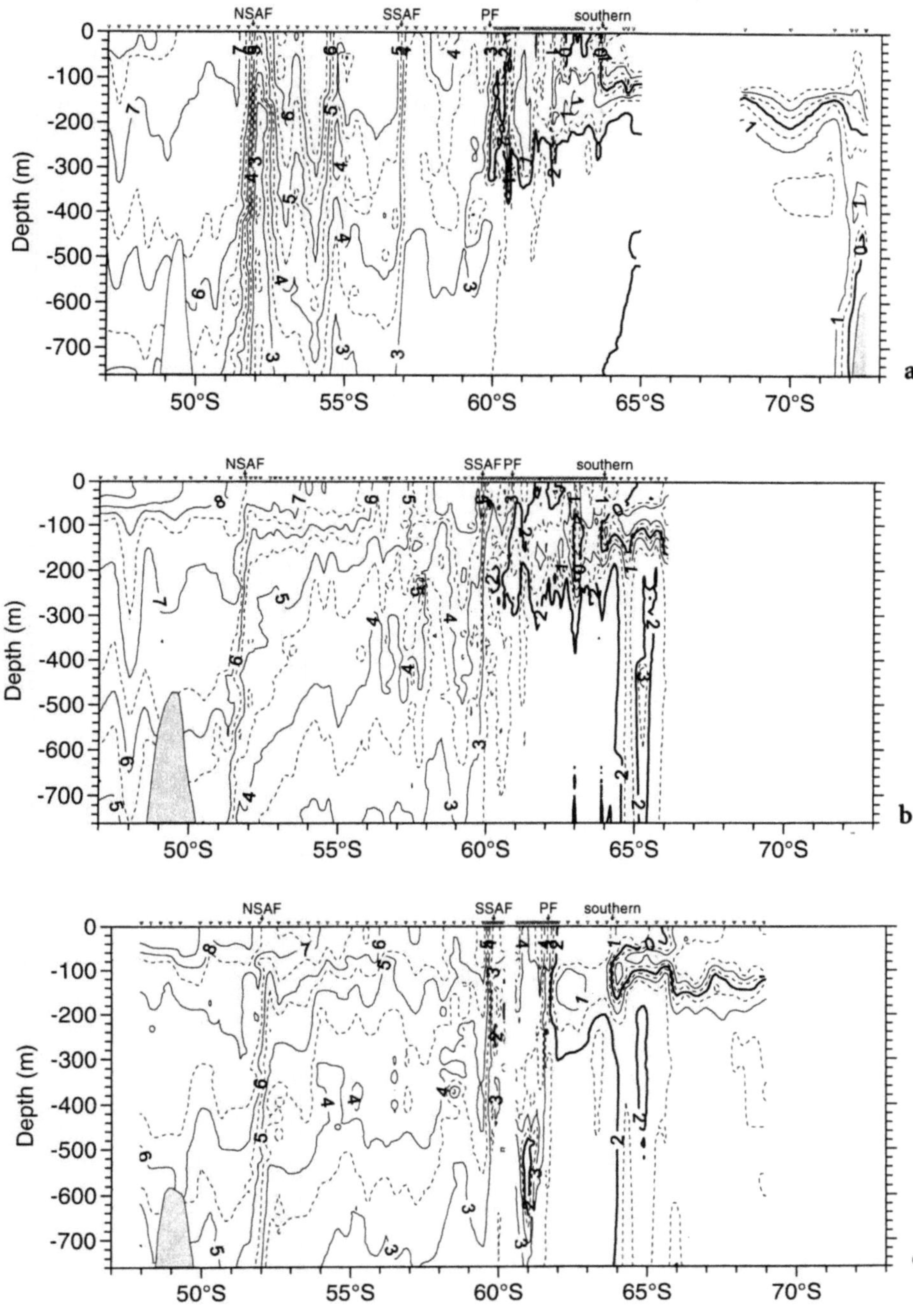

Fig. 2. Isotherms along the six sections: 3-11 November 1994 (**a**), 28 December 1994-1 January 1995 (**b**), 6-11 January 1995 (**c**), 25 February-2 March 1995 (**d**), 6-12 January 1996 (**e**) and 13-17 February 1996 (**f**); contour interval 0.5 °C. The location of the observed fronts is reported on the top axis of each section

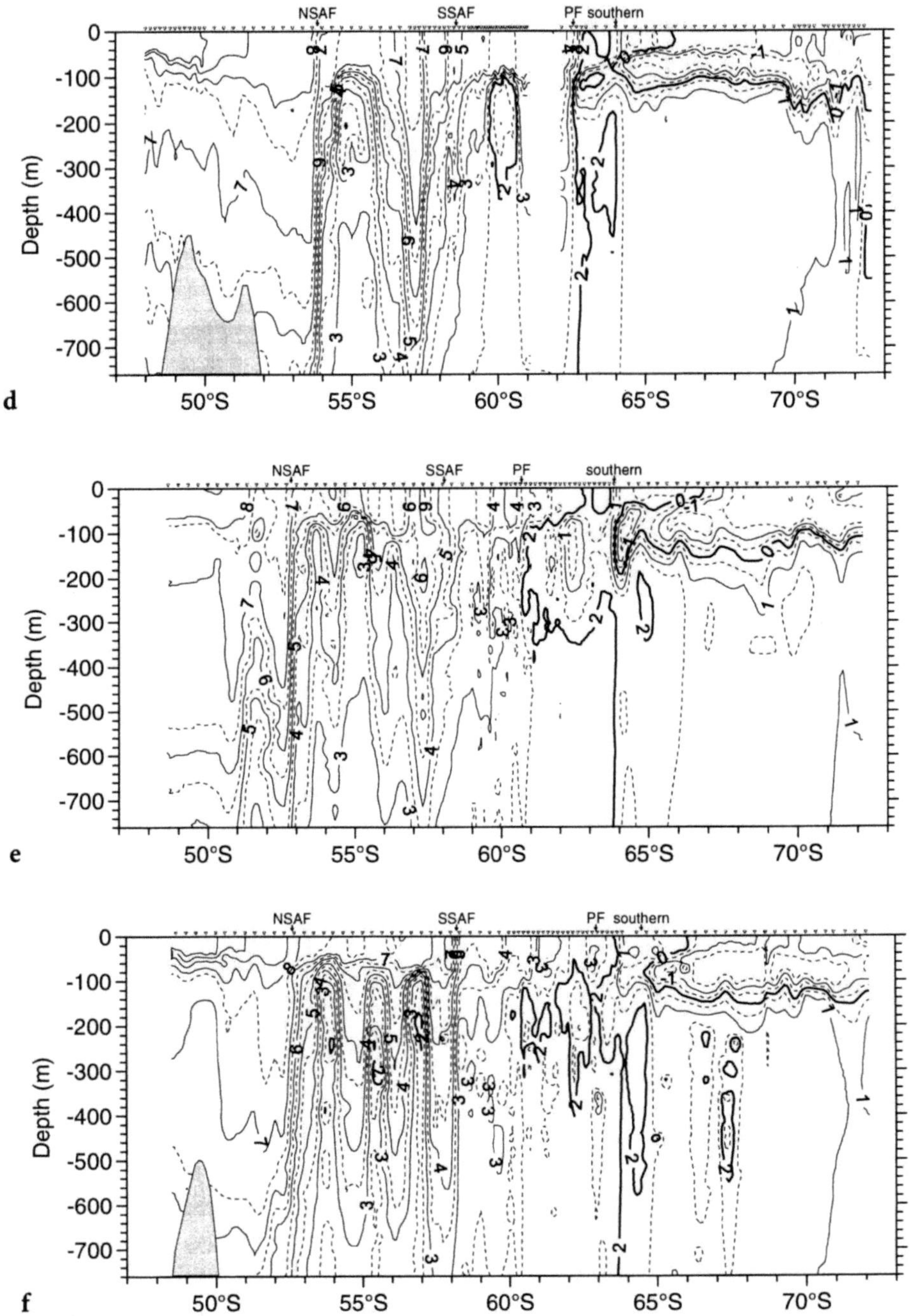

Fig. 2. Continued

Table 1. Latitude of the observed fronts

Section	Period of measures	NSAF	SSAF	PF	Southern front
9495-1I	3-11 November 1994	51° 50' S	56° 50' S	59° 50' S	63° 40' S
9495-2I	28 December 1994– 1 January 1995	51° 50' S	59° 50' S	60° 50' S	64° 00' S
9495-3I	6-11 January 1995	52° 00' S	59° 50' S	61° 40' S	63° 50' S
9495-4I	25 February–2 March 1995	53° 50' S	58°-59° S	62° 30' S	64° 00' S
9596-1I	6-12 January 1996	52° 50' S	58° 00' S	60° 40' S	63° 50' S
9596-2I	13-17 February 1996	52° 40' S	58° 10' S	63° 00' S	64° 30' S

NSAF, Northern SubAntarctic Front; *SSAF*, Southern SubAntarctic Front; *PF*, Polar Front

surface layer. Moreover, the PF is observed at its northernmost position, about 60° S. The high resolution (10-km) XBT launches in the PF area of November and December 1994 reveal that the area between the southern front and the PF could be subjected to intense dynamics and mixing, as can be inferred by the tormented temperature structure of the first 300 m; the same characteristic is not present in the area between the PF and the "southern" SAF where the high resolution launches of January and February 1995 were performed. Just south of the NSAF, we observe a cold feature, about 100 km wide at 400-m depth, which reaches the sea surface.

The second (Fig. 2.b) and third (Fig. 2.c) sections are very close in time (from 28 December 1994 to 1 January 1995), and are quite similar in their northern part. No relevant eddy or meander is detectable in these two sections. It can be observed that the surface layer warming extends more southward, the 5.5 °C isotherms being at 54° 20' S in the first section, while it was at 56° 40' S in the second one and at 58° S in the third one.

The fourth section (25 February–2 March 1995; Fig. 2.d) shows a cold eddy just south of the NSAF, wider than the November cold feature (about 200 km at 400 m). The eddy has a temperature minimum at 200 m and is intense until 80 m from the surface, becoming much weaker in the surface layer due to the solar heating effect. This cold eddy appears to be coupled to a warm one, almost of the same dimensions, evident from the surface until the bottom of the section. Further south, we found another cold feature, about 100 km wide, more recently detached from the PF, with characteristics similar to the northern one (minimum temperature between 100 and 300 m, weak signal in the first 80 m).

The two sections in 6-11 January (Fig. 2.e) and 13-17 February 1996 (Fig. 2.f) show an interesting triple cold core feature located between the NSAF and the SSAF. These eddies or meanders are just forming in January, and 1 month later they are observed almost at the same position fully developed, incorporating nuclea of Antarctic Surface Water (AASW). In fact, each of the three cold cores now reaches a temperature of less than 2 °C, and this is the minimum temperature that we observed in our six sections at this latitude. The most evident part of this signal occurs 60-80 m from the surface, but a weaker signature can be observed also in the surface layer. Just north of the PF, other water bodies are separating from it, more evidently in February. Comparing these two sections with the two corresponding 1995 ones, we observe that the T_{min} layer in January and February 1996 is thicker than in January and February 1995, probably due to a minor surface warming during summer 1996.

5 Discussion and Conclusions

Main fronts between New Zealand and Antarctica have been individuated on the basis of thermal structure, with the exception of STF (too close to the New Zealand coast in this area) and Bdy (difficult to locate using temperature data only). According to Orsi et al. [3], the Bdy marks the separation between the ACC and the subpolar regime and can be individuated as the southern limit of the Upper Circumpolar Deep Water (UCDW); in this area, the maps of Orsi et al. report the Bdy very close to the southern front. We detected the southern front on the northern flanks of the Southeast Indian and Pacific-Antarctic Ridges, a position in good agreement with Orsi et al., so we can consider these two ridges as the northern limit of the subpolar area.

We observed that the SAF is not uniquely individuated in our sections, so, following a definition of a similar situation observed sporadically in the Tasmanian sector [4], we hypothesize in the New Zealand area a splitting of the SAF into a "northern" and a "southern" part. These two fronts being persistent, at least for what concerning our six sections, we propose the names Northern SubAntarctic Front (NSAF) and Southern SubAntarctic Front (SSAF) respectively. The NSAF is defined by a temperature variation between 6.5 and 4.5 °C at 400 m and the SSAF by a temperature variation between 4.5 and 3.5 °C at the same depth.

We observed NSAF just south of the Campbell Plateau, while SSAF is located around 58° S, arriving at 56° 50' S in November 1994 and at 59° 50' S in December 1994–January 1995, where the PF was in November. The PF appears to retreat its position southwards, going from austral spring to summer, presumably related to the surface layer warming caused by the increasing solar heating. The southern front shows the most stable location, being always on the northern flanks of the Pacific-Antarctic and Southeast Indian Ridges.

We observe many cold features between the NSAF and PF, and most of them could be detached from the PF at around 165° S, where the Southeast Indian Ridge and the Campbell Plateau form a relatively narrow passage in which, according to the Belkin and Gordon fronts location map, PF and SAF are very close. Then these cold eddies and/or meanders would continue to propagate toward north-east, some of them along the Campbell Plateau continental slope. The hypothesis that these features were detached from the PF is supported by the evident subsurface temperature minima observed between 200 and 300 m in most of them (fourth and sixth sections in particular, where they show up to 1000 km north of the PF). These mesoscale phenomena transfer cold water in the upper layer toward lower latitudes and could play a relevant role in the global heat budget.

XBT launches during the "R/V Italica" route tracks between New Zealand and the Ross Sea are continuing to be executed, so it will be possible to have information on the interannual variability of the upper ocean temperature in this area.

Acknowledgements. This work was carried out in the framework of the Italian National Programme for Antartic Research (PNRA), whose support is gratefully acknowledged. Thanks are due to the many persons who collaborated in the XBT launching from the "R/V Italica".

References

1. Peterson RG, Stramma L (1991) Upper-level circulation in the South Atlantic Ocean. Progress in Oceanography 26:1-73
2. Belkin IM, Gordon AL (1996) Southern Ocean fronts from the Greenwich Meridian to Tasmania. J Geoph Res 101:3675-3696
3. Orsi AH, Whitworth III T and Nowlin W (1995) On the meridional extent and fronts of the Antarctic Circumpolar Current. Deep Sea Res 42:641-673
4. Rintoul SR, Donguy JR, Roemmich DH (1997) Seasonal evolution of upper ocean thermal structure between Tasmania and Antarctica. Deep Sea Res 44:1185-1202

The Large-Scale Thermohaline Structure of the Ross Gyre

V. Gouretski

WHP Special Analysis Centre (jointly operated by Max-Planck-Institut für Meteorologie und Bundesamt für Seeschiffahrt und Hydrographie), Bernhard-Nocht Str. 78, 20359 Hamburg, Germany

Abstract

This study describes the large-scale thermohaline structure and circulation patterns of the Ross Gyre and is based on a quality controlled data set including about 7,000 historical and modern hydrographic stations. A new regional climatology was obtained through the objective analysys of the data on neutral surfaces. Steric height anomaly maps portray Ross Gyre as a depression south of the mid-ocean ridge between about 170°E and 140°W. The centre of the gyre shifts with depth from about 68°S 164°W in near surface levels to about 63oS 150oW at 1500 meters. Maximum estimate of the geostrophic transport within the gyre is 8.5 Sv near 150°W, a factor of three lower than corresponding estimates for the Weddell Gyre. Data from a quasi-synoptic survey made by three former U.S.S.R vessels in 1985-86 are used for a more detailed description of the gyre's northern, eastern and southern boundaries. Similar to the Weddell Gyre the location of the eastern boundary is controlled by the southward extension of the Antarctic Circumpolar Current core flowing through the Udintsev Fracture Zone with the geostrophic transport of 56 Sv. The Circumpolar Deep Water spreads westwards along the southern limb of the gyre with potential temperatures still as high as 1.6 °C at 75oS, 150°W.

1 Introduction

The area of the Southern Ocean south of the Antarctic Polar Front (APF) is characterized by a net heat loss to the atmosphere. This loss of heat is compensated by the poleward heat transport in the ocean [1]. A cyclonic wind-stress field exerted on the ocean surface south of the Antarctic Circumpolar Current (ACC) in combination with the bottom topography creates two large-scale cyclonic circulations in the Southern Ocean: the Weddell Gyre in the Atlantic sector and the Ross Gyre in the Pacific sector. These cyclonic circulations facilitate southward transport of the Circumpolar Deep Water (CDW) from the southern margins of the ACC to the Antarctic continental shelves, thus providing a large-scale mechanism for the southward heat and salt transports. The CDW, entrained along the eastern limbs of the gyres, is modified due to strong interaction with shelf and slope waters, thus producing Antarctic Bottom water (AABW). Some of this cold and dense water escapes to the north, filling large areas of the deep ocean in the lower latitudes. Whereas a large number of studies have been conducted for the Weddell Gyre, the Ross Gyre (with the exception of the offshore areas of the Ross Sea) still remains one of the least sampled in the Southern Ocean.

The first maps indicative of a cyclonic circulation pattern south of the Antarctic Polar Front (APF) in the Pacific sector of the Southern Ocean were compiled by Treshnikov [2]. Reid and Mantyla [3] produced dynamic height maps, which implied a single cyclonic gyre from about 160°E to 130°W or a number of closed circulations between these longitudes. Reid [4] hypothesised an extension of the Ross Gyre to the Antarctic Peninsula, though his property maps were based on only 1780 stations for the whole South Pacific. Levitus [5] and later Olbers et al. [6] used much larger datasets to produce new maps of the Southern Ocean, which indicate the eastern boundary of the Gyre at about 140°W. The first quasi-synoptic description of the Ross Gyre was given by Antipov et al. [7] and Maslennikov [8]. It was based on quasi-synoptic data from nine meridional sections between 167.5°E and 132.5°W. The authors put the eastern boundary of the Gyre at about 140°W and give detailes of the northern boundary. A descriptive study of the physical oceanography of the South Pacific is given by Patterson and Whitworth [9]. Locarnini [10] presents a comprehensive analysis of the Ross Gyre and its environs. He also locates the eastern end of the Ross Gyre near 140°W in the upper and mid-depth levels.

The goal of this study is to improve the climatological description of the Ross Gyre. In a deviation from earlier studies our results are based on a much larger dataset of modern and historical hydrographic data. These quality controlled data with accounts made for inter-cruise offsets for salinity, oxygen and nutrients allow a detailed description of the water masses, fronts and circulation within the offshore area of the Ross Gyre.

2 Data and Methods

A set of hydrographic data for the area between 130°E and 100°W is used here to infer large-scale thermohaline and circulation patterns of the Ross Gyre. Most of the data came from the World Ocean Atlas [11], Southern Ocean Hydrographic Atlas Data Base [6] and hydrographic lines S4, P16A and P17A, occupied during the World Ocean Circulation Experiment (WOCE). Figure 1 shows the location of 7164 hydrographic stations south of 40°S used for this study. South of 57°S (taken crudely as a mean position of the APF) the dataset consists of 3597 stations. Our collection differs from the previous studies in the inclusion of additional data collected in the Ross Gyre area in the 1980s and 1990s. Thus, Russian expeditions from 1985 to 1986 [7, 12] contributed with 272 profiles, and WOCE Pacific lines S4, P16A and P17A provided 338 profiles. Since the focus of this study is on the deep ocean we used only bottle and low-resolution CTD casts obtained over depths greater than 200 m with up to six parameters given at observed levels: temperature, salinity, oxygen, silicate, nitrate and phosphate. Despite the inclusion of the additional data the data coverage is still insufficient south of 70°S and east of 150°W. A survey in early 1994 made

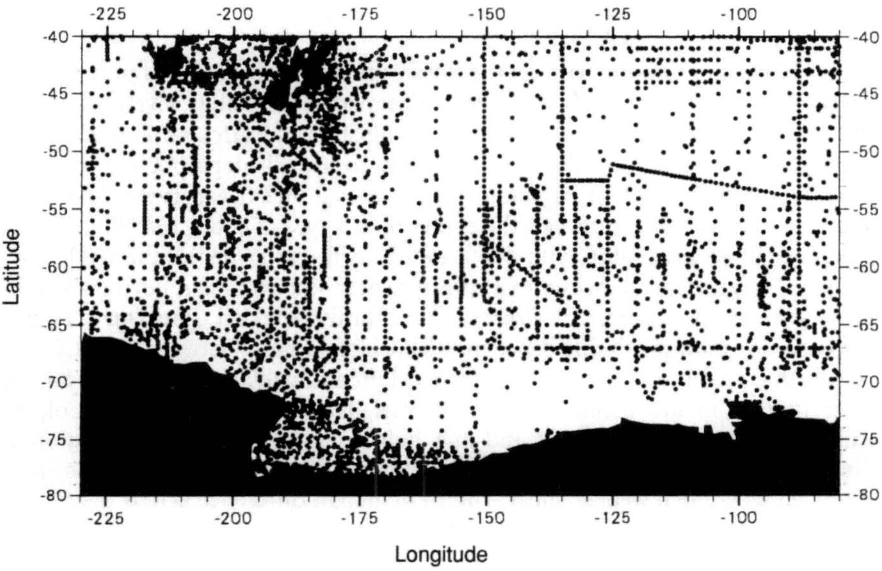

Fig. 1. Locations of 7164 hydrographic stations south of 40°S used in this study

aboard R/V "Nathaniel B. Palmer" [13] provided the first hydrographic data for this area, but these were unavailable to us during the preparation of this manuscript.

Our dataset also differs from earlier studies by a more stringent quality control procedure. Gouretski and Jancke [14] report on a very inhomogeneous quality of the historical data in the South Pacific. The validation procedure applied to our dataset was based on an experimental fact that relationships between parameters in the deep ocean are relatively stable [15]. The quality control was done by comparing individual observations with locally averaged neutral density-parameter relations. Since some parts of the study area are only marginally sampled, we used all historical data, spanning the time period of about 60 years. These data were acquired by different instruments and methods, that often lead to large systematic offsets between the cruises [14]. We determined the inter-cruise offsets by comparing individual profiles with mean potential temperature-parameter relationships in the deep ocean, calculated on the basis of the high-quality (WOCE) data. The validated data were analyzed objectively on the isopycnal surfaces and re-interpolated back on 45 standard levels between 0 and 5000 m. As demonstrated by Lozier et al. [16] such procedure mimics the isopycnal mixing in the real ocean and thus guarantees against the formation of artificial water masses, which may occur in regions of steeply sloping isopycnals when interpolating (averaging) the data on the isobaric surfaces. To model spatial correlations between the observations a gaussian function with a decorrelation scale of 400 km was used to get enough data points for the objective analysis in the data-poor areas. The climatological fields refer to the annual mean climatic state of the ocean, with a strong bias to the period of austral summer.

3 Geographical Position of the Ross Gyre

To define dynamic boundaries for the Ross Gyre we consider geostrophic circulations (Fig. 2) referenced to a number of assumed zero-velocity levels. Choice of the levels is subjective, with the intention to illustrate the circulation of the near-surface, intermediate and deep waters respectively. All maps portray the Ross Gyre as a dynamic height depression south of the mid-ocean ridge. The steric anomaly map 100/500 m reveals the center of the gyre at about 164°W, 68°S. The gyre's center seems to shift with depth to the north-east, whereas the 1500/3000-m map (Fig. 2c) depicts the center of the gyre near 150°W, 63°S. A northward shift of the gyre axis is reported by Orsi et al. [17] for the Weddell Gyre. Our averaged maps do not show much difference in the horizontal shear between the ACC and the northern limb of the gyre. The steric height anomaly isoline 0.155 m for the layer 100/500 m lies within the zone of high lateral geostrophic shear of the Antarctic Circumpolar Current and may be chosen to represent the location of the gyre's northern boundary. In contrast, the southern limb of the gyre is characterized by a much weaker shear. The steric height difference across the gyre is about 5 cm for the layer 100/500 m, whereas below 500 m it is only about 3 cm, indicating a decrease of the vertical shear with depth.

At all levels isolines of steric height are aligned quasi-meridionally between 150 and 135°W, indicating the eastern boundary of the gyre at about 140°W. Compared with the Weddell Gyre no pronounced western boundary is revealed by the dynamic topography at all levels. This may be explained by a quasi-zonal orientation of the continental slope in the area. In contrast, the Antarctic Peninsula in the Weddell Sea provides a meridional barrier along which a northward boundary current with a total transport of about 28 Sv is developed [18].

Integral volume transports within the study area based on the gridded density fields are shown in Fig. 3. The ocean bottom was used as a zero velocity level. Accumulated westward transports were calculated for the area between the Antarctic continent and 65°S (north of this latitude the direction of the water transport is to the east everywhere). The maximum westward transport of 8.5 Sv is found near 150°W, with transport magnitudes exceeding 5 Sv between 165 and 145°W. These estimates include both transports within the narrow Antarctic Slope Front described by Jacobs [19] and within the southern limb of the Ross Gyre. The non-zero westward transports are found only between 179 and 135°W, thus implying the virtual abscence of the Antarctic Slope Current west of 190°W and east of 135°W.

Geostrophic transports of 24-26 Sv relative to the bottom are reported by Fahrbach et al. [20] for the Weddell Gyre at 6°W. Thus, in terms of geostrophic transports the Ross Gyre is about a factor of 3-4 weaker than its Weddell Sea counterpart. At all levels depicted in Fig. 2 the ACC is characterized by a strong horizontal shear north of the mid-ocean ridge crest. Being a deep-reaching current, the ACC is strongly controlled by the bottom topography. Two deep gaps provide the conduits for two main high-speed cores of the ACC. The Antarctic Polar Front core flows through the Udintsev Fracture Zone near 143°W. It is well pronounced on the 100/500-m steric height anomaly map (Fig. 2a) as an intensive jet between the mid-ocean ridge and 120°W. The current core, associated with the

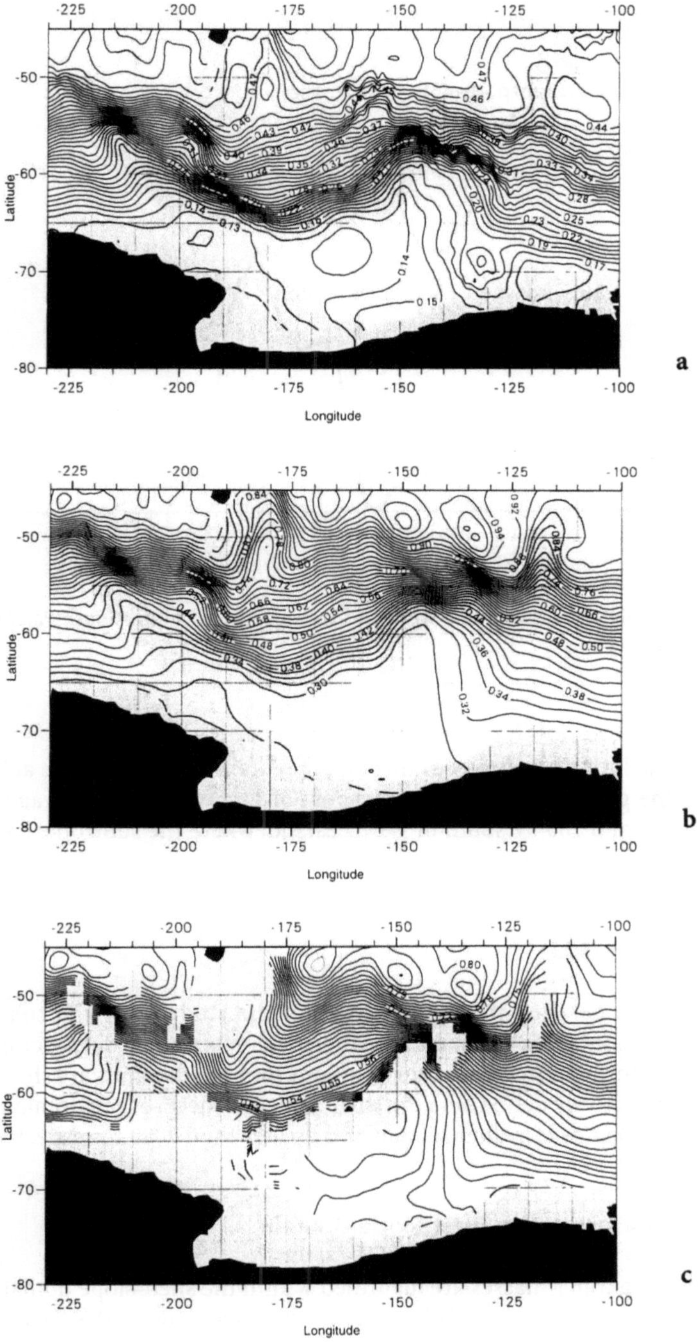

Fig. 2. Steric height anomaly (in meters) for the layers; **a** 100/500 m, **b** 500/1500 m and **c** 1500/3000 m

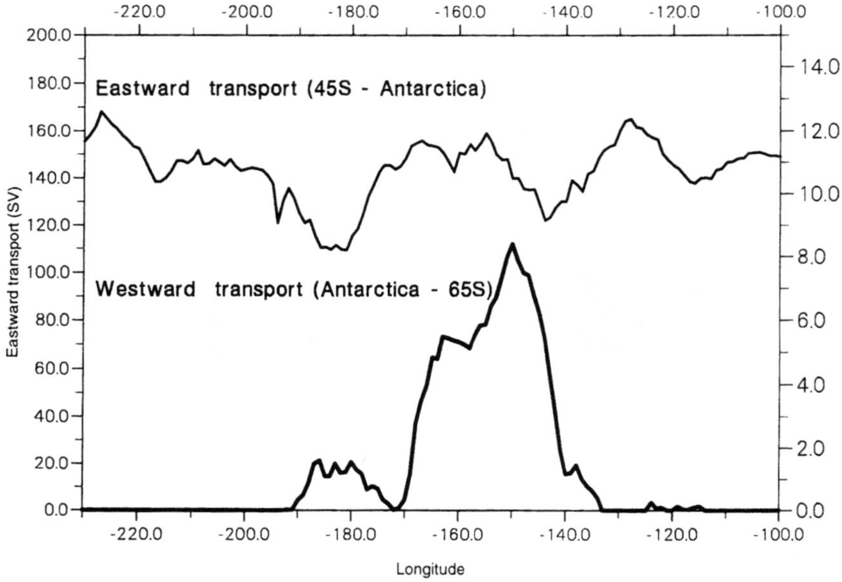

Fig. 3. Accumulated eastward and westward transports (Sv) in the South Pacific based on gridded density data

Subantarctic Front (SAF), passes through the Eltanin Fracture Zone at 130°W. We estimate the geostrophic volume transport of the ACC as the average of 110 Sv between 185 and 180°W south of New Zealand, where the subtropical gyre is not crossed by the meridional section.

4 Water Masses

To distinguish between the water masses in the study area it is convenient to divide the area into two domains: a shelf-slope domain and a deep-ocean domain. The temperature-parameter scatter diagrams for both domains are shown in Figs. 4-5. (We subjectively define here the shelf-slope and deep-ocean domains as those with bottom depths less than 2000 m and greater than 2000 m respectively.)

4.1 Water Masses of the Shelf/Slope Domain

The Following water masses are identified within the shelf-slope domain (Fig. 4): Antarctic Surface Water (AASW), High Salinity Shelf Water (HSSW), Low Salinity Shelf Water (LSSW), Ice Shelf Water (ISW) and Modified Circumpolar Deep Water (MCDW.) The *Antarctic Surface Water* occupies a layer of about 50 m thick, found throughout the whole shelf area in summer. The layer has temperatures up to sev-

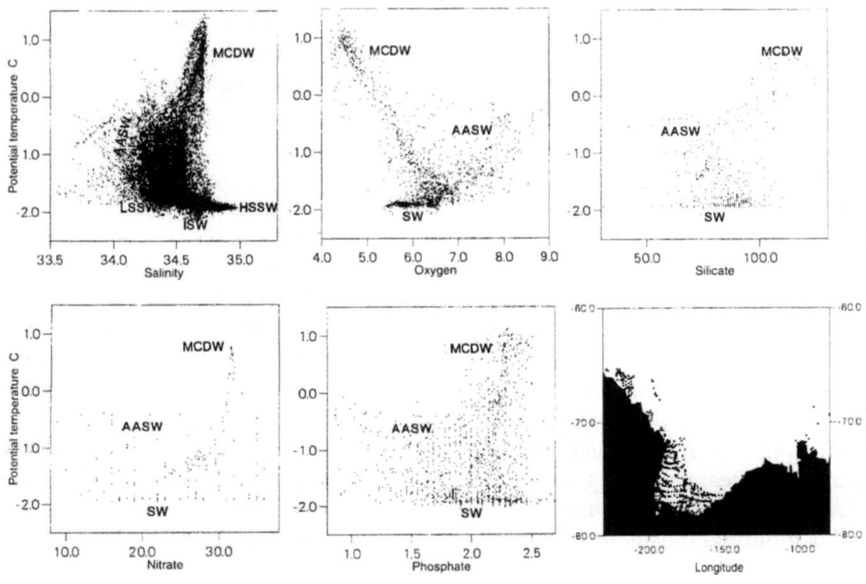

Fig. 4. Potential temperature-parameter scatter diagrams for the shelf-slope domain (bottom depth < 2000 m). *AASW* Antarctic Surface water; *HSSW* High Salinity Shelf Water; *ISW* Ice Shelf Water; *LSSW* Low Salinity Shelf Water; *MCDW* Modified Circumpolar Deep Water; *SW* Shelf Water. Based on data below 30 m

eral degrees above the freezing point and a wide range of salinities (33.5–34.5). Shelf waters represent dense waters observed over the continental shelf (below the surface layer). *High Salinity Shelf Water*, the saltiest (at –1.9 °C, S>34.75) and densest water found in the Southern Ocean, is formed over the continental shelf in the north-western Ross Sea. The high salinity of HSSW is explained by the intensive process of ice formation in quasi-permanent polynyas along the coast [21]. This water is involved in the formation of the high-salinity modification of the bottom water. The *Low Salinity Shelf Water* (at > –1.9 °C, S<34.60) is found in the eastern Ross Sea [22] and also along the Adelie coast [23]. Carmack and Killworth [24] argue that mixtures of LSSW with circumpolar deep water may ventilate the water column at intermediate depths. The *Ice Shelf Water* represents a water mass with temperatures below the sea surface freezing point – an indication of the interaction with the floating part of the Ross Ice Shelf at depths. Two types of ISW are known to exist in the study area. Jacobs et al. [22] describe a shallow type of the ISW, found at depths between 50 and 250 m near the Ross Ice Shelf edge. This water is formed by the interaction of MCDW with the Ice Shelf. Mac Ayeal [25] describes a tongue of ISW at 300–600 m. This (Deep) ISW is formed, presumably, when the HSSW interacts with the base of the Ice Shelf. Deep waters of the oceanic domain penetrate the shelf in a form of the *Modified Circumpolar Deep Water*. It is warmer than about –1.6 °C and its salinity is greater than 34.50.

Typical depth range occupied by this water mass is between 200 and 300 m. This water mass is produced by the isopycnal mixing between Lower Circumpolar Deep Water (see below) and shelf waters. As noted by Locarnini [10] this mixing results in a subsurface temperature maximum and dissolved oxygen minimum.

4.2 Water Masses of the Oceanic Domain

The oceanic domain is occupied by the following water masses (Fig. 5): Antarctic Surface Water (AASW), Circumpolar Deep Water (CDW) and Antarctic Bottom Water (AABW). *Antarctic Surface Water* has its northern limit at the Antarctic Polar front. This relatively cold (−1.9 to 2.0 °C) and low saline (34.0–34.3) layer of about 200 meter depth is formed by convective mixing in winter. Warming of the top of this layer in summer produces a well pronounced temperature minimum on the vertical temperature profiles.

Circumpolar Deep Water is by far the most voluminous water mass of the Southern Ocean, deriving its properties from the North Atlantic Deep Water (NADW), which is introduced into the Southern Ocean within the Atlantic sector and is modified during its circumpolar eastward flow with the ACC. Gordon [26] was the first to make a distinction between the two types of the CDW, which derive their characteristics from two different northern sources. The *Lower CDW* (LCDW) bears the imprint of NADW by having a deep maximum in salinity (34.75 at 1.7 °C) and minima in nitrates (31 μmol/kg) and phosphates (about 2.15 μmol/kg). *The Upper CDW* (UCDW) is marked by a pronounced minimum in oxygen (3.8–3.9 ml/l at 2.2 °C) and silicate (65 μmol/kg) and maxima in concentrations of nitrate (35 μmol/kg) and phosphates (2.5 μmol/kg). The oxygen minimum of UCDW is acquired due to the lateral exchange with the Indian and Pacific deep waters, both having much lower oxygen concentrations and higher nutrient content compared with those of NADW. Scatter diagrams for oxygen, nitrate and phosphate also display extrema near 1.3 °C: the oxygen concentrations fall below 4.1 ml/l, whereas nitrates and phosphates increase to 33 μmol/kg and 2.35 μmol/kg respectively. This water type, which we call here Ross Deep Water (RDW), is found in the center of the Ross Gyre (Fig. 6). As in the Weddell Gyre, this oxygen minimum is obviously due to the oxygen consumption by biological processes during a prolonged isolation of the deep water within the inner recirculation area, as noted earlier by Locarnini [10]. Both modifications of CDW occupy most of the water column within the ocean domain.

It is generally accepted that the Weddell Sea contributes mainly to the production of the *Antarctic Bottom Water* (AABW). The next important source of the bottom water is located in the south-western part of the Pacific sector. Shown in Fig. 5b are scatter diagrams below 300 m for the stations from the deep-ocean domain. All diagrams display two distinct segments within the deeper part of the water column. The upper segment (0.5–0.0 °C) indicates mixing between the deep waters (0.5 °C, S=34.705) and bottom waters (0.0 °C, S=34.700). The latter indices characterize the old AABW [9]. Its most pronounced characteristics are maxima in silicate, nitrate and phosphate concentrations (150, 31 and 2.2 μmol/kg respectively). Within the lower segment, below the inflection point at 0.0 °C the scatter

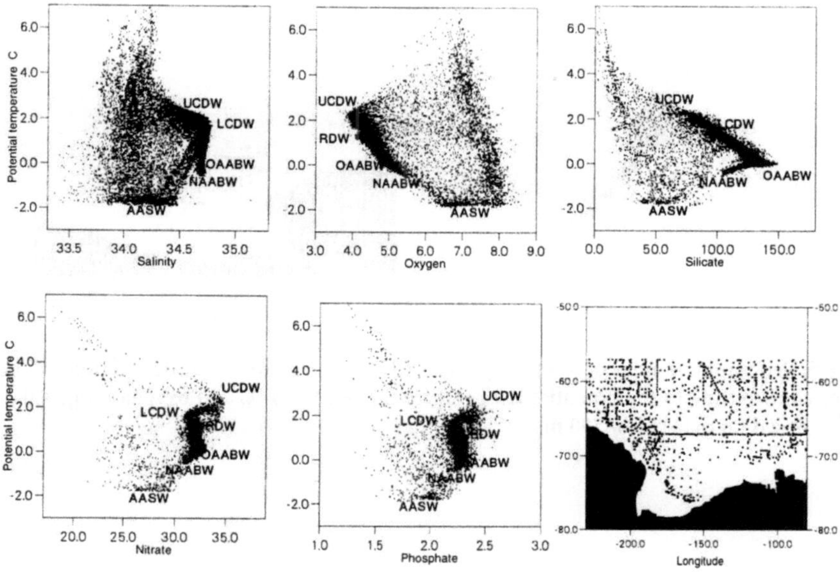

Fig. 5a. Potential temperature-parameter scatter diagrams for the deep-ocean domain (bottom depth > 2000 m). *AASW* Antarctic Surface Water; *LCDW* Lower Circumpolar Deep Water; *NAABW* New Antarctic Bottom Water; *OAABW* Old Antarctic BottomWater; *RDW* Ross Deep Water; *UCDW* Upper Circumpolar Deep Water. Based on data below 30 m

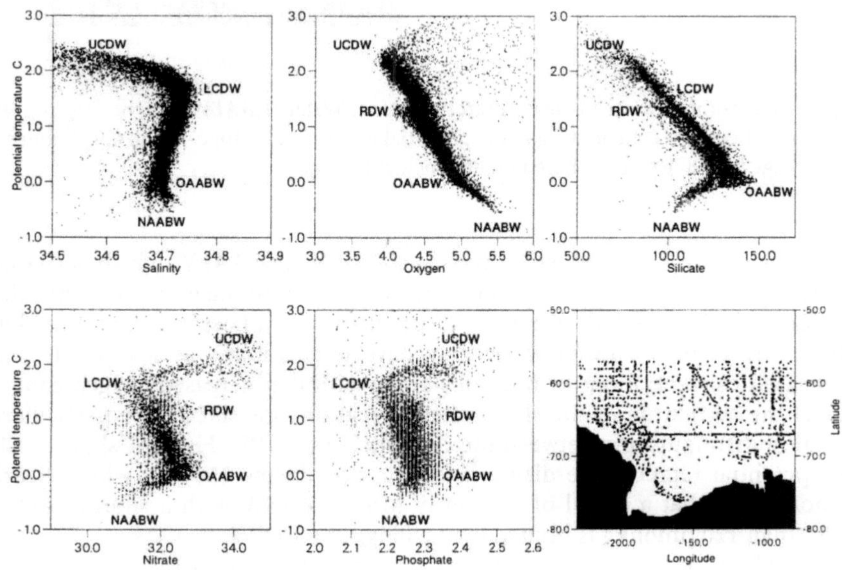

Fig. 5b. Same as Fig. 5a but for data below 300 m

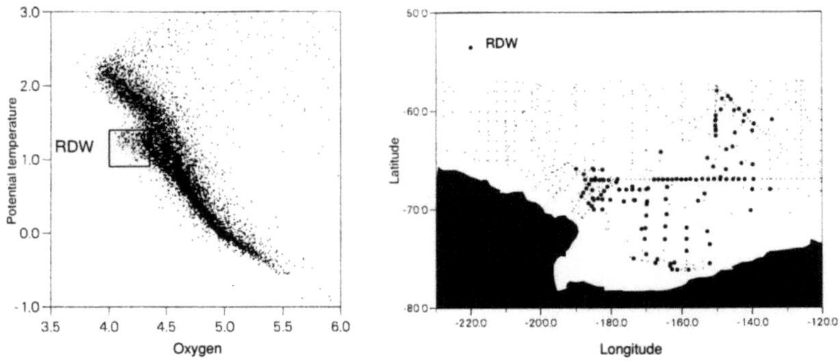

Fig. 6. Geographical areas of Ross Deep Water *(RDW)*. *Large dots* indicate locations of profiles contributing to the potential temperature range 0.9-1.4 °C and oxygen range 4.0-4.3 ml/l. Based on data below 300 m

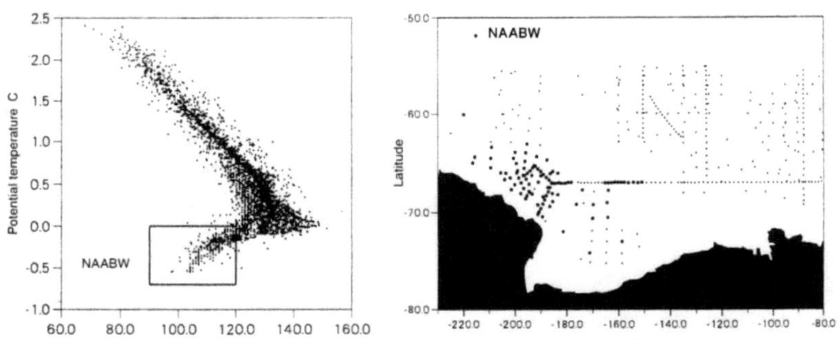

Fig. 7. Geographical areas of New Antarctic Bottom Water *(NAABW)*. *Large dots* indicate locations of profiles contributing to the potential temperature range –0.7 to 0.0 °C and silicate range 90-120 μmol/kg. Based on data below 1500 m

diagrams indicate another water mass, a so-called New AABW, which is characterized by the oxygen maximum (up to 5.5 ml/l) and minima in the concentration of silicate (105 μmol/kg), nitrate (30.5 μmol/kg) and phosphate (2.2 μmol/kg). The newly formed bottom water may have temperatures as low as –0.5 °C). Decreased concentrations of nutrients along with the increased oxygen content reflect the admixture of the shelf waters. Diagrams for silicate are particularly useful in distinguishing between the Old and New AABW. Figure 7 shows a silicate-potential temperature diagram for the data below 1500 m, which clearly demonstrates that a signal of the newly formed AABW with a silicate content lower than 120 μmol/kg is limited by the longitude 150°W.

According to the potential temperature salinity diagram the mean bottom water salinity is about 34.700 at < 0.0 °C. Unlike the other parameters displayed in Fig. 5b, salinity depicts two varieties of the bottom water. The *High Salinity*

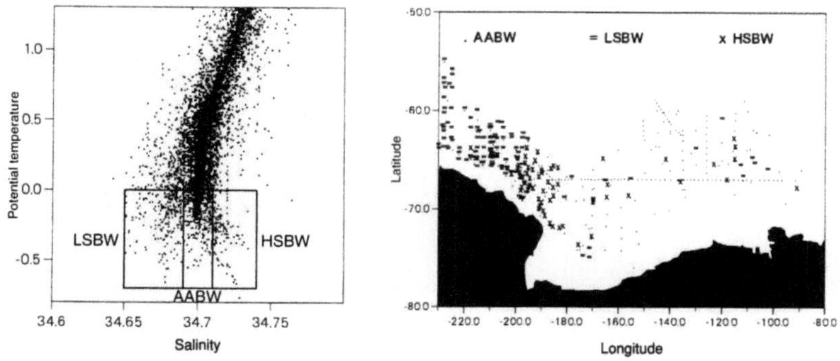

Fig. 8. Geographical areas of Low Salinity Bottom Water *(LSBW)* (S<34.69), High Salinity Bottom Water *(HSBW)* (S>34.71) and bottom water with median salinity (34.69<S<34.71). Based on data below 1500 m within the deep-ocean domain (bottom depth > 2000 m)

Bottom Water (HSBW) (S>34.71) has its source in the north-western Ross Sea, formed of a mixture of HSSW and CDW. The area of this water seems to be confined to the sector 200-185°W, extending from the continental slope of the western Ross Sea to about 64°S (Fig. 8). Some historical stations indicate the eastward spreading of HSBW with the northern limb of the Ross Gyre (as far as 115°W), though these are probably an artifact of erroneous salinity observations. High-quality data from the WOCE S-4 line give no evidence of this water type along 68°S. The westernmost stations showing the presence of HSBW are located near 225°W.

The *Low Salinity Bottom Water* (S<34.685) is formed off the Adelie coast [23, 27]. Its areal is located west of 205°W. The eastern Ross Sea also seems to contribute to the formation of LSBW near 170-174°W (Fig. 8). According to Jacobs et al. [21] both shelf and ice shelf waters are the components for the formation of LSBW in this area. According to our refined dataset there is little overlap between the areas occupied by HSBW and LSBW, with the boundary between the two areas being between 195 and 200°W. Only three stations indicative of HSBW are found west of the Balleny Islands.

5 Property Distributions on Neutral Surfaces

The meridional structure of the Ross Gyre along 150°W, where the maximum geostrophic transport is observed according to our calculations, is shown in Fig. 9. This section is based on the gridded data (a product of the objective analysis of the non-synoptic data) because no synoptic transects with a complete set of parameters across the whole gyre are available east of about 180°W in our data set. Isopleths for all parameters indicate a rather sharp boundary between the ACC and the Ross Gyre. An inflow of CDW into the gyre is indicated by the maxima in potential temperature and salinity near 70-72°S. A pronounced silicate maximum

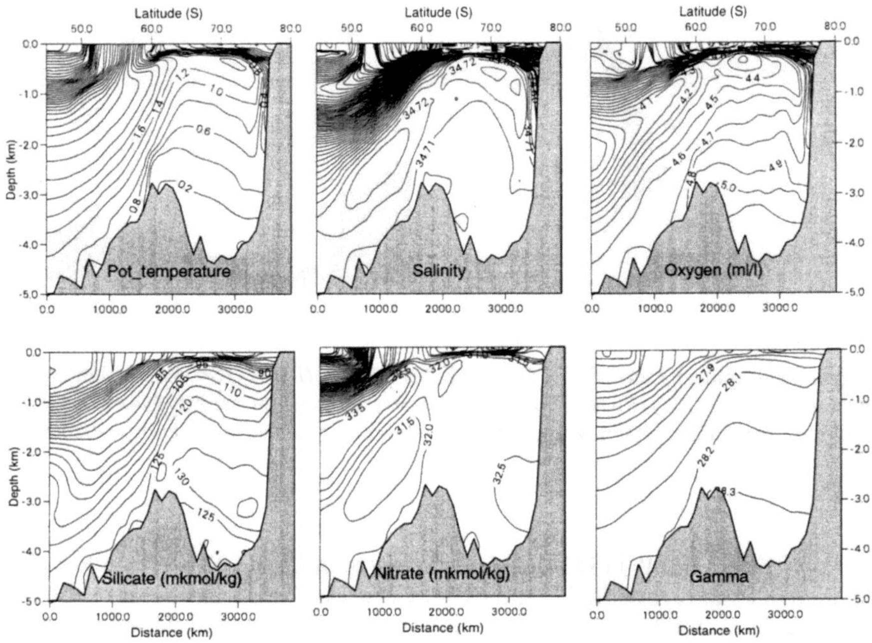

Fig. 9. Annual mean property distributions on the vertical section along 150°W obtained by optimal interpolation of profile data on neutral surfaces

is a signal of the inflowing silica-rich deep-water from the south-east Pacific, whereas the near-bottom maximum is due to a newly formed bottom water. Property distribution at a number of selected neutral surfaces [28] is used to illuminate the circulation and thermohaline structure of the Ross Gyre within characteristic wâter masses.

5.1 Neutral Density γ^n=27.90 (Fig. 10a)

The isopycnal surface of γ^n=27.90 coincides with the lower boundary of AASW, which typically occupies the layer up to 500 meters thick near the APF and up to 100 m in the center of the gyre. The property fields exhibit wide ranges in water mass characteristics. A zone of intense temperature, salinity and oxygen gradients extends over the crest of the mid-ocean ridge, delineating a sharp boundary between the ACC and Ross Gyre. This boundary curves abruptly southward near 145°W, with property isolines exhibiting a 12-degree latitudinal shift. The influence of the cold, fresh and highly oxygenated shelf waters is apparent in the interior of the gyre. Extreme values are observed near 70°S in the center of a tongue between 170 and/at 160°W (<−1.4 °C, S<34.35, O_2>6.6 ml/l, SiO_2<60 µmol/kg).

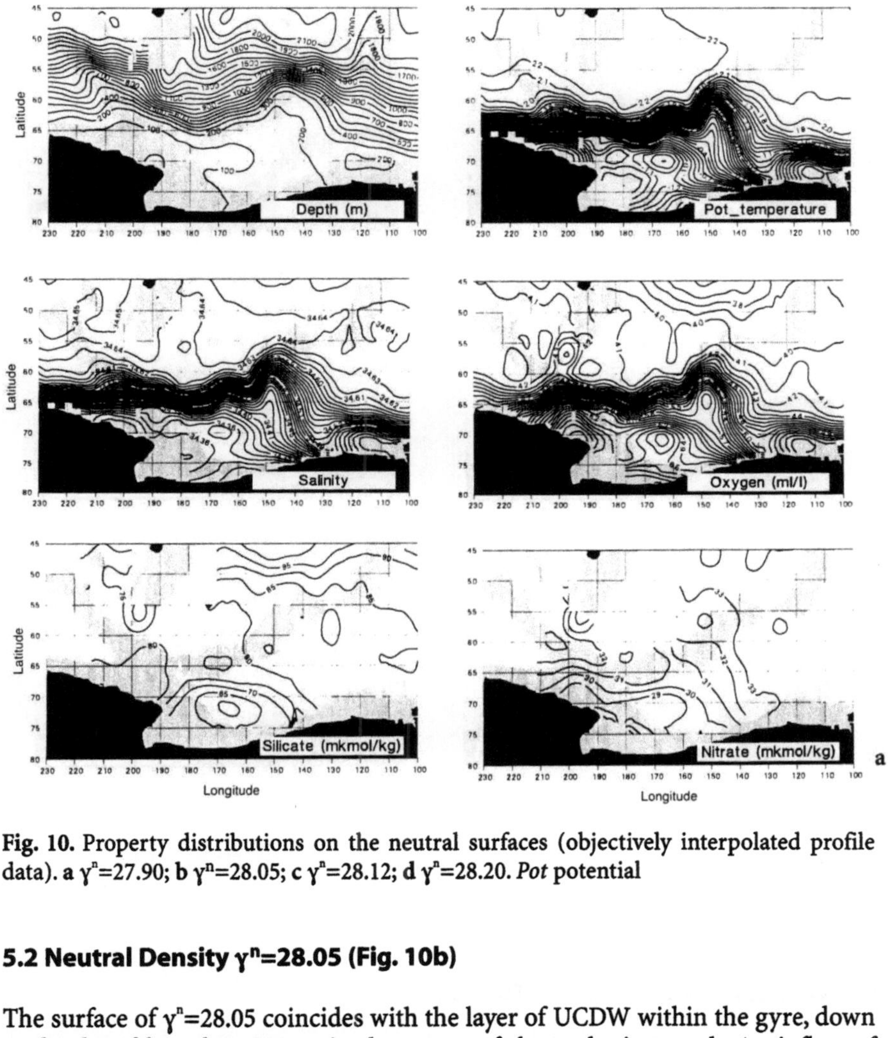

Fig. 10. Property distributions on the neutral surfaces (objectively interpolated profile data). **a** γ^n=27.90; **b** γ^n=28.05; **c** γ^n=28.12; **d** γ^n=28.20. *Pot* potential

5.2 Neutral Density γ^n=28.05 (Fig. 10b)

The surface of γ^n=28.05 coincides with the layer of UCDW within the gyre, down to depths of less than 300 m in the center of the cyclonic trough. An inflow of UCDW with the southern limb is well shown by the depth of the isopycnal surface. The characteristic warm and salty signal spreads westward from the eastern boundary of the gyre (~140°W) to about 165°W. This signal weakens abruptly further to the west, indicating strong modification of the inflowing water at 170-175°W. A cold and fresh tongue of the modified outflowing water appears east of 170°W, spreading to the north-east. Whereas temperature and salinity exhibit very similar patterns, the oxygen field for this surface is marked by a minimum centered at 66°S, 157°W. This feature may be attributed to biological processes, reducing the oxygen content in the gyre's interior. Both temperature and salinity fields show two separate cells within the gyre's interior, centered near 164°W and 150°W respectively. Since both cells coincide with the location of similar features on the previous neutral surface, they may indicate in-situ convection in this vicinity.

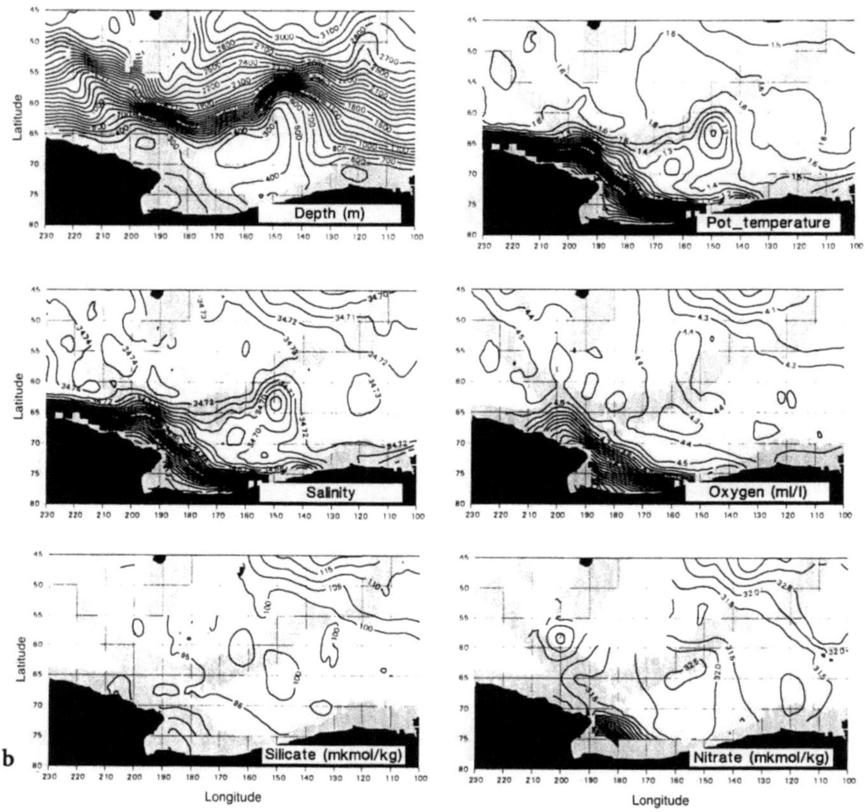

b

Fig. 10. Continued

5.3 Neutral Density γ^n=28.12 (Fig. 10c)

The surface of γ^n=28.12 depicts the distribution of LCDW; its core lies typically at depths of 500-800 m in the gyre. Whereas the depth of this surface still shows a sharp ACC–Ross Gyre boundary, other parameters do not show any front associated with the northern boundary. The inflow of LCDW along the southern limb is, however, seen in the salinity field, with the 34.73 isohaline marking the westward penetration of LCDW, consistent with the pattern of UCDW. Compared with the upper surfaces, the range of water characteristics north of the continental slope is considerably reduced, so that some patchiness observed in the salinity field may be due to undersampling and observational errors. Both Fig. 9 and 10c show maximum salinity values within LCDW exceeding 34.73 along the southern limb of the Ross Gyre. These values are at least 0.03 higher than maximum LCDW salinities in the southern limb of the Weddell Gyre, which eventually cause higher salinities of the bottom water formed in the study area.

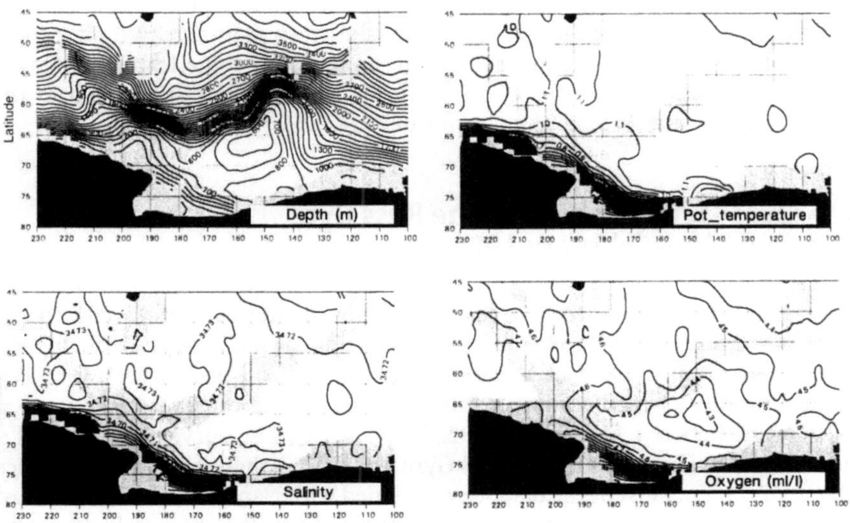

c

Fig. 10. Continued

5.4 Neutral Density γ^n=28.20 (Fig. 10d)

The surface of γ^n=28.25 characterizes distribution of the deep water in the gyre. The isopycnal depth distribution does not show much difference from the γ^n 28.05 surface, except that the center of the gyre shifts nearer to the mid-ocean ridge.

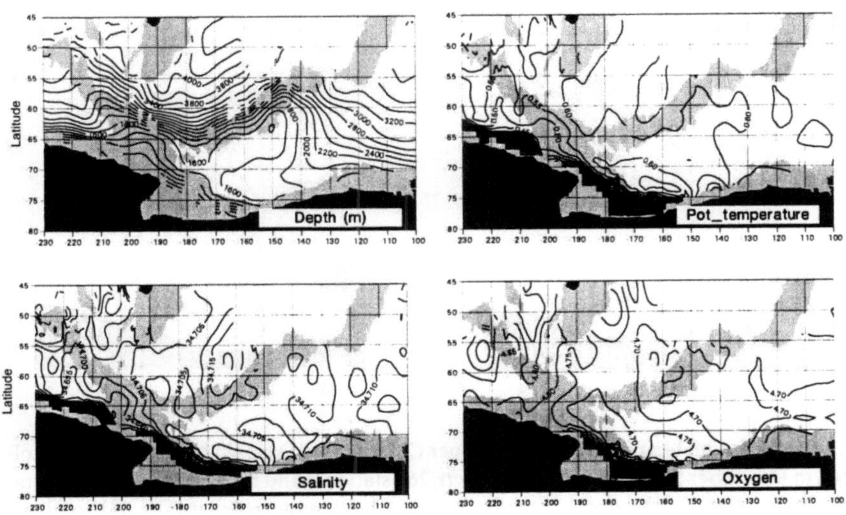

d

Fig. 10. Continued

The range of parameters on this surface north of the continental slope is reduced to only 0.2 °C for temperature and 0.02 for salinity, so that increased patchiness due to observational errors makes the interpretation of some features ambiguous. However, an intrusion of LCDW with the southern limb is evident in the higher temperatures (>0.65°) and salinities (>34.715).

6 A Quasi-synoptic View of the Ross Gyre in July 1985–February 1986

Hydrographic observations were made aboard Russian research vessels "Professor Zubov" and "Professor Vieze" over most of the Ross Gyre during the time period from the end of December 1985 to the end of February 1986 [7]. Vertical profiles of temperature, salinity, oxygen and silicate were measured at 260 stations between 167.5°E and 132.5°W (Fig. 11). The survey consisted of 9 meridional sections crossing the Ross Gyre from its northern boundary to the ice edge at 67-71.5°S. Most of the data were obtained with Nansen bottles at standard levels. Additional 1000-m-deep casts were obtained within the frontal zones with a temperature-conductivity probe what reduced station spacing to 33 km. Our validation procedure revealed some deficiencies in salinity and oxygen, so that only the temperature data are used for the analysis.

6.1 Northern Boundary

Orsi et al. [29] defined a deep-reaching Southern Front (SF) observed to the south of the APF as a circumpolar feature. Further to the south they find a sharp boundary (BDY), which divides the poleward edge of UCDW within the ACC

a 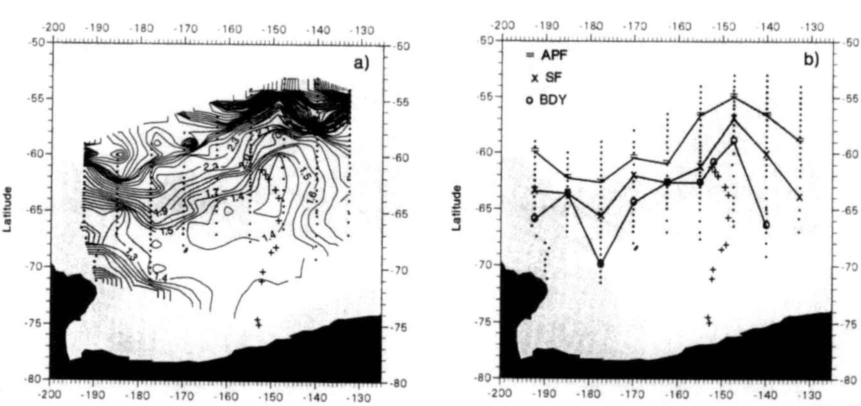 b

Fig. 11. a Maximum temperature within Upper Circumpolar Deep Water based on data collected in December 1985-February 1996 (*dots*, 260 stations) and in July 1985 (*pluses*, 12 stations). **b** Locations of the Antarctic Polar Front *(APF)*, Southern Front *(SF)* and southern ACC Boundary *(BDY)* from the same dataset from the weakly stratified waters of the Antarctic zone. The boundary is also associated with an increased geostrophic shear contributing to the eastward water transport. In general, this scheme is confirmed by our analysis of temperature distributions on the sections. As noted by Maslennikov [8] distribution of temperature within the UCDW temperature maximum provides a good characteristic of the upper Ross Gyre and its boundaries. Figure 11a gives a quasi-synoptic view of the temperature field within the temperature maximum layer and Figure 11b shows locations of the APF, SF and BDY during the survey. The location of the Antarctic Polar Front is marked by a zone of strong meridional gradients within the water warmer than 2.3–2.5 °C. Farther to the south, two other frontal discontinuities are found within the temperature range 2.0–1.4 °C. We identify these as the Southern Front and the ACC subpolar regime boundary (BDY). On all sections east of 140°W the APF is located just north of the northern slope of the mid-ocean ridge. The SF follows approximately the ridge crest, whereas the BDY lies south of the ridge crest.

Figure 12 shows the potential temperature distribution on the section along 177.5°W. A pronounced intrusion of the waters from the southern margin of the

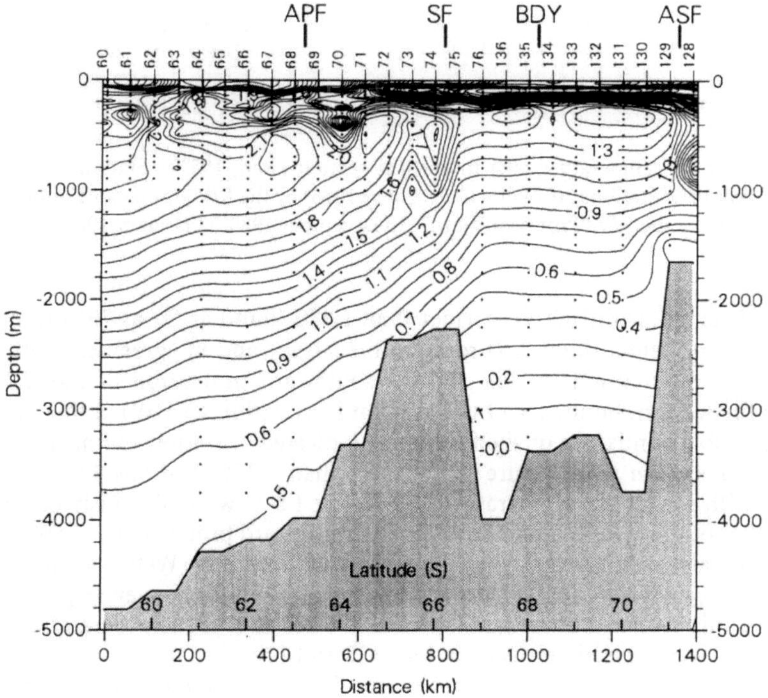

Fig. 12. Potential temperature (°C) in vertical section along 177.5°W. Stations *60-76* by the R/V "Professor Zubov", (8-9 February 1986); stations *128-136* by the R/V "Professor Vieze" (20-24 January 1986)

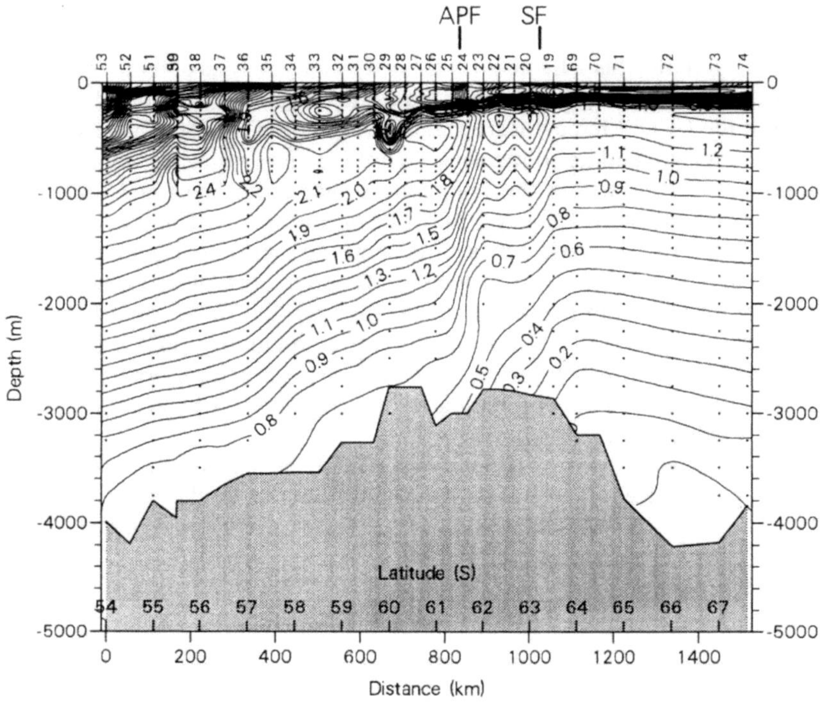

Fig. 13. Potential temperature (°C) in vertical section along 155°W. Stations *51-53* (27 January 1986) and stations *70-74* (15-17 February 1986) by the R/V "Professor Zubov"; stations 128-136 by the R/V "Professor Vieze" (30 December 1985-02 January 1986)

ACC is observed, with UCDW warmer than 1.5 °C found as far as 70 °S near the Antarctic continental slope. According to the survey and in agreement with climatological distributions in Fig. 10, this southward protrusion of the UCDW defines the western boundary of the gyre (at least within UCDW) at this meridian. A possible admixture of shelf waters is indicated at station 134 near 68.5 °S, whereas maximum temperature falls to less than 1.3 °C. Stations 133-130 (68-70.5 °S) with maximum temperatures exceeding 1.5 °C were obviously occupied within the southern limb of the Ross Gyre. A southward protrusion of CDW from the north is also confirmed by the high-quality data from the WOCE S-4 line: neither Ross Deep Water (Fig. 6) nor a newly formed bottom water (Fig. 7) were found on this section between 177 °W and 170 °W at 67 °S. The Antarctic Slope Front (ASF) was crossed between stations 128 and 129.

The section along 155 °W (Fig. 13) reveals both the pronounced SF at 61.5 °S and the BDY at 63.2 °S. Similar to other sections, the temperature distribution indicates formation of eddies or meanders between the SF and the southern terminus of UCDW, seen as warm cells with typical temperatures in the core of at least 1.5 °C. These structures are similar to the warm cells described by Gouretski and Danilov [30] for the boundary of the Weddell Gyre.

6.2 Eastern Boundary

Comparing the circulation patterns at different levels and surfaces with the bottom topography of the study area suggests a close link between the bottom topography and the eastward extension of the Ross Gyre. The situation particularly comparable with the eastern Ross Gyre is found in the Atlantic sector of the Southern Ocean where the ACC crosses the Southwest Indian Ridge. Here, the ACC core associated with the Antarctic Polar Front flows through a gap near 25 °E and turns southward on the eastern side of the ridge [29]. The southward deflection of the flow occurs due to the vorticity conservation requirements. The section along 140 °W (Fig. 14) crosses a very intense core of the ACC at the APF near 57.5 °S. This current core passes through the Udintsev Fracture Zone and is especially pronounced east from this gap in the upper levels (see Fig. 2a). A temperature front is developed throughout the whole water column with a temperature drop of about 0.9 °C between stations 37 and 35 only 110 km apart. The temperature front is associated with a strong salinity front (not shown), so that the integral eastward transport relative to the bottom between these two stations amounts to 56.4 Sv (about half of the baroclinic transport of the ACC between Antarctica and New Zealand). A well-pronounced mesoscale feature (an eddy or

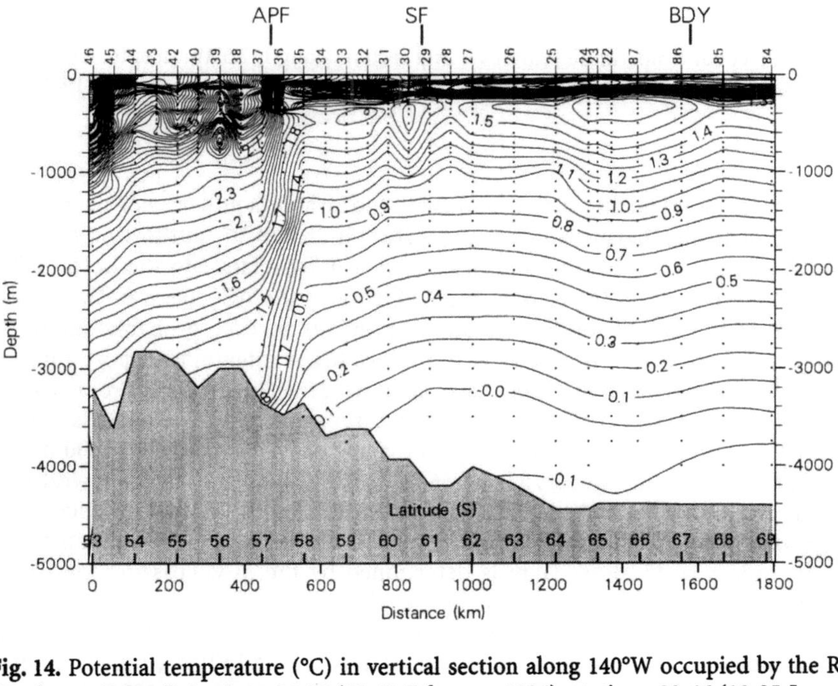

Fig. 14. Potential temperature (°C) in vertical section along 140°W occupied by the R/V "Professor Zubov". Stations *84-87* (21-22 February 1986); stations *22-46* (18-25 January 1986)

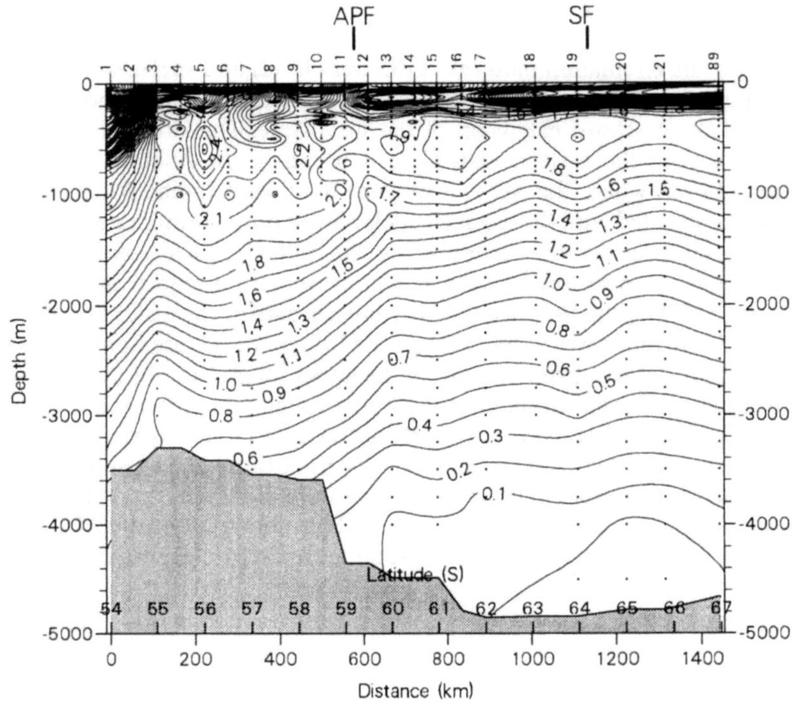

Fig. 15. Potential temperature (°C) in vertical section along 132.5°W occupied by the R/V "Professor Zubov". Station *89* (23 February 1986), stations *1-21* (13-18 January 1986)

meander) is found farther to the south, with the SF at its southern side near 60.5 °S. Doming of the isotherms south of the feature indicates the outflowing northern branch of the gyre with the total transport of 8 Sv relative to the bottom. The 132.5 °W section (Fig. 15) crosses the APF core of the ACC near 60 °S, with the front being not so pronounced as on the previous section, probably due to the oblique angle of the section relative to the front. The integral transport amounts to 37.4 Sv. The northern ACC core associated with the Subantarctic Front is found between stations 44 and 46, with the integral transport of 44 Sv. The SF is implied near 64 °S, with the associated integral transport of 7.2 Sv relative to 3000 m. The lowest CDW temperature (1.75 °C at 66 °S) is 0.3 °C higher than the lowest temperature at 140 °W (1.45 °C at 61.5 °S), thus indicating the virtual absence of the deep water typical of the interior of the gyre. Both sections suggest that the eastern boundary of the Ross Gyre is determined by the ACC crossing the Pacific Antarctic ridge, as noted earlier by Read et al. [31].

The easternmost oceanographic section within the inflow area available to us was occupied by the Russian R/V "Mihail Somov" along 150 °W (Fig. 16). Nansen hydrographic casts were obtained between 26 July and 11 August 1985, thus being probably the only winter time measurements within the Ross Gyre [12] (station

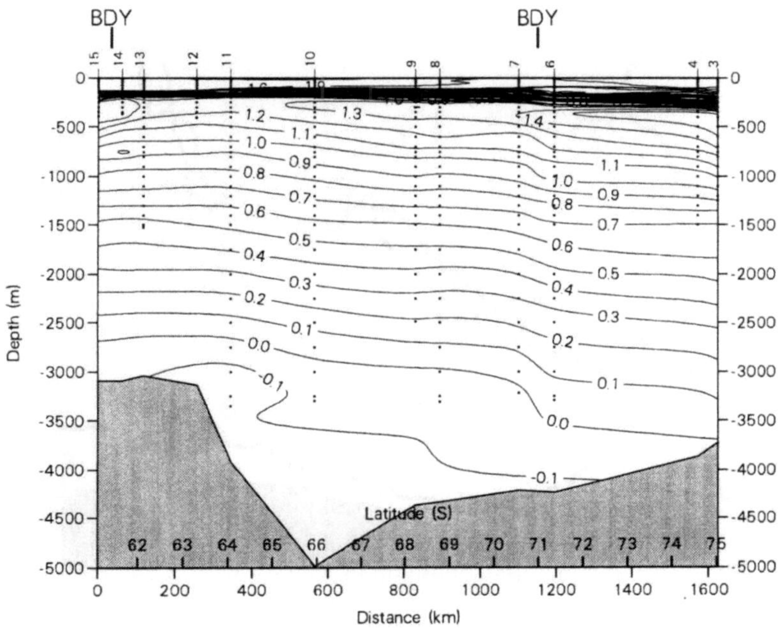

Fig. 16. Potential temperature (°C) in vertical section nominally along 150 °W occupied by the R/V "Mihail Somov" in July 1985

locations are shown in Fig. 10). AASW occupies to upper 150–200 m layer, with rather constant temperatures of about −1.9 °C due to vigorous winter cooling throughout the whole transect. Below the thermocline the section reveals more structure. Highest potential temperatures within the CDW (1.61 °C at 400 m) are observed at the southern most station 3 at 75 °S. The lowest deep-water temperature was measured at 63 °S (1.28 °C at 250 m, station 12) in agreement with the location of a cold UCDW pattern in Fig. 9b. Slope of the isotherm implies eastward flow north of station 11 (64°S) and westward flow south of it. A slight intensification of the westward flow is indicated by the increased slope of the isotherms between stations 6 and 7 near 73 °S.

Evidence of the mesoscale variability is observed on most of the sections in the form of warm-core cells. Intense mesoscale eddies embedding the water of the southern ACC margin were found to be important for the property transports at the eastern boundary of the Weddell Gyre [32]. The rms sea surface height (SSH) variability calculated from the GEOSAT altimeter data at crossover points [32] is shown in Fig. 17 for the sector 190-100 °W. The SSH variability pattern at the eastern boundary of the Ross Gyre seems to be different from that of the Weddell Gyre, where a meridionally oriented high variability area coincides with gyre's dynamic eastern boundary. The area with variability exceeding 8 cm is found to the north of the SF, east of the Eltanin Fracture Zone (see Fig. 2a). Within the gyre and in the immediate vicinity of its eastern boundary (135–140 °W, 60–64 °S) the

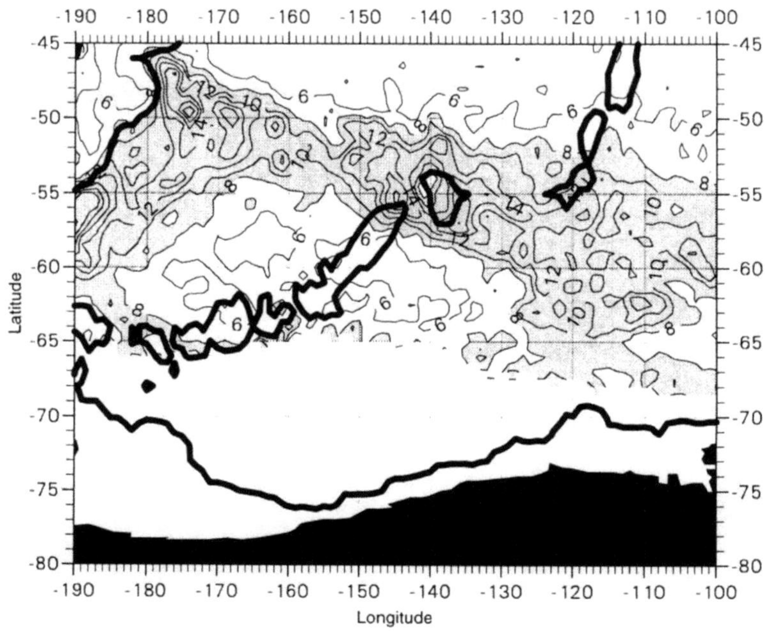

Fig.17. Root mean square sea surface height variability (cm) from GEOSAT altimeter data. The 3000-m isobath is shown by the *solid line*

rms SSH variability is relatively low (< 6 cm). The time-latitude SSH diagrams based on the GEOSAT altimetry (not shown) do not indicate persistent southward movement of isolated ACC eddies similar to those documented by the altimetry and hydrography at the eastern boundary of the Weddell Gyre. An increased variability west of 130 °W is observed south of 65 °S (a maximum of about 10 cm near 133 °W, 66 °S). It may imply an increased role of eddy transports within the inflowing branch of the gyre. However, the altimeter data here are more noisy due to the proximity to the is seasonal sea ice edge.

Acknowledgements. This work was supported by the German WOCE Programme (grant No. 03F0157A) from the Bundesministerium für Bildung, Wissenschaft, Forschung und Technologie. We are grateful to K.-P. Koltermann and S. Dobroljubov for helpful comments. Our thanks are also extended to I.V. Chuguy who made the data obtained by the R/V "Mihail Somov" available to us.

References

1. Gordon AL (1981) Seasonality of the Southern Ocean sea ice. J Geophys Res 86:4193-4197
2. Treshnikov AF (1964) Surface water circulation in the Antarctic Ocean. Information Bulletin of the Soviet Antarctic Expedition 5:81-83 (English translation)

3. Reid JL, Mantyla A (1971) Antarctic work of the Aries expedition. Antarctic J United States 6:111-113
4. Reid JL (1986) On the total geostrophic circulation of the South Pacific Ocean: flow patterns, tracers and transports. Prog Oceanogr 16:1-61
5. Levitus S (1982) Climatological atlas of the World Ocean. NOAA, Professional Paper 13, Rockville
6. Olbers D, Gouretski VV, Seiss G, Schroeter J (1992) Hydrographic atlas of the Southern Ocean Bremerhaven
7. Antipov NN, Maslennikov VV, Pryamikov SM (1987) Location and structure of the Polar Frontal zone in the western part of the Pacific sector of the Southern Ocean (in Russian). In: Biological oceanographic investigations in the Pacific sector of Antarctica. VNIRO, Moscow, pp 19-32
8. Maslennikov VV (1987) Secondary frontal discontinuities in the western part of the Antarctic Pacific sector (in Russian). In: Biological oceanographic investigations in the Pacific sector of Antarctica. VNIRO, Moscow, pp 32-41
9. Patterson SL, Whitworth T (1990) Physical oceanography. In: Antarctic sector of the Pacific. Elsevier, New York, pp 55-93
10. Locarnini RA (1994) Water masses and circulation in the Ross Gyre and environments. PhD thesis, Texas A&M University, College Station (unpubl.)
11. World Ocean Atlas (1994) National Oceanographic Data Center CD-ROM Data Sets. US Department of Commerce, NOAA, National Environmental Satellite, Data, and Information Service, National Oceanographic Data Center, Ocean Climate Laboratory, Washington, DC, 11 disks
12. Botnikov VN, Chuguy IV (1989) Main peculirities of the circulation and spatial distribution of the hydrological elements in the Ross Gyre Inf Bull Sov Ant Exped 112:87-99
13. Jacobs S, Hellmer H (1996) Antarctic ice sheet melting in the Southeast Pacific. Geoph Res Letters, 23:957-960
14. Gouretski VV, Jancke K (1996) A new hydrographic data set for the South Pacific: synthesis of WOCE and historical data. WHP Special Analysis Centre Technical Report No. 2, WOCE Report No. 143/96, Hamburg (unpubli.)
15. Saunders PM (1986) The accuracy of measurements of salinity, oxygen and temperature in the deep ocean. J Phys Oceanog 16:189-195
16. Lozier MS, McCartney MS, Owens WB (1994) Anomalous anomalies in averaged hydrographic data. J Phys Oceanogr 24:2624-2638
17. Orsi AH, Nowlin WD Jr, Whitworth T (1993) On the circulation and stratification of the Weddell Gyre. Deep Sea Res 40:169-203
18. Muench RD, Gordon AL (1995) Circulation and transport of water along the western Weddell Sea margin. J Geophys Res 100:18503-18515
19. Jacobs SS (1991) On the nature and significance of the Antarctic Slope Front. Marine Chem 35:9-24
20. Fahrbach E, Schroeter M, Klepikov AV (1994) Circulation and water masses in the Weddell Sea. A contribution to the summer school in Savolinna, 1994 (unpubl.)
21. Jacobs SS, Fairbanks RG, Horbie Y (1985) Origin and evolution of water masses near the Antarctic continental margin: evidence from ratios in seawater. Antarctic Res Ser 43:59-85
22. Jacobs SS, Gordon AL, Ardai JL (1979) Circulation and melting beneath the Ross Ice Shelf. Science 203:439-443
23. Gordon AL, Tchernia P (1972) Waters of the continental margin off Adelie Coast, Antarctica. Antarctic Res Ser 19:59-69
24. Carmack EC, Killworth PD (1978) Formation and interleaving of abyssal water masses off Wilkes Land, Antarctica. Deep Sea Res 25:357-369
25. MacAyeal DR (1984) Thermohaline circulation below the Ross Ice shelf: a consequence of tidally induced vertical mixing and basal melting. J Geophys Res 89:597-606
26. Gordon AL (1967) Structure of Antarctic waters between 20W and 170W. Antarctic Map Folio Series 6:1-10
27. Carmack EC (1977) Water characteristics of the Southern Ocean south of the Polar Front. In: Angel M (Ed.) A voyage of doiscovery George Deacon 70th Anniversary Vol, Suppl to Deep Sea Res Pergamon, Oxford, pp 15-41
28. Jackett DR, McDougall T (1997) A neutral density variable for the World's Ocean. J Phys Oceanogr 27:237-263

29. Orsi AH, Whitworth T, Nowlin WD (1995) On the meridional extent of the Antarctic Circumpolar Current. Deep Sea Res 42:641-673
30. Gouretski VV, Danilov AI (1993) Weddell Gyre: Structure of the eastern boundary. Deep-Sea Res 40:561-582
31. Gouretski VV, Danilov AI (1994) Characteristics of warm rings in the African sector of the Antarctic Circumpolar Current. Deep Sea Res 41:1131-1157
32. Gouretski VV, Seiss G (1993) Calculations of gridded sea surface height anomaly fields from GEOSAT altimeter data south of 40oS. Tech Rep Berichte aus dem Fachbereich Physik 42, 27 pp (unpubl.)

Water Masses and Dynamics of the Ross Sea Continental Shelf

Temporal Variability of Currents in the Ross Sea (Antarctica)

P. Picco[1], L. Amici[2], R. Meloni[2], L. Langone[3] and M. Ravaioli[3]

[1] ENEA CRAM, P.O. Box 316, 19100 La Spezia, Italy
[2] CNR Istituto di Oceanografia Fisica, P.O. Box 316, 19100 La Spezia, Italy
[3] CNR Istituto di Geologia Marina, Via Gobetti 110, 40129 Bologna, Italy

Abstract

Some results from the analysis of about one year long time series of currents, temperature and salinity data collected in the central-western part of the Ross Sea during 1995 are here presented. Temporal evolution and seasonal variability of currents at intermediate and deep layers in different locations are described. The role of surface seasonal processes and of the advection of different water masses on temperature and salinity variations is investigated. Spectral analysis, focusing on low frequency oscillations, evidences that kinetic energy is mainly concentrated on mesoscale phenomena time scale at which barotropic movements prevail. At higher frequencies, the tidal component strongly dominates, while the inertial oscillation, whose frequency is the same band, is less important. The ice formation and melting processes affect also the intermediate level as evidenced in temperature and salinity time series, but do not seem to modify significantly the circulation in this layer. Apart from a weakeness in kinetic energy observed in most of the locations from late April to early June, the general circulation does not present an important seasonal variability, suggesting that the open ocean Antarctic Circumpolar Current acts as a prevailing forcing with respect to the thermoaline forcings.

1 The Experiment and Data Treatment

Six moorings composed of current meters, temperature and conductivity sensors, sediment traps and turbidimeters were deployed in the central-western part of the Ross Sea Shelf by the end of December 1994 during the oceanographic expedition of N/O "Italica", in the framework of the Italian National Programme for Antarctic Research [1]. Three of the moorings (D, F, H) were deployed in the framework of CLIMA (Climatic Long-term Interaction for the Mass balance in Antarctic) Project; the others (A, B, C) were managed by the CNR Istituto di Geologia Marina of Bologna to support the studies of biogeochemical processes in the Ross Sea (ROSSMIZE Project) involved in the seasonal retreat of the sea–ice margin. All the moorings were recovered, maintenanced, and deployed again the following year, during the 1995/1996 expedition [2].

Mooring F was located 2 miles in front of the edge of the Ross Sea Ice Shelf, at the latitude of 177°W. H was located 2° north of mooring F, at about the same lon-

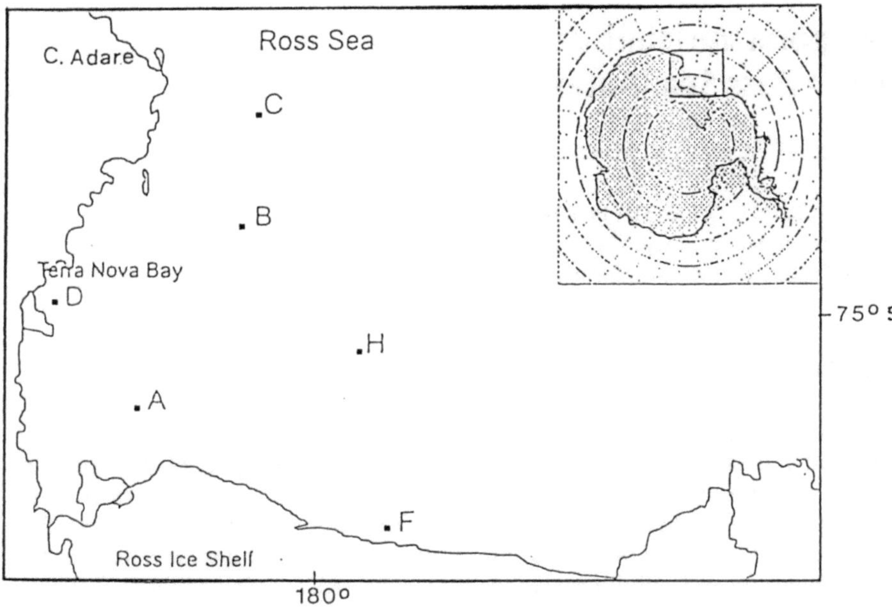

Fig. 1. The Ross Sea with location of moorings

gitude, bounded west by the Ross Bank; mooring D was deployed in Terra Nova Bay, at a depth of 912 m, in an area of recurrent polynya [3, 4] with the aim of investigating the physical processes of the High Salinity Shelf Water formation. Mooring C was at the northwestern edge of the shelf, in front of Cape Adare, where the shelf becomes narrow and the effects of the open sea circulation are more intense. B was about 90 miles south of C at a depth of 555 m, and A was on the western corner of the Ross Ice Shelf barrier, north of Ross Island (Fig. 1).

Current meters were Aanderaa RCM7, or RCM8, equipped with temperature and conductivity sensors; other temperature and salinity sensors, which allow better accuracies, were SeaCat SBE3 and SBE4. Moorings A, B, and C also had strain-gauge pressure sensors at the top to detect the vertical stability. Except for mooring D, which had the upper current meter at 55-m depth [4], all the moorings had the upper current meter at a depth around 240 m in order to avoid damage from drifting icebergs [5], and, one (at least) at about 40 m near the bottom. The analysis is thus limited to the intermediate and deep level. Apart from current meters at mooring F, which stopped working on September 3 and 7, 1995, at 244 and 579 m respectively and on October 10 1995 at 391 m, the others regularly operated until they were recovered, obtaining time series of about 1 year or more. One conductivity sensor was found broken and a few of them did not provide reliable results, so some of the temperature and salinity time series are not complete (Table 1). Data were preliminarily checked for errors and some small omissions of data were reconstructed by means of cubic splines. Correction for the magnetic declination for each mooring was applied. Due to the different sampling interval, 1 or half an

Table 1. Location of moorings (A-H), period of measurements and depth of current meters

	A	B	C
Latitude	76°41.03' S	74°01.50' S	72°29.55' S
Longitude	169°0.23' E	175°0.55' E	175°08.10' E
Sea depth (m)	815	588	456
Deployment	22/01/95	11/12/94	09/12/94
Recovering	19/01/96	12/01/96	12/01/96
Current meters depths (m)	220	220	230
	780	555	425
	D	F	H
Latitude	75°06.10' S	77°59.99' S	75°56.20' S
Longitude	164°13.04' E.	177°01.62' W	177°35.98' W
Sea depth (m)	912	602	625
Deployment	17/02/95	27/01/95	04/02/95
Recovering	14/01/96	21/01/96	23/01/96
Current meters depth (m)	55	244	283
	140	390	450
	402	579	503
	748		597
	882		

hour, all the time series were reduced to hourly values and, from them, daily mean values were obtained. Filtering and resampling were applied to eliminate high frequency variations and to better evidence long-term trends. Unless specified, reported data are daily mean values; temperature are in-situ values.

2 Statistical and General Description

Currents in front of the Ice Shelf barrier at 177°W longitude (mooring F) show a mean northwestward pattern and a northward deviation from mid April, at all the examined levels. About 50% of daily mean direction is grouped in the range over 270–360 °N, 15% is between 225 and 270 °N and the remaining, which is spread over the wide range between 0 and 225 °N, can be explained by the observed meandering. Velocities are quite high: more than 40% of daily mean values being higher than 10 cm/s, with peaks of more than 30 cm/s, especially at the bottom level in January and July. Even though the flow is mainly barotropic (the first barotropic mode accounts for more than 79% of variance, and the second for 15%), the analysis of temperature and salinity time series reveals more detailed features such as the alternative signature, at the depth of 398 m, of different water masses. From hydrological measurements taken during the austral summer campaign, it was observed that the current meter at 391 m depth was just at the interface between two well-defined water masses whose "cores" were found at about the same depth: the Ice Shelf Water (ISW) at west and the warmer Modified

Circumpolar Antarctic Water (MCAW) at east. Salinity at this level is around 34.46–34.48 for most of the year (78% of the registered daily mean values are below 34.5), with corresponding temperature ranging from a maximum of −1.1 °C in mid April to −1.8 °C during July-August. This indicates the presence of the warm Modified Circumpolar Antarctic Waters circulating in the shelf; the high variability in temperature could be ascribed to mixing with the colder waters above and to possible displacements of the core of this water mass. The examined time series also evidenced episodes lasting from a few days up to more than a week, with salinity peaks of more than 34.52 and temperature lower than −2.1 °C, typical of the Ice Shelf Waters formed below the Ross Ice Shelf [6, 7]. They appeared more pronounced, with salinity higher than 34.62, after August. In correspondence with these events, an intensification of the northern component of the currents is expected, indicating an outflow from below the Ice Shelf, but such correlation was not clearly found. In addition, on 255 days of current measurements, 157 daily mean data were in the northern sector (from 270 to 360°N and from 0 to 90°N) and 98 had a southern component; during the same period, the signature of ISW, defined by salinity higher than 36.5, was observed only in 38 cases, while the other 217 identified the presence of MCAW (Fig. 2). This suggests that, at this longitude, both ISW and MCAW mainly flow along the Ice Shelf barrier, meandering, and only in some cases being directed inside the cavity. In terms of transport per square meter for the whole period of measurements, it was com-

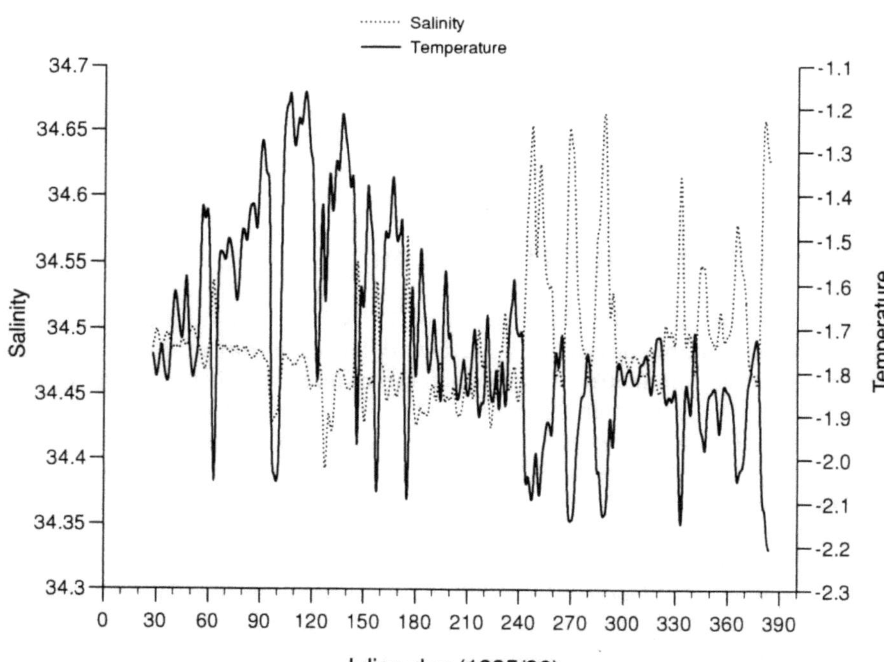

Fig. 2. Daily mean data of temperature and salinity at 390 m from mooring F reported in a temperature-salinity diagram

puted respectively as 1.45 x 10^6 m^3 northward and 0.75 x 10^6 m^3 southward; resulting net transport is 0.11 x 10^6 m^3 directed 337°N.

Moving northward to the shelf break, the signature of the core of Ice Shelf Water disappears at about 77.5 °S, where it supposedly deviates west; this is why such a cold water is never detected by the sensors of mooring H. Currents and temperature and salinity data from mooring H indicate waters leaving the Ross Shelf to the open ocean after having circulated in the western side. Prevailing directions of currents are north-northeast and southeast with respectively 45 and 38% of daily mean data. Southeastern currents, with lower intensity, are mainly observed in April and May, while in December and January they are higher and northward. Daily mean velocities are rather high: more than 25% higher than 10 cm/s, with peaks of more than 35 cm/s. The flow is strongly barotropic: from empirical orthogonal function (EOF) analysis the first mode reaches 93.8%, while the second is only 4.4%. Meandering, oscillations and complete anticlockwise rotations occurring in July and January 1996 can be observed simultaneously at all depths (Fig. 3a). Resulting annual mean velocity at all depths is directed around 65°N with an intensity of 3-4 cm/s (Table 2).

Table 2. Mean current intensity with standard deviations and maximum; mean velocity components with standard deviations and resulting module and direction. Velocities are in centimeters per second; direction in °N. The whole period of measurements was considered for each mooring (A-H)

	A		B		C		F		
Depth (m)	220	780	220	555	230	425	244	391	579
Mean speed	6.4	6.9	5.0	5.3	9.6	9.2	9.7	10.0	9.9
σ	3.3	3.1	3.1	2.4	4.1	3.5	5.5	5.7	6.3
Max speed	22.2	18.7	15.6	15.5	20.7	21.8	26.8	27.1	31.7
Vx	0.6	1.5	0.4	2.8	-2.3	-5.9	-3.5	-2.2	2.0
σx	6.0	5.3	4.1	3.2	4.9.	4.2	8.0	8.5	7.4
Vy	-0.5	-0.7	-0.6	0.1	6.0	5.8	2.5	2.0	4.6
σy	3.6	5.2	4.4	4.1	4.2	3.1	6.6	7.2	7.7
Module	0.7	1.3	0.6	2.3	8.0	8.4	4.3	3.0	5.0
Direction	129	127	147	89	343	316	306	312	337

	D					H			
Depth (m)	55	140	402	748	885	283	450	503	597
Mean speed	14.1	9.3	7.1	6.7	8.0	7.9	7.7	6.7	6.5
σ	9.0	5.4	4.4	4.9	5.7	4.9	4.7	4.4	3.9
Max speed	48.2	27.5	24.7	24.0	26.3	28.2	26.6	26.5	22.9
Vx	-0.8	-1.1	0.6	2.6	4.9	3.2	3.8	2.7	2.7
σx	9.6	7.2	5.7	5.4	5.8	5.2	5.0	4.1	3.9
Vy	7.7	3.0	2.1	0.9	0.4	1.7	1.5	1.5	0.9
σy	11.3	7.3	5.7	5.8	6.2	6.8	6.3	6.2	5.9
Module	7.8	3.2	2.1	42.8	4.9	3.6	4.1	3.1	2.9
Direction	354	340	15	71	86	62	69	61	72

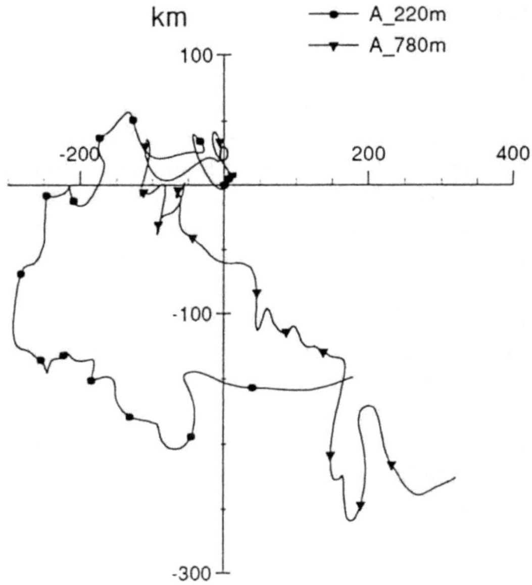

a

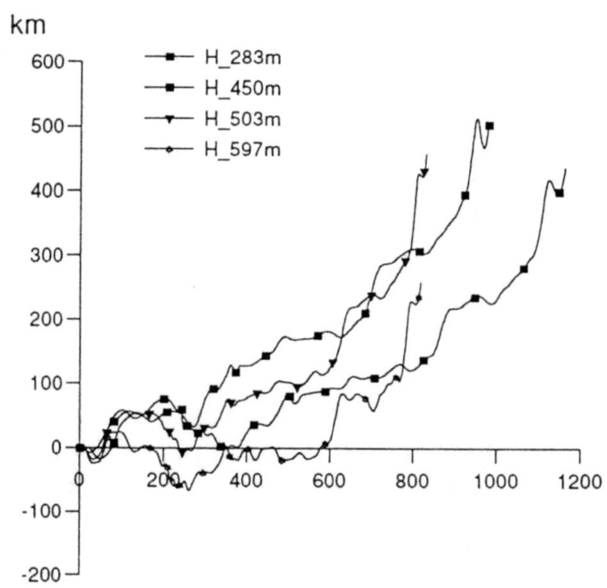

b

Fig. 3. Progressive vectors from mooring A (**a**) and from mooring H (**b**). Data are smoothed and markers indicate 30-day intervals

Currents observed at mooring A are characterised by a high variability with frequent inversions and rotations; daily mean velocities are generally less than 10 cm/s except during a few events in May and July when they reach more than 20 cm/s and are directed southeast at both levels. Daily mean data of direction show a clear bimodal distribution more accentuated at 780 m with 32% in the range 46–90°N and 25% in the range 180–225 °N. Currents are directed west during the period February–July; after then they turn southeast, indicating cold and salty waters (temperature below -1.9 °C and salinity around 34.72) probably coming from the polynya region entering the Ross Ice Shelf or recirculating in the shelf at a depth of around 250 m. Resulting annual mean values, which account for the water masses transport, are extremely small: less than 1 cm/s directed southeast. Correlation between correspondent velocity components at the intermediate and near-bottom levels is around 0.75. Warming at the depth of 220 m occurs from March to late April with maximum temperature (–1.54 °C) followed by a long period (from June to January) with temperature below –1.9 °C.

Flow at 220 m at mooring B is characterised by meandering, frequent and rapid complete rotations both clockwise and anticlockwise and slow speeds with 65% of daily mean values below 5 cm/s (Fig. 3b). Main direction is southwest until mid March; then it inverts, going northeast for about 6 months. From November it turns southeast with a more regular pattern. Thirty five percent of daily mean currents is directed in the range 45–135 °N while the remaining is almost uniformly distributed. Temperature in this level is generally warmer than in the inner shelf, ranging from about –0.4 °C at the end of March to –1.84 °C in mid December. An episode of rapid warming with temperature higher than –0.7 °C and a peak of –0.5 °C, lasting about 20 days is observed in March; after decreasing to 1.4 °C, temperature still increases, remaining higher than –1.2 °C until early August. This discrepancy from the expected seasonal cycle, associated with the high variability of the currents, could be ascribed to intrusions into the shelf of warmer waters from the open sea. Deeper current (550 m) patterns differ from the upper: flow is mainly directed east (55% of daily mean direction being in the range 45–135°N), with some north-south oscillations of about 2 month periods clearly evident during austral summer months; velocities are extremely low (75% of values less than 5 cm/s), especially during the period April-August.

Currents in the northwestern edge of the Ross Sea Shelf (mooring C) are rather constant in direction, due to the topographic effects such as the narrowing of the shelf. Daily mean direction is concentrated in the range 315–45°N (80%) at 230 m and 270–360°N (88%) at 425 m (Fig. 4). Forty percent of velocities are higher than 10 cm/s with peaks of 25 and hourly mean intensity reaching more than 90 cm/s. Resulting annual mean velocities are respectively 8 and 8.4 cm/s directed 343 and 316°N. Temperature is higher than elsewhere in the shelf, being around 0.5 °C at 22 m and –0.7 °C at 425 m probably owing to the stronger interaction with the general open sea circulation.

Main feature and temporal variations of currents in Terra Nova Bay are deduced by the analysis of data from mooring D. A more detailed description is found in [4]. A two-layer system, persistent all year long, characterises the circulation in this area. Currents at 55 and 140 m show strong northern alongshore components; peaks of more than 50 cm/s are frequent at the upper level, partly

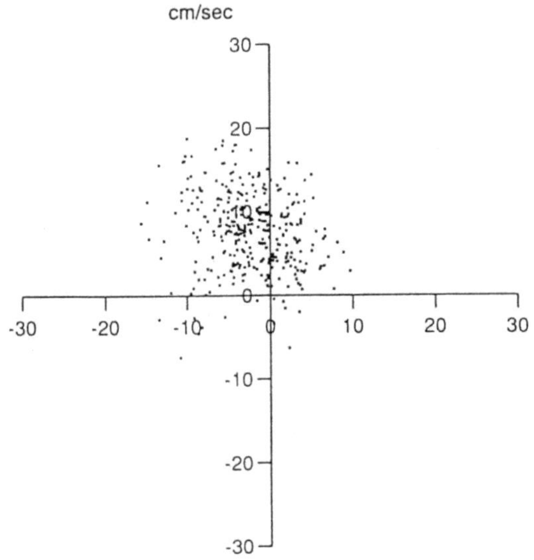

Fig. 4. Scatter plot form for mooring C at 230 m

explained by the Ekman transport due to the intense offshore winds. At the depth of 402 m, northern flux is still prevailing, but at higher depths a clear eastward turning is observed. Meandering is observed at all levels. Results from EOF analysis confirm these observations: the first mode accounts for 78% of the variability, but the second is also important (about 17%).

3 Spectral Analysis

Data from two current meters for each mooring (one around 240 m and one near the bottom) were analysed to outline the horizontal and vertical characterisation of frequency pattern above the Ross Sea. Frequency analysis was performed separately for the "main" tidal range and for a mesoscale range which includes periods of up to about 15 days. Tidal and high frequency movements are important in the polar oceans as they produce breaks in the sea ice pack, favouring the melting processes; low frequency currents account for the transport of particles. In the first case, power spectra of velocity magnitude were obtained by FFT-based algorithms after separating velocity module vectors into groups of 1024 data (for 30-min rate sampled series) or 512 data (for hourly sampled series, namely from mooring H). This yielded a frequency distribution of kinetic energy, apart from a constant factor. For the mesoscale, half-day mean values were obtained from velocity magnitude series before processing them with a non-FFT spectrum-evaluating method (via autocovariance): this was because series after resampling resulted too short to apply the FFT routines. The rate of resampling was chosen

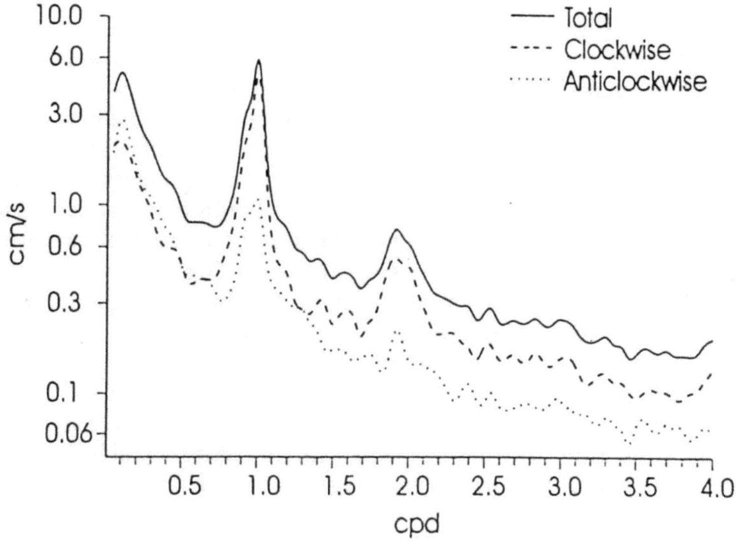

Fig. 5. Rotary spectra of currents at 283 m from mooring H

to allow periods of around 14 days to be distinguishable, which might occur for meteorological and astronomic forcing.

3.1 Tidal and Inertial Currents

Inertia (Coriolis) current periods at mooring latitudes vary from 12.23 to 12.55 h respectively 1.91 and 1.93 cycles per day (c.p.d.), thus falling into the semidiurnal band (M2 frequency being 1.93 c.p.d.) and making it quite difficult to detect distinct peaks. Only removal of principal tidal constituents after harmonic analysis would allow us to detect possible inertia circle events (by inspection of residual time series) as well as seiche-like frequency packets, which, at present, might have been partially masked by principal lunar and solar tide constituents.

Following Godin [8], a rotary-component method was used to separate clockwise and anticlockwise parts of the complex Fourier transform of the velocity vector: as in the southern hemisphere, the inertia cycle is clockwise and comparison between the two components around the investigated frequency should allow us to distinguish the signal, if it is not too weak or if resolution is not too poor. Time series relative to about 7 (mooring F) to 13 (mooring C) months of measurements were divided into pieces covering about 21 days; for each subset the complex vector of velocity, the real part of which was the west-east component series while the imaginary part was the north-south one, was de-averaged and detrended, and then Fourier transformed; from the transformed vectors, anticlockwise, clockwise and total frequency components were derived according to Gonella [9]; finally, an average of the sections was performed.

In spite of these attempts, no clear result could be achieved, except that in the semidiurnal band, the clockwise signal is generally stronger than the anticlockwise one, indicating a little weight of inertia current phenomenum during the overall period, which does not exclude short lasting events (Fig. 5). Spectral analysis revealed a strong predominance of diurnal components with respect to semidiurnal ones, which is in complete agreement with previous results as outlined by Lutjeharms et al. [10], who particularly quoted several sea-level measurements performed in the Ross Sea since the beginning of the twentieth century up to 1979. Analogous outcome was determined by Stocchino and Manzella [11], who analysed currents in the area of Terra Nova Bay.

In the present analysis only current meters on mooring B showed a large semidiurnal signal at both depths of 230 and 550 m, with a substantially equal percentage of kinetic energy in both diurnal (0.8–1.2 c.p.d.) and semidiurnal (1.8–2.2 c.p.d.) bands, from now on shortened to Di and SDi (Table 3); the other instruments measured Di energy percentages as great as 1.67 (mooring D at 55 m) and SDi ones as great as 9.9 (mooring F at 579 m).

Mooring D, which showed a relatively strong SDi component, is rather near to the site in the Gerlache Inlet in which Stocchino and Manzella found [11] a predominant SDi signal in power spectra. Moorings A and C, where the ratio between Di and SDi has intermediate values, are relatively nearer the coastline than moorings H and F (the latter are very close to the Ross Ice Shelf), where the largest ratios were detected. Depth of measurement seems not to be relevant for the magnitude of the cited ratio.

For each mooring, energy distribution among main frequency bands does not differ significantly between the two examined depths. With the exception of moorings C, where signals are by far stronger than elsewhere, power spectra from other sites showed signals of comparable magnitude in both Di and SDi bands. The lowest peak in the dominant band occurred at site A, a slightly one at F and

Table 3. Kinetic energy percentage in the bands Di = 0.8 c.p.d. and SDi =1.8 c.p.d. for moorings A-H

	A		B		C	
Depth (m)	220	780	220	555	230	425
% Di	14.54	17.45	29.80	21.96	10.80	18.39
% SDi	4.27	2.87	27.27	30.61	3.88	5.12
Di/SDi	3.41	6.08	1.09	0.72	2.78	3.59
	D		F		H	
Depth (m)	55	748	244	579	283	503
% Di	4.12	5.55	13.04	13.56	37.55	34.90
% SDi	2.47	2.28	2.74	1.37	4.43	5.53
Di/SDi	1.67	2.43	5.01	9.90	8.48	6.31

D, and progressively higher at H, B and C. Tidal amplitudes seem to increase from region closer to the coasts or to the Ross Ice Shelf barrier to offshore ones.

3.2 Low Frequency Currents

On lower frequency bands, two well-known tidal periods, lunar fortnightly (Mf; 13.66 days) and lunar-solar fortnightly (MSf; 14.76 days) have been found in datasets from B (both instruments), C (only Mf at 420 m), D (MSf at 55 m) and H (Mf from both instruments). The peaks seen in the spectrum were particularly high and neat for C, H and B, even if the two components were in general not well resolved, in agreement with relevance of signal noticed for other bands above (Fig. 6).

Components with periods of ~9 and 11 days, typical of mesoscale activity, and others spread in the range of a few days, were found in most spectra, with peculiar strength and evidence for the 11 days component of mooring H. Mesoscale activity interests the whole examined water column: important meandering and rotations are observed simultaneously in all the current metres in the same mooring. A noticeable feature cannot be ignored in the C 420-m spectrum: at 6.8 days, i.e. about half the Mf period, there was a signal as high as the Mf one, while it scarcely and weakly emerged in the 230-m spectrum from the same mooring. Other peaks occurred at lower frequencies, particularly between 16 and 33 days, but rather irregularly; they were very neat for A meters, but with low energetic content.

4 Discussion

Seasonal cycle is mainly determined by the ice formation and melting processes occurring at the surface. Ice formation in the Ross Sea Shelf generally starts by the end of February, with observed interannual variability in the initiation [12], and rapidly increases in concentration, reaching maximum coverage after about 5 weeks. In this area ice growth starts early in the eastern part, advancing west and, from the coast, advancing southward, so that the northwestern sector of the shelf can be considered the one with lower ice concentration. Ice concentration on the continental shelf was found lower than in the open ocean, probably due to the newly formed ice exportation from strong continental winds. Ice decay starts in December; summer minima are found from the end of January to the first 2 weeks of February [13].

A seasonal signal due to these processes, detectable by temperature and salinity variations, also reaches the intermediate level with a certain delay, as a result of strong vertical convective movements which mix the water column above. Rapid temperature decreasing is observed in March in Terra Nova Bay polynya (mooring D) where the absence of total ice coverage favours an intense exchange with the strong atmospheric forcings, in May near the Ice Shelf barrier (moorings A and F) and later moving north (B and H). Minimums of temperature, ranging from −1.7 to −1.9 °C, are reached at moorings A, F and H in the period from late

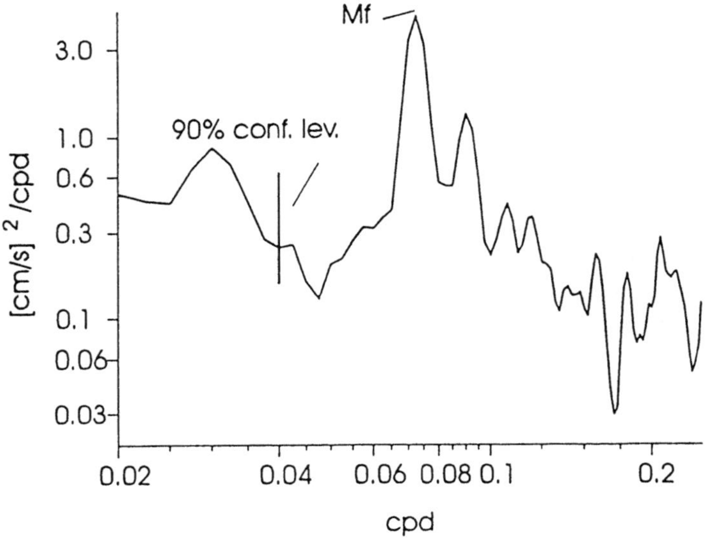

Fig. 6a. Power spectra of currents at 503 m from mooring H

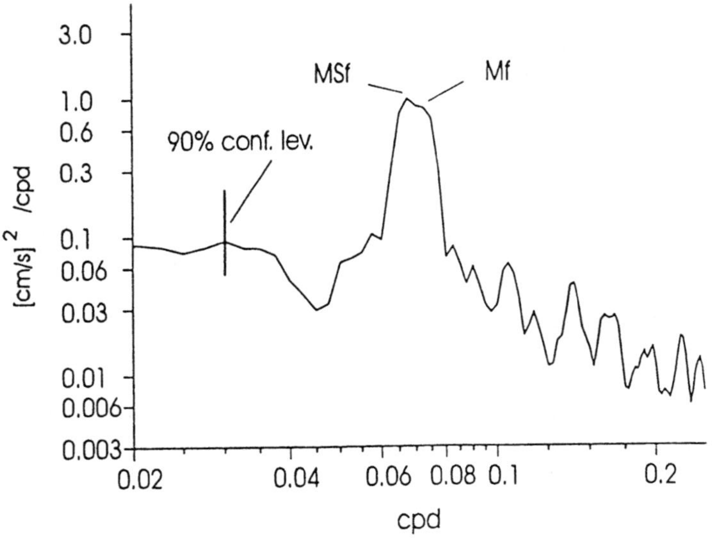

Fig. 6b. Power spectra of currents at 555 m from mooring B

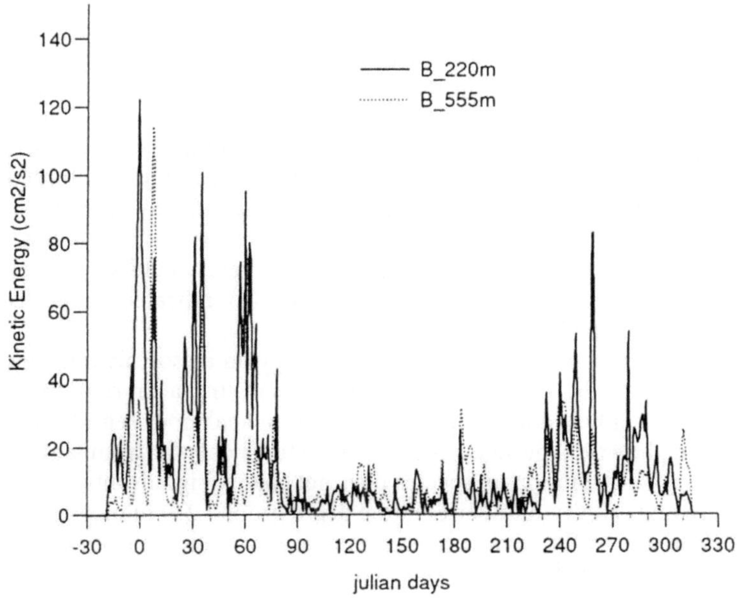

Fig. 7. Kinetic energy from mooring B

June and mid July, remaining almost constant for the rest of the year, until the summer warming, which, at this level, occurs from February. In correspondence with water cooling, salinity starts to increase more slowly, reaching its maximum not before September.

The response of the circulation in the examined layers to these thermodynamic forcings, resulting in a clear seasonal variability, was not evidenced in temporal series of currents. Nevertheless, from early April, when the ice coverage processes are completed, to late June, a weakeness in kinetic energy can be observed in some of the time series, in particular in moorings A, B and H (Fig. 7).

Sparse long-term current measurements in the same area have been performed in the past and reported by different authors. Pillsbury and Jacobs [14] analyzed current meter data taken during 1978 and 1983 in front of the Ross Sea Ice Shelf, between 172 and 180°W. They observed an inflow of Modified Circumpolar Antarctic Water below the Ice Shelf in the eastern mooring, a prevailing westward flow around the latitude 175°W and a weaker northern outflow of Ice Shelf Water at about 180°, with no remarkable seasonal differences in mean direction. The main pattern did not differ significantly from that found in mooring F. Strong and constant northward currents in front of Cape Adare, in correspondence with mooring C, were also observed by Jaeger et al. [15] during 1990. Currents in a location near mooring A, taken from February 1990 to February 1992 and reported in the same paper, showed important differences in the mean direction from one year of measurements to the others and did not show correlation between the intermediate and deep layer as described in this work.

5 Conclusions

The analysis of more than 1 year of temporal series of currents, temperature and salinity in six different sites of the central western part of the Ross Sea Shelf has permitted us to outline some main characteristics of the circulation in the area and its temporal variability. In spite of the presence of several water masses in the Ross Shelf, the strong air-sea-ice interactions and the complicated topography, the circulation in the layer below 240 m is mainly barotropic with a limited seasonal variability. This suggests that the important current systems out of the shelf break (such as the Antarctic Circumpolar Current) could act as a prevailing force for the circulation in the whole shelf, as also indicated by results from numerical simulations [16].

Spectral analysis shows a large amount of energy on mesoscale frequencies in all the moorings. Tidal currents are important, showing a diurnal component prevalence, sided only in one case (mooring B) by an analogous semidiurnal component. Fortnightly tidal periods are relevant features only far from land and the ice shelf. Energy distribution on spectra does not change significantly with depth. Temperature and salinity time series at the depth around 240 m display a limited seasonal cycle due to the surface processes of ice formation and retreat, but, in some cases, are modified by the intrusion of different water masses. Higher salinity is found in polynya areas, decreasing moving east and south. Near-bottom temperature and salinity do not present any seasonal variation, indicating the constant presence of the so-called High Salinity Shelf Water, with a temperature of −1.9 °C and salinity around 34.8, spreading on the lower levels of the whole shelf.

The resulting annual mean pattern of circulation in the central western sector of the Ross Sea Shelf describes a clockwise gyre with waters moving mainly northwest along the coasts and the Ice Shelf barrier and recirculating southeast along the northern part; some slight modifications due to the topography are observed in the near-bottom level. Currents in front of Cape Adare seem to be scarcely representative of the the shelf area, being more directly linked to the open sea circulation.

Acknowledgements. This work was carried out in the framework of the Italian National Programme for Antarctic Research.

References

1. PNRA (1995) Rapporto sulla Campagna Antartica Estate Australe 1994/95 X Spedizione. Progetto Antartide ANT 95/02
2. PNRA (1996) Rapporto sulla Campagna Antartica Estate Australe 1995/96 XI Spedizione. Progetto Antartide ANT 96/07
3. Kurtz DD, Bromwich DH (1985) A recurring atmospherically forced polynya in Terra Nova Bay. In: Budillon et al. Oceanology of the Antarctic Continental Shelf. Antarct Res Ser 43:177-202, AGU, Washington, DC
4. Manzella GMR, Meloni R, and Picco P (1998) Current, temperature and salinity observations in the Terra Nova Bay polynya area (This Volume)

5. Keys JR, Jacobs SS, and Barnett D (1990) The calving and drift of iceberg B-9 in the Ross Sea, Antarctica. Antarctic Sci 2(3):243-257
6. Jacobs SS, Amos AF, Bruchhausen PM (1970) Ross Sea oceanography and Antarctic bottom water formation. Deep Sea Res 17:935-962
7. Jacobs SS, Fairbanks RG and Horibe Y (1985) Origin and evolution of water masses near the Antarctic continental margin: evidence from $H_2\,^{18}O/\,H_2\,^{16}O$ ratio in sea water. In: Oceanology of the Antarctic Continental Shelf. Antarct Res Ser 43:87-107, AGU, Washington, DC
8. Godin G (1972) The analysis of tides. Liverpool University Press, Liverpool
9. Gonella J (1972) A rotary-component method for analysing meteorological and oceanographic vector time series. Deep Sea Res 19:833-846
10. Lutjeharms JRE, Stavropoulos CC, and Koltermann KP (1985) Tidal measurements along the Antarctic coastline. In: Oceanology of the Antartic Continental Shelf. Antarct Res Ser 43:273-290, AGU, Washington, DC
11. Stocchino C, Manzella GMR (1991) Analisi delle correnti marine di Baia Terra Nova (Mare di Ross, Antartide). Consiglio Nazionale delle Ricerche, Genova
12. Comiso JC, Maynard NG, Smith WO Jr, Sullivan CW (1990) Satellite ocean colour studies of Antartic ice edges in summer and autumn. J Geophys Res 95(C6):9481-9496
13. Jacobs SS, Comiso JC (1989) Sea Ice and oceanic processes on the Ross Sea continental shelf. J Geophys Res 94(C12):18195-19211
14. Pillsbury RD, and Jacobs SS (1985) Preliminary observation from long-term current meter moorings near the Ross Ice Shelf, Antartica. In: Oceanology of the Antartic Continental Shelf Antarct. Res Ser 43:87-107, AGU, Washington, DC
15. Jaeger JM, Nittrouer CA, De Master DJ, Klechner C and Dunbar RB (1996) Lateral transport of settling particles in the Ross Sea and implication for the fate of biogenic material. J Geophys Res 101(C8):18479-18488
16. Bergamasco A, Carniel S, Caldesi Valeri L (1998) Reconstructing the general circulation of the Ross Sea using a robust diagnostic model (This Volume)

Reconstructing the General Circulation of the Ross Sea (Antarctica) Using a Robust Diagnostic Model

A. Bergamasco, S. Carniel and L. Caldesi Valeri

CNR-ISDGM, San Polo 1364, 30125 Venice, Italy

Abstract

A robust-diagnostic numerical model for the general circulation of the Ross Sea is presented. The model is forced by open boundary velocities and wind field, while temperature and salinity values have been objectively analyzed from data acquired during the Xth Italian expedition in Antarctic. Even if model assumptions could be improved (e.g. enhance the description of the sea-ice interaction, use available wind time series and updated topography), comparisons between current meters moorings data averages and reconstructed fields shows that temperature, salinity and velocity fields obtained are in good agreement with experimental observations and theoretical studies, even if velocity modules have been under estimated.

1 Introduction

The Ross Sea is an interesting site where different important aspects can be observed and studied. Water formation processes, water mixing, melting and spreading from the Ross Sea and from below the Ross Ice Shelf (RIS), the southern part of the Sea almost permanently covered by hundreds of meters of ice, and the consequent interaction with the circumpolar current are major topics to be explored. The water formation below the RIS and near its edge is not completely understood: it involves interactions between the RIS and both the relatively warm cores of Modified Circumpolar deep Water (MCDW), incoming from the Pacific, and High Salinity Shelf Water (HSSW), formed inside the basin.

Another interesting aspect of the region is the water circulation itself, which can broadly recirculate in the Ross Sea continental shelf break before deepening into the Pacific intermediate and deep layers. Its residence time is an important factor that can trigger the heat exchange between the Southern Ocean and mid-latitude areas, allowing the Ross Gyre to be one of the engines of the circumpolar current.

Last, but not least, the Ross Sea is located in a very critical position near one of the earth's Poles; it can be extremely sensitive to climatic variations. Therefore a good understanding of the mechanisms that drive the general circulation can be an important step towards the understanding of the possible global change and its implications.

At the moment, the hypothesis about the circulation of the area is inferred by different temperature-salinity data sets, mainly vertical profiles, and by a few time series of current meter moorings displaced in the area. The mean circulation of the Ross Sea is mainly west-north-westward; in the western part, waters seem to describe a cyclonic gyre, with a flow directed southward, which later on turns back equatorward and closes the gyre itself.

In the eastern part of the Ross Sea, waters along the coast seem partially directed southward beneath the coast and the Ross Island, entering the RIS, and partially moving equatorward until Cape Adare, thus originating two separate gyres [1, 2]. The development of a numerical model could be seen as a tool to gain more information about the general circulation of the examined area using dynamic constrains, and to check these results against the patterns up until now proposed, based merely on T–S distribution.

This work refers to the general circulation of the Ross Sea during January-February 1995, when the second leg of the tenth Italian expedition in Antarctica took place. Cast profiles explore the major areas; the first one along the RIS edge, the second one going from mooring F to mooring H, cutting the Ross Sea from south to north, and the third one from the continental shelf to Terra Nova Bay and the Drygalsky basin. Unfortunately, the bad weather conditions did not allow us to completely cover the south-west region, and this will reflect using traditional objective analysis methods, which could have some reliability problems during the reconstruction of the variable fields.

Therefore, a numerical model used for a dynamic adjustment of the interpolated fields can give estimations of T and S consistent with the dynamics involved. Comparisons between data averages of current meter moorings and reconstructed current fields give us first validation of the model to reach a complete vision of the quasi-steady summer circulation of this Antarctic region.

Phenomenologically speaking, the tenth Antarctic campaign confirmed the presence of the locally most common water masses: the Antarctic Surface Water (ASW), the Low Salinity Shelf Water (LSSW), the Deep Ice Water (DIW) and the High Salinity Shelf Water (HSSW). Figure 1 shows the ensemble plot of all 152 T–S density casts, measured during the cruise.

From the density plot we can easily see two different bottom water masses, which are the dense Ross Sea basin water and the lighter circumpolar one. A clear differentiation is also visible for temperatures, which are clustered in two major groups, with an interesting and important temperature homogeneity for inside basin waters. Salinity signatures show the HSSW (34.8 PSU, Practical Salinity Unit) even in the slope, at depths of about 1200 m, while the typical ocean basin salinity near the Ross Sea is about 34.67 PSU. The surface salinity high variability in the upper 200–m layer is a confirmation of the summer melting processes.

In the following sections, a brief description of the numerical model characteristics and implementation adopted is given; then a section is dedicated to the description of forcings factors used. Finally, the discussion of numerical results and some conclusions and suggestions for future activity are given.

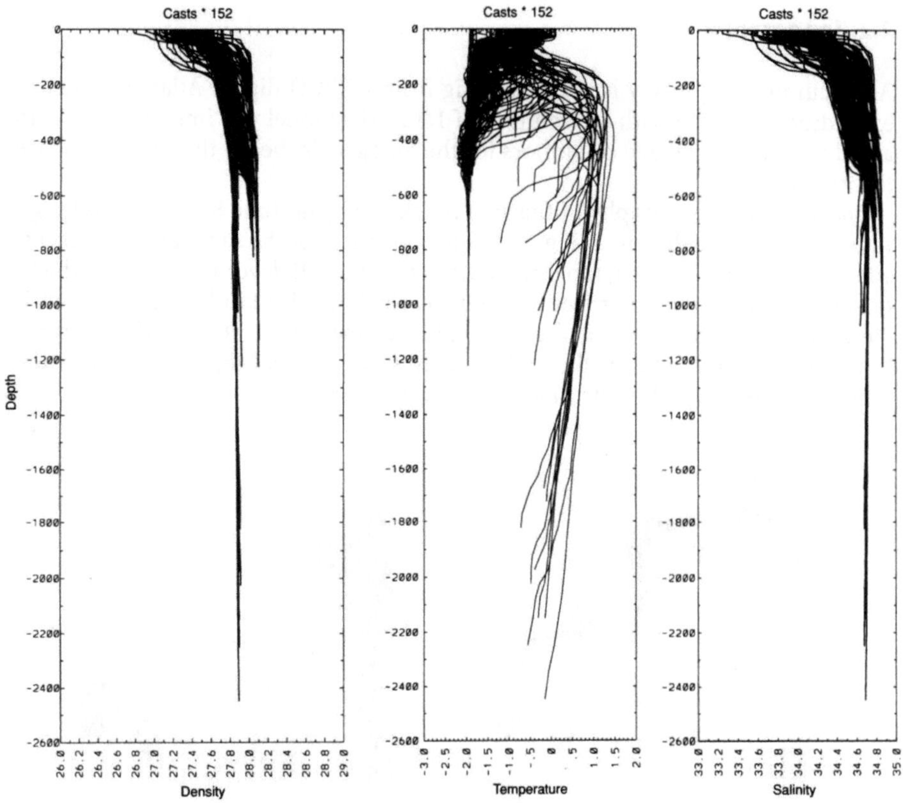

Fig. 1. Plot of 152 cast values for density anomaly (kg dm^{-3}), temperature (°C) density and salinity (PSU) versus depht (m)

2 Methodology

2.1 The Numerical Model

The hydrodynamic circulation model used is a modified version of the Princeton Ocean Model (POM) [3], using a grid configuration of *imax*=56 points along the x axis and *jmax*=73 along the y one. Along the vertical one, *kmax*=30 levels is used within a sigma reference system. The grid size dimensions are variable, with *DX(i,j)* ranging from 43.7 km at 67°S to 9.9 km at 80°S, and a constant *DY(i,j)* equal to 27.8 km, in order to cope with the earth's curvature.

The model implementation does not consider water deeper than 1500 m and therefore behaves as if there were an homogenous virtual topography beyond the continental slope. This allows us to capture all the desired interactions inside the Ross Sea and within the shelf break, neglecting only the oceanic abyssal ones which, in the present work, we are not interested in. Moreover, they should not contribute significantly to the time scales examined.

2.2. Topography

An accurate bathymetry is used. Starting from GEBCO digital Atlas bathymetry extending until 78°S with a resolution of 1/12 °, the model was implemented with an ad hoc one digitalized by authors for the topography below the RIS and the ice thickness [1].

In Fig. 2a the topography of the area bracketed by latitude 85–67 °S and longitude 161–206 °E is shown; in Fig. 2b the ice thickness of the same area is presented. As can clearly be seen, moving to the south, the RIS layer increases its thickness and reduces the water free to circulate. Northward, a kind of plateau is visible at about 600 m: after that, the bottom line deepens until 1000 m, then abruptly points up again and, in the end, quickly sinks to 3000 m; this shows as the topography has been heavily modelled by glacier erosion.

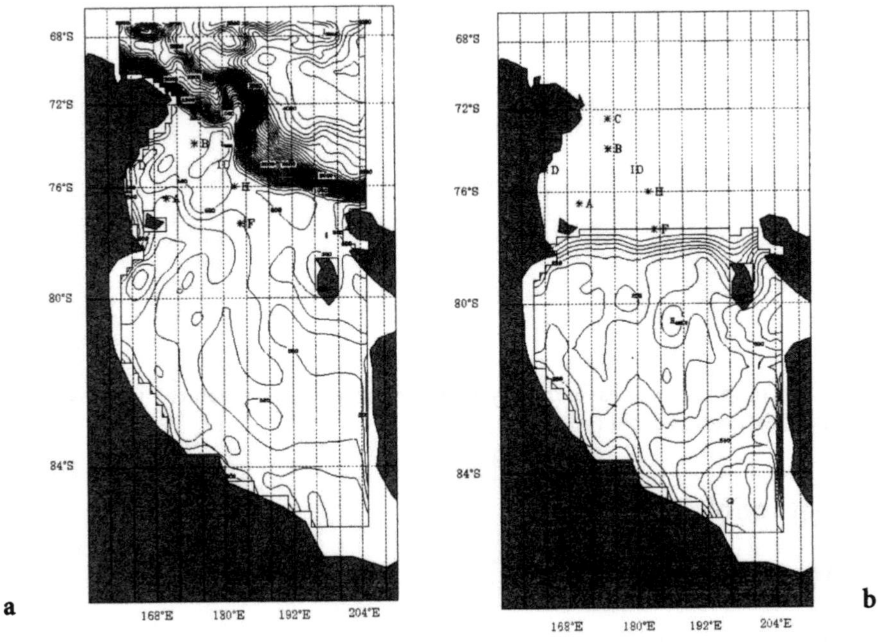

Fig. 2. a Topography (m) of the area between latitude 85 and 67 °S and longitude 162 and 204 °E. **b** Ice thickness (m) in the same area

2.3 Temperature and Salinity Initial Conditions

In order to provide the initial conditions to the model, the T–S data collected in the austral summer 1995 survey have been processed using statistical interpolation techniques (objective analysis). The procedure has been carried out for the whole 3D domain of the model, using values trending to Levitus climatology in the oceanic areas where no data were available. Below the RIS, the present configuration of the model uses values of sea surface elevation and horizontal velocity is set to zero, while temperature and salinity values are computed according to the T–S vertical profile [4]. We used a temperature linearly decreasing from the ice melting pressure-depending one to −1.85 °C at the bottom, while salinity linearly ranges from 34.4 immediately under the ice cover to 34.8 in the bottom. The initial condition sets have been used in runs lasting 2 months.

In the numerical experiments a robust diagnostic integration has been performed. This means that the prognostic values of T and S, adjusted to the velocity field, have been relaxed to their original values, with a relaxation time of $\gamma=1/\tau=0.1$, corresponding to actual time of hours. Different values have been used in order to test the sensitivity of the adjustment: large values of γ almost prevented any change in the T–S distribution, while small values allow the system to adjust better under the baroclinic forcings.

2.4 Open Boundary Velocities

In order to reproduce the forcing caused by the Ross Gyre current, the eastern inflow and the western outflow have been simulated prescribing velocity values in four grid points in the east and west zone (about 100 km). Both vertical averaged velocities and those related to the single level have been imposed, using a decreasing profile.

These values intend to simulate the Ross Gyre current, having direction west-east and entering the Ross Sea area in the eastern part following the continental shelf break, and outflowing the domain near Cape Adare. The computational points where the velocities are imposed have a depth of 1500 m, and are completely ice-free.

A sensitivity analysis of the inflow points shows that the general circulation is slightly sensitive to the forcing points spatial variation, reproducing results quite close even if input points are moved 2–3 degrees. In these preliminary numerical tests, the topography was expanded beyond the examined limits (202 °E, 70 °S), thus creating a numerical buffer zone which allowed the local gyre to close and bend to the output area.

Model forced only with these velocities at the open boundaries produced very small transport values inside the Ross Sea, confirming wind to be the major forcing factor in driving the general circulation. After these numerical evidences, in order to save computing time, the authors decided to reduce the extension of the numerical buffer near the open boundary.

2.5 Wind Stress

A synthetic reconstruction of the prevalent summer winds has been forwarded, in order to reproduce their direction and intensity over the whole basin. It is the authors' intention to analyse and use real wind data, but up until now a wind field has been synthesized on the basis of Buzzi et al. [5]. In Fig. 3 the adopted wind field pattern is shown; maximum velocities are about 10 m/s.

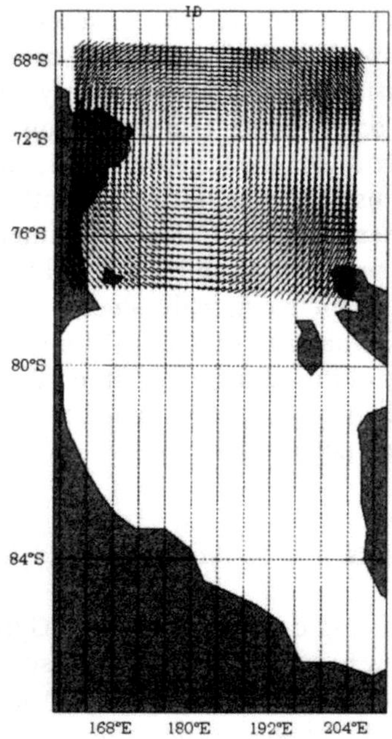

Fig. 3. Adopted wind field pattern. (Reconstructed after [5])

3 Numerical Results

Figure 4 shows the 30-day averaged vertically integrated total transport field, where points A, B, C, D, F, and H are the PNRA mooring locations. A strong topographic control is noted, clearly presented in Fig. 4 by the west-east flux along the steep bathymetric gradient, besides a cyclonic circulation in the proximity of Cape Adare. There is evidence of a transport east-west, closer to the RIS. This situation starts another gyre around the west coast which later on ties up with the major transport flowing to the outflow area. The wind stress, as we expected, acts as a relevant forcing for the large-scale circulation. More attention will be devoted to the acquisition of wind time series and to its climatological analysis. The overall circulation is in good agreement with Locarnini hypothesis.

Table 1 presents a comparison between yearly averaged current meter data [6] and modelled ones at the adjustment. As moorings data have been acquired at different depths, the comparison is carried out at intermediate levels, representative of the sea column slightly affected by boundary effects. For the same reason, modelled velocities shown in the Table are computed as average over the whole column depth. Considering the approximations and the synthetic wind stress adopted, model results are substantially in good agreement with measured ones, even if velocity modules are slightly underestimated.

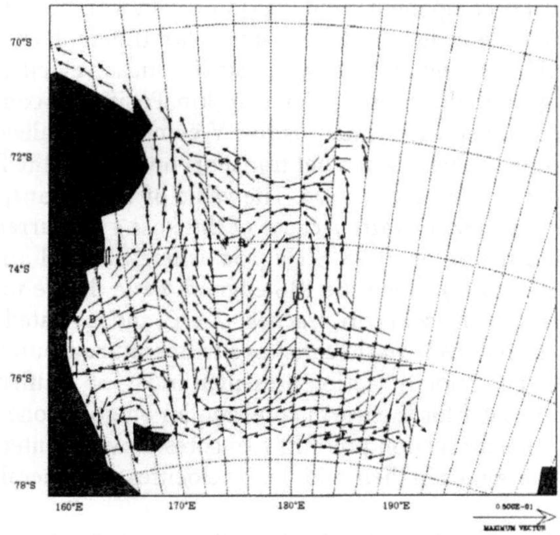

Fig. 4. Model of 30-day averaged total transports (m²/s), forced by open boundary velocities and wind stress

Table 1. Yearly mean velocities and directions measured during 1995 [6] and modelled ones

Mooring	Yearly mean measured velocity (cm/s)	Modelled mean velocity (cm/s)	Yearly mean measured direction (°N)	Modelled mean direction (°N)
A	0.51	0.69	348	329
B	0.65	0.57	129	328
C	7.99	4.87	343	288
D	2.32	0.79	340	29
F	4.27	2.03	306	278
H	3.62	1.10	62	307

Table 2. U- and V-components of integrated transports (Sv)

Latitude (°S)	Transport across x direction (Sv)	Transport across y direction (Sv)
78	-18.9	3.4
77.5	-28.6	8.9
77	-28.8	22.4
76.5	-9.5	29.0
76	-0.7	30.0

In Table 2, zonal (U-component) and meridional (V-component) modelled transports, integrated along longitude, are given for different latitude values. Values are computed by adding the net transports (i.e. mean velocities multiplied by depth) for every grid point, covering about 900 km. Positive U-component values are east-directed transports, while positive V-component values are north-directed ones. Integrated values show a net transport outflowing the RIS along its edge with a prevalent westward direction. The value of the y-transport component is increasing going equatorward because of the Ross gyre current injection.

The temperature distribution at 78 °S (Fig. 5a, where blanked areas near the surface are ice-covered ones), along the RIS edge, shows a dipole warm–cold of interacting water masses. The cold core reaches −2.1 °C, and is related to isohaline of 34.65 PSU, while the relatively warm core has a temperature around −1.5 °C and salinity of 34.45 PSU. The velocity section presented in Fig. 5b confirms an entering velocity under the shelf for the warm core, and an outgoing one for the cold one. Particularly, V velocity section presents two cores of water entering the RIS, while the outgoing one stands in their middle; U velocities are generally westward directed.

The cold core water is saltier than the warm one; in order to produce such a water, warm core water entering below the RIS must interact with HSSW which can be produced during events of strong evaporation associated with cold wind. Such events could be related to polynya areas that appeared more frequently in the west area of the Ross Sea during winter. This kind of water can be stored below the RIS along its western part, constrained by deeper topography.

The T–S field along the west-east transect (again 161–190 °E) at parallel 77 °S (Fig. 6) confirms the water masses movements: the cold core is quite clear; if compared with Fig. 5a, a portion of it has been able to move to the north, while a part of it will be later constrained by the sea mount visible at its west, probably mixing vertically. The signature of the intermediate minimum is lost, while the warm core, entering the Ross Sea, is still evident, with higher maximum values. Figure 6 reveals another characteristic aspect of the Ross Sea, that is a salinity decreasing almost linearly going from west to east, and increasingly from surface to bottom.

We can now take into consideration some vertical sections of the area below the RIS edge latitude, which could be, even if preliminary, of relevant interest in order to gain some useful information about the process simulations under the RIS itself. For instance, checking the temperature plot of Fig. 5a against Fig. 7 (representing T–S field along a west-east transect, 161–190 °E, at parallel 78.5 °S), we can observe that the cold core originated closer to the RIS base (Fig. 7) and after that sinked to deeper depths (Fig. 5a).

Velocity plots (not presented here) demonstrate a dominant east-directed (positive values) U velocity in the western and eastern part of the cross. V-component values are mainly directed into the RIS in the western region; still, water masses with opposite velocities are present along the section. This activity is also confirmed by the U-V field along a west-east transect, again 161–190 °E, at parallel 79 °S, even if in a reduced way. Under the RIS, a slow cyclonic circulation establishes, probably moving the saltier water closer to the RIS bottom base, where warm-cold interaction processes take place. The major activity areas may extend for about 40–60 km along the RIS. Considering Fig. 8, the warm core signal

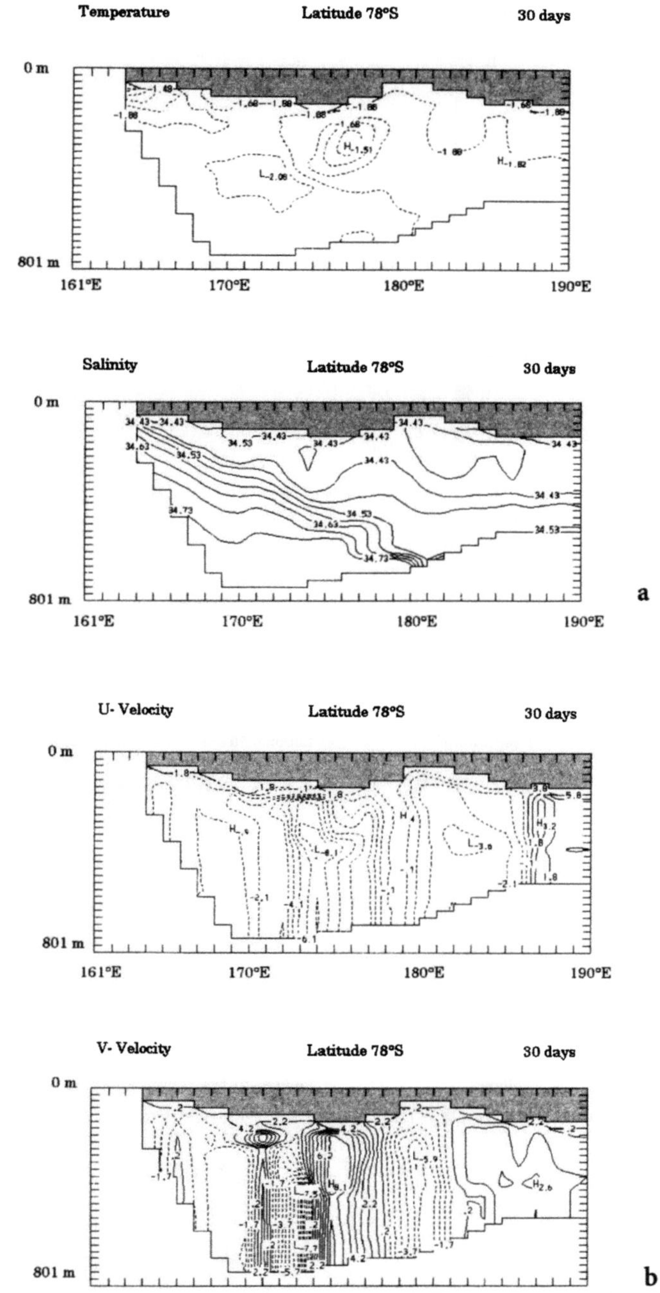

Fig. 5. a Temperature (°C) and salinity (PSU) section west-east, latitude 78 °S, longitude 161–190 °E. **b** U- and V- velocities in the same area (cm/s). Model adjustment after 30 days

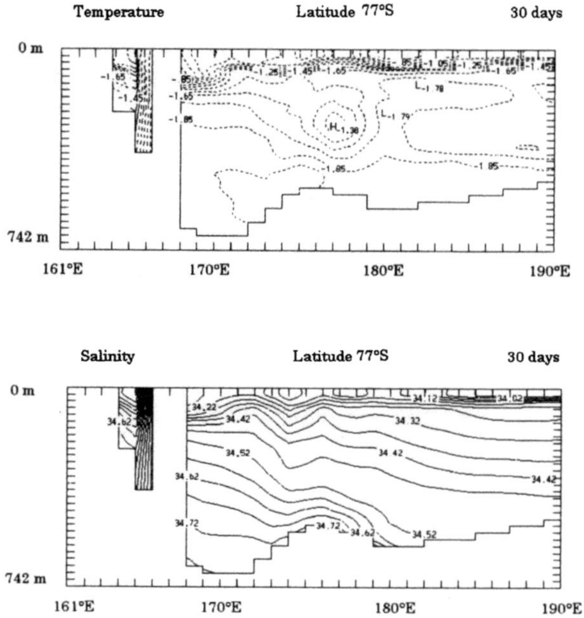

Fig. 6. Temperature (°C) and salinity (PSU) section west-east, latitude 77 °S, longitude 161–190 °E

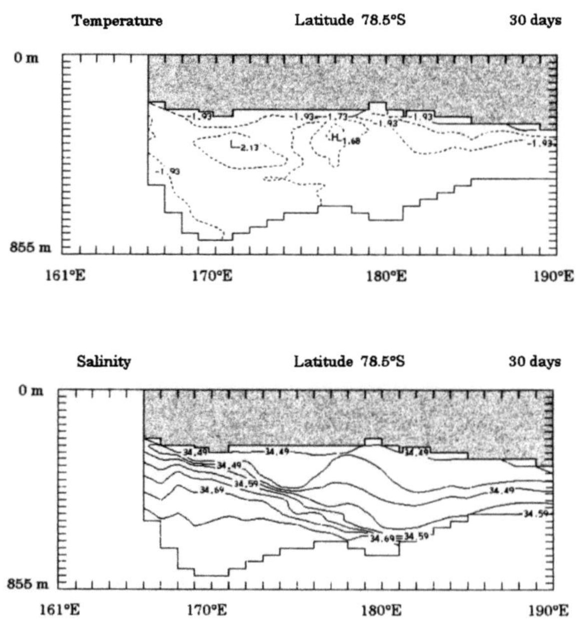

Fig. 7. Temperature (°C) and salinity (PSU) section west-east, latitude 78.5 °S, longitude 161–190 °E

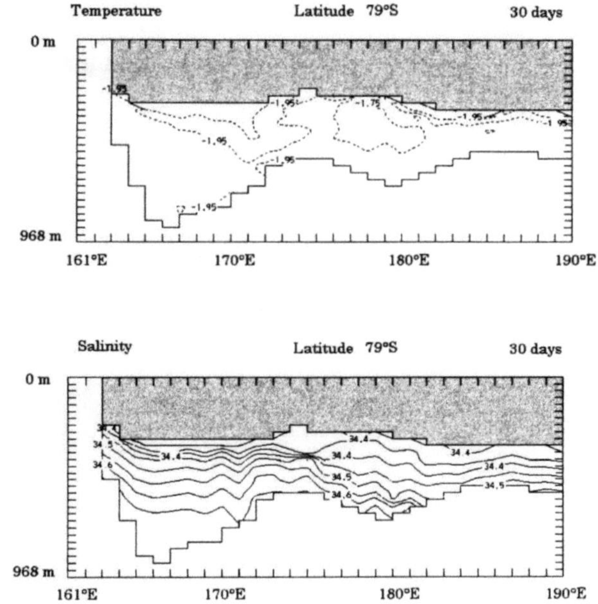

Fig. 8. Temperature (°C) and salinity (PSU) section west-east, latitude 79 °S, longitude 161–190 °E

appears to be the very weak, and mixing processes seem to be the prevailing ones. Salinity values in Fig. 8 reveal the salinity reservoir in the deeper topography of the western area.

A couple of south-north sections give the behaviour of T-S fields, more or less in the middle of the basin. Figure 9a shows a temperature (°C) and salinity (PSU) vertical section south-north (80 °–72 °S) and longitude 183 °E, again at the end of the model adjustment. The section, approximately cutting the Ross Sea in the middle, shows a strong thermal and haline front between the warmer and deeper Modified Circumpolar Deep Water and the Ross Shelf water. It also shows the signature of the cold core water with T<-1.9 °C, flowing out from the bottom of the Ross Ice Shelf cavity, as confirmed by Fig. 9b. This figure also demonstrates a strong exchange activity around the RIS edge, clearly shown by alternating positive-negative U velocities. The barotropic velocity field is still evident outside the RIS, while under the RIS some baroclinic eddies caused by melting processes appear.

At a latitude of about 80 °S, closer to the shelf, the melting slightly dilutes the water, increasing the distance between the isohalines. On the bottom, close to the thermal front, we find a small haline intrusion, probably an entrainment due to diffusion, which modifies the circumpolar current and allows the existence of super cold water almost until 77 °S. The presence of the front in the northern

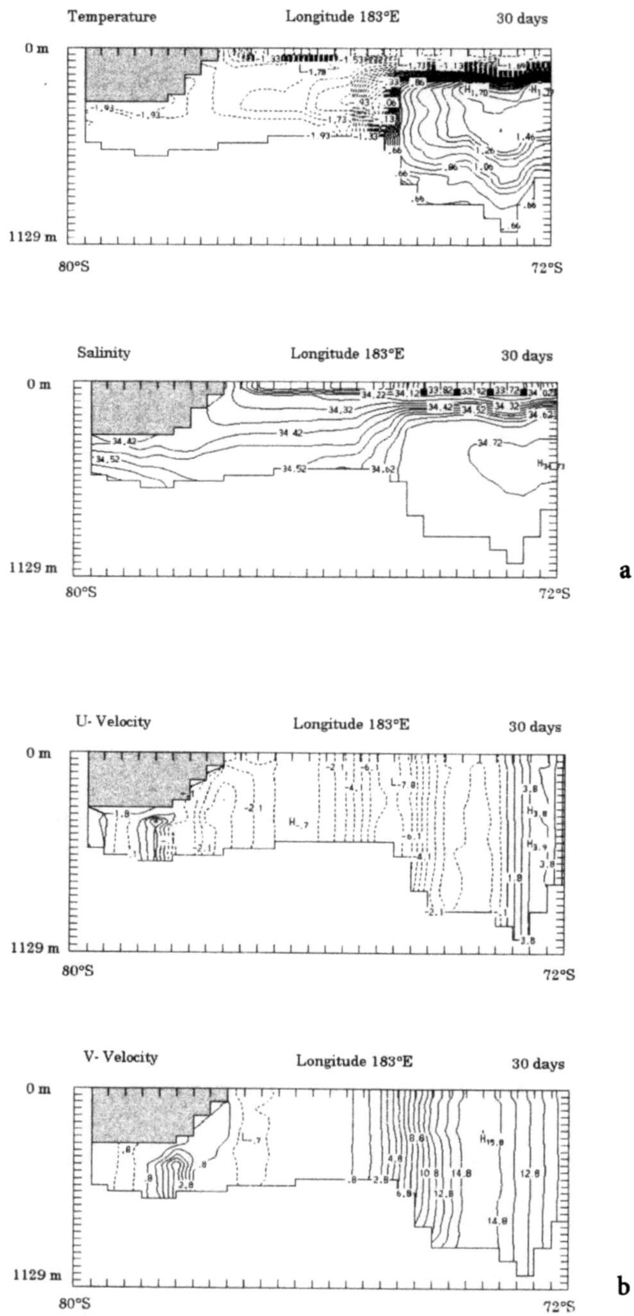

Fig. 9. a Temperature (°C) and salinity (PSU) section south-north, latitude 80–72 °S, longitude 183 °E. **b** U- and V- velocities in the same area (cm/s). Model adjustment after 30 days

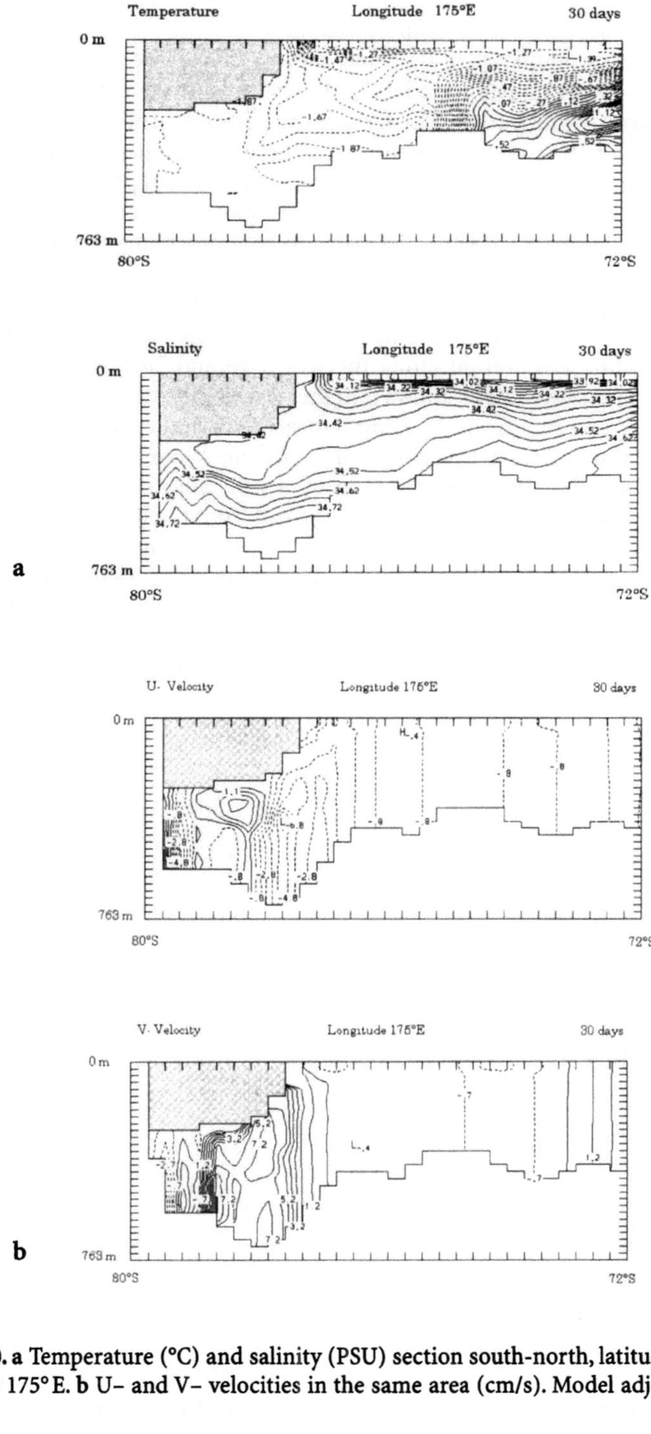

Fig. 10. a Temperature (°C) and salinity (PSU) section south-north, latitude 80–72 °S, longitude 175° E. **b** U– and V– velocities in the same area (cm/s). Model adjustment after 30 days

part, and of a sea mount in the western area, constrains the cold core water flow, and thermal mixing surely induces an upwelling.

Figure 10a shows temperature (°C) and salinity (PSU) for a vertical section south-north (80–72 °S) and longitude 175 °E, resulting from the model adjustment after 30 days of integration. The temperature front is still present, even if it is not as clear as in the previous section, and there is evidence of an upwelling induced by topography. The salinity signature has almost disappeared and clearly shows the eastern reservoir of HSSW. Figure 10b shows again some eddies originating at the RIS base, which allow the water to cool down due to the RIS melting but also to mix with the HSSW reservoir.

Figure 11 presents the horizontal section of the basin after 30 days of model adjustment, at 100 m of depth, for temperature values. In this sub-surface layer, temperature minima are present along the RIS edge and Drygalsky area. (The salinity plot, not presented here, shows a higher variability signal, with a tongue of fresher MCDW entering from mooring B directed to mooring A.

A saltier one, starting from mooring D, moves along Victoria Land.)

Figure 12 shows the situation at 300 m of depth. The thermal front of the MCDW is clear, as well as the cool water flowing out from the RIS near mooring F. The pool around mooring A appears as an almost homogenous one, almost as cold as the one near mooring F. (Salinity maximums appear along Victoria Land, with a tongue going back under RIS, mixing to form the ISW with higher salinity.)

The temperature plot presented in Fig. 13, relatively to 400 m, is quite close to the 300-m one. Topographic control is the reason for the meandering of the fronts around Pennell Bank.

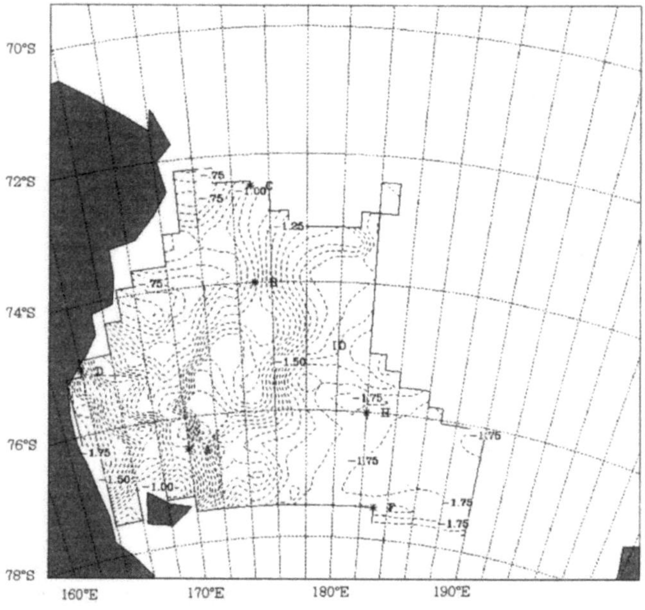

Fig. 11. Temperature (°C) horizontal section, 100-m depth. Model adjustment after 30 days

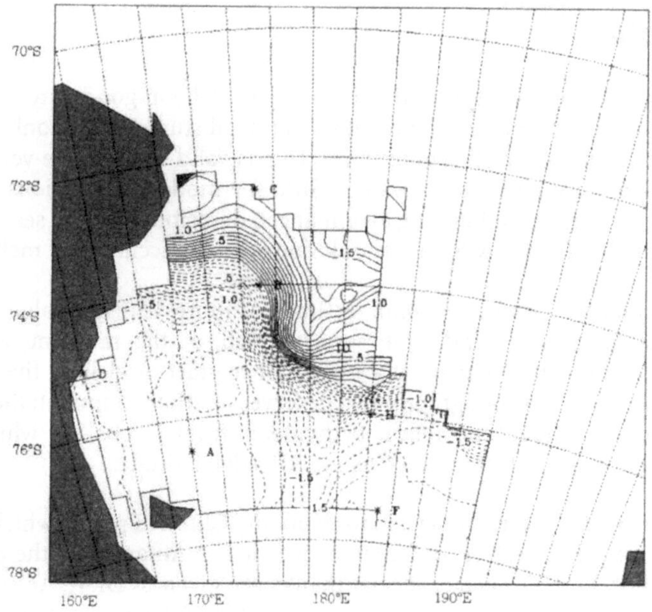

Fig. 12. Temperature (°C) horizontal section, 300-m depth. Model adjustment after 30 days

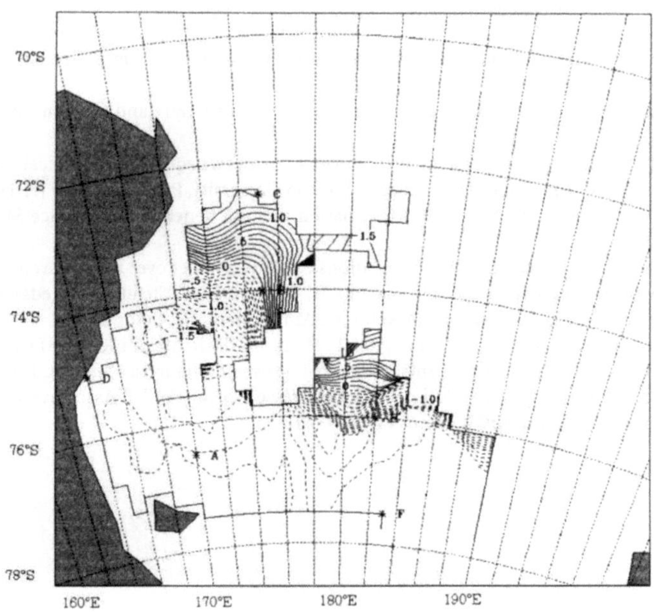

Fig. 13. Temperature (°C) horizontal section, 400-m depth. Model adjustment after 30 days

4 Conclusions

The velocity fields obtained from the model seem to be in good agreement both with experimental observations and with theoretical studies commonly accepted by the international scientific community. This is valid also for the vertical sections of temperature and salinity values. Still, the model assumptions could be improved. For example, embedding an enhanced description of the sea-ice interaction, especially near the RIS area, should improve the accuracy of melting phenomena.

Velocity modules have been underestimated; more realistic values may be probably reached using wind time series, which, at the moment, are being acquired by authors: this is mandatory in order to start modelling the Ross Sea seasonal variability. The topography also deserves another comment; the authors received a more detailed and recent bathymetry from Jacobs [1], which could change some important factors, such as the RIS edge latitude.

Acknowledgements. The authors wish to thank the CLIMA Project, which operated in the second leg of the tenth Italian expedition in Antarctic in the person of the coordinator, Prof. G. Spezie. Special thanks are given to Dr. A. Artegiani and Dr. G. Budillon, for their helpful collaboration in the CTD data acquisition, to Ing. R. Meloni and Dr. M. Ravaioli for the mooring data availability, and to Mr. G. Arcari and Dr. F. De Biasio for supporting the data elaboration.

References

1. Jacobs SS (1989) Marine controls on modern sedimentation on the Antarctic continental shelf. Marine Geol 85:121-153
2. Locarnini RA (1994) Water masses and circulation in the Ross Gyre and environs. PhD dissertation, Texas A&M University
3. Mellor GL (1991) User's guide for a three dimensional, primitive equation, numerical ocean model. Progress in atmosphere and ocean science. Princeton University, Princeton, New Jersey, p 35
4. Jacobs SS, Gordon AL, Ardai JL (1979) Circulation melting beneath the Ross Ice Shelf. Science 203:439-443
5. Buzzi A, Malguzzi P, Cadelli R (1996) Simulations of a low level jet over the Antarctic region: 3-D diagnostics of potential vorticity. In: Colacino M, Giovannelli G, Stefanutti L (eds), Conference Proceedings Vol 51, Italian Research on Antarctic Atmosphere, SIF, Bologna
6. Picco P, Bergamasco A, Demicheli L, Langone L, Manzella GMR, Meloni R (1998) Current meter, temperature and salinity measurements in the Ross Sea shelf (Antarctica) during 1995/1996: data presentation. In: Picazzo M (ed) Atti del XII Congresso Nazionale dell'Associazione Italiana di Oceanologia e Limnologia-AIOL, Genova II:105-120

A Wind and Boundary Driven Circulation Model of the Ross Sea

V. Commodari[1] and S. Pierini[2]

[1] Istituto di Meteorologia e Oceanografia, Istituto Universitario Navale, Via Amm. Acton 38, 80133 Napoli, Italy
[2] Dipartimento di Fisica, Università dell'Aquila, Via Vetoio, 67100 Coppito (AQ), Italy

Abstract

The application of a barotropic primitive equation model to the Ross Sea has allowed to determine the main features of the vertically integrated transport as induced locally by the wind and forced by the external action of the East Wind Drift during the ice-free season. Furthermore, the implementation of the same model with higher resolution to a coastal area including Terra Nova Bay has provided information on the local circulation regime. The circulation model of the Ross Sea is implemented in a domain including the Ross Sea and an external zone which allows for the formation of the Ross Sea gyre. A regular grid of 20 km resolution is defined on a adapted Gauss-Boaga projection which reduces notably the deformation of the distances. The model is first forced by an idealized wind system representing schematically the local atmospheric circulation on a global scale and the runs are carried out up to the steady state. The Ross Sea gyre and an overall cyclonic circulation in the interior of the Ross Sea locally shaped by the topography are observed. On the other hand, the response inside the Ross Sea to the remote effect of a boundary forcing given by the Antarctic Circumpolar Current and of the East Wind Drift in the absence of winds is found to be very weak. This puts in evidence that the wind is by far the most energetic forcing of the local barotropic circulation. These numerical results are found to be in good agreement with classical observational data and with current meter measurements taken within the Italian P.N.R.A.. The model is then applied to a coastal zone around Terra nova Bay with a 2 km resolution. This model is forced by an idealized wind and nested with the coarse resolution model of the Ross Sea. Information are thus obtained on the local circulation.

1 Introduction

The Antarctic continent exerts a profound influence on both the global atmospheric circulation and the oceanic thermohaline and dynamic structure, thus interfering with marine and terrestrial ecosystems. The oceanic circulation surrounding the continent can be roughly represented by two mainly wind-driven currents, the Antarctic Circumpolar Current (ACC) flowing eastward and a coastal countercurrent, the East Wind Drift (EWD), which is strongly influenced by the local coastal and bathymetric morphology. At the longitudes corresponding to the Ross Sea, this two-current system gives rise to a recirculating cyclonic dynamic structure known as the Ross Sea gyre, extending well beyond the Ross Sea itself into the Southern Ocean [1].

In this context and within the framework of the Italian National Programme for Antarctic Research (PNRA) we have developed a process study aimed at analyzing the vertically integrated transport of the Ross Sea gyre and of the circulation within the Ross Sea as induced by an idealized but fairly realistic wind system during the ice-free season. The effect of the Ross Sea gyre on the circulation in the Ross Sea through topographic coupling is also investigated. Moreover a high resolution model of the small region enclosing Terra Nova Bay (where the Italian base of the PNRA is located) is nested with the large-scale circulation model in order to obtain information on the local barotropic circulation.

In Section 2 the circulation model is presented, along with the projection used to represent the area of interest on a plane Cartesian system. The domain of integration and the wind forcing is described; moreover, a boundary forcing is introduced in order to investigate the direct influence of the Ross Sea gyre on the circulation inside the Ross Sea. In Section 3 the nearly steady state obtained asymptotically by resolving the initial-boundary value problem forced by the wind shows two well-defined cyclonic systems: a weaker one inside the Ross Sea and a stronger one yielding a closed recirculation outside (the Ross Sea gyre). When the system is forced only through a boundary forcing representing schematically the Ross Sea gyre, the internal circulation is found to be very weak so that, as a consequence, the circulation in the Ross Sea turns out to be mainly locally wind-driven. Moreover, our numerical results are found in good agreement with current meter measurements taken in the framework of the PNRA. Finally in Section 4 the numerical results concerning the circulation in Terra Nova Bay are presented.

2 The Circulation Model and Wind and Boundary Forcings

The wind stress at the air-sea interface transfers to the ocean energy and momentum in the form of surface Ekman currents. Their divergence generates a surface topography and corresponding pressure-driven geostrophic currents. Under the assumption of negligible nonlinear and JEBAR (Joint Effect of Baroclinicity and Relief) (e.g. [2]) terms it can be shown (e.g. [3]) that the equations of motion for the barotropic component of the velocity including both the Ekman and geostrophic currents in a vertically integrated way reduce to the so-called shallow water equations (e.g. [3, 4]):

$$\begin{cases} \mathbf{u}_t + (\mathbf{u} \cdot \nabla)\mathbf{u} + f\mathbf{k} \times \mathbf{u} = -g\nabla\eta + \dfrac{(\boldsymbol{\tau}_w - \boldsymbol{\tau}_b)}{\rho_o H} + A_H \nabla^2\mathbf{u}, \\ \eta_t + \nabla(H\mathbf{u}) = 0 \end{cases} \tag{1}$$

where $\mathbf{u}(\mathbf{x},t) = \frac{1}{H}\int_{-H}^{0} U(x,z,t)dz$ is the depth-averaged horizontal velocity [$\mathbf{u}=(u,v)$, $x=(x,y)$, U is the current velocity], η the surface elevation, ρ_o a reference density, f the Coriolis parameter, $\mathbf{k}=(0,0,1)$, g the acceleration of gravity, A_H the horizontal

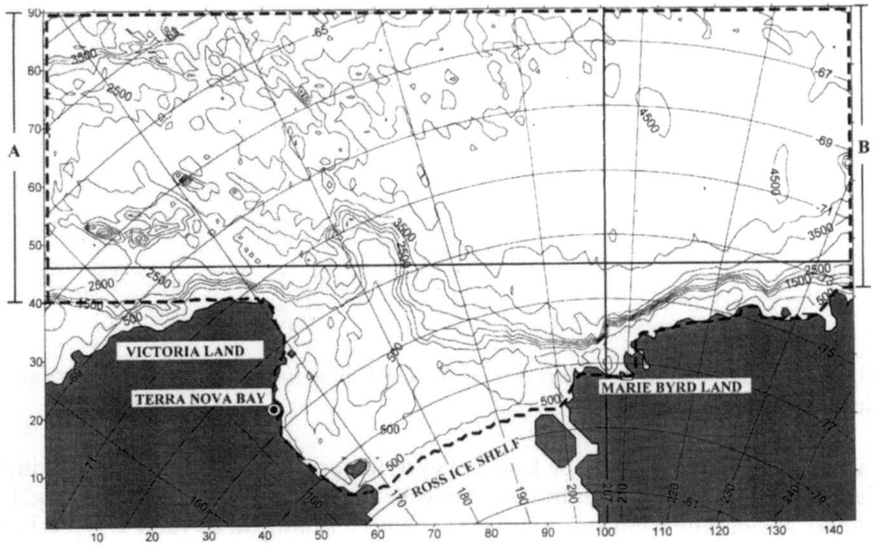

Fig. 1. Domain of integration (enclosed by *dashed line*) with the bathymetry

eddy viscosity coefficient, $\boldsymbol{\tau}_w$ the wind stress, $\boldsymbol{\tau}_b = \rho\, C_{Db}|\mathbf{u}|\mathbf{u}$ the bottom stress and $H=D+\eta-d$ [where D is the mean water depth, d(**x**) the bottom topography and η(**x**,t) the free surface displacement]. We will make use of a finite-difference model for Eq. (1) on the C-grid previously implemented for circulation studies in the Mediterranean [5, 6].

In a process-oriented study like the present one it is acceptable to project the region of interest onto a plane. The classical β-plane approximation is, however, not applicable in this peculiar geographical context. We have therefore applied a cylindrical projection similar to the Mercator one but tangent to a maximum circle that, rather than being the equator, crosses the area of interest normal to the meridian of longitude 153°W (Fig. 1). The Coriolis parameter has then been evaluated at each point of the regular grid. The domain of integration is the region delimited by the dashed line in Fig. 1, where the bathymetry is also shown. It contains the Ross Sea (the lower boundary corresponds to the edge of the Ross Ice shelf) and an external zone sufficiently vast to allow for the formation of the Ross Sea gyre. A regular grid of nominal spacing $\Delta x=\Delta y=20$ km was adopted. The following parameters have been used: $\Delta t=8$ s, $A_H=10\,000$ m^2 s^{-1}, $C_{Db}=0.002$. Since the aim of this study is to obtain circulation patterns indicative of a climatological situation during the ice-free season, we solve Eq. (1) with vanishing initial conditions forced by steady forcings (see below). In such a way, after a transient phase during which eddy viscosity and – to a lesser extent – bottom friction dissipate high frequency, mainly gravitational motions, a virtually stationary circulation is achieved.

The main forcing used in the numerical simulation is a wind system as shown in Fig. 2a. It represents schematically the typical atmospheric circulation at sea level at these latitudes (e.g. [7, 8]). Although the climatological conditions in the Ross Sea area are characterized more precisely by a low atmospheric pressure system with an

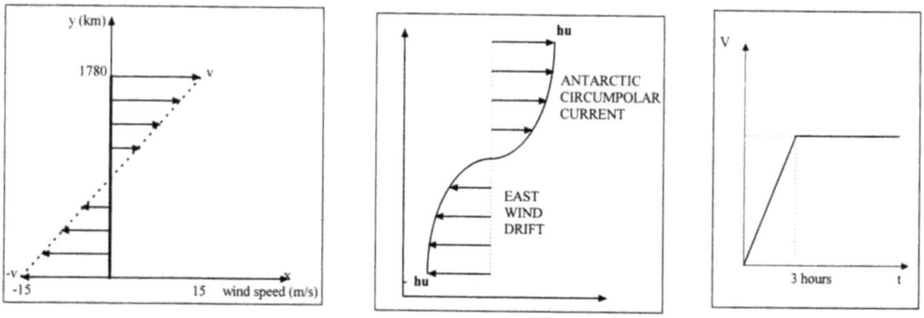

a b c

Fig. 2. a Wind forcing. **b** Boundary forcing imposed along *lines A* and *B* of Fig. 1. **c** Time dependence of forcings **a** and **b**

extension corresponding to the Ross Sea gyre, in our case a one-dimensional wind field such as that of Fig. 2a is more suitable for our domain of integration of Fig. 1. The corresponding wind stress curl is spatially constant and its value is typical of the local atmospheric circulation. On the other hand, the lateral boundaries A and B provide the closure of the streamlines and the establishment of an idealized Ross Sea gyre.

In order to assess the importance of the local winds in the circulation inside the Ross Sea compared with the effect of the Ross Sea gyre that, through topographic coupling, also contributes to the internal circulation, we used (in the absence of winds) the boundary forcing of Fig. 2b, representing schematically the Ross Sea gyre structure. The transport profile is imposed along both the A and B boundaries (note the fictitious straight coastline introduced west of Victoria Land, whose presence is, however, unimportant for the Ross Sea circulation). Finally, the forcings were given a time dependence as shown in Fig. 2c in order to reduce undesirable transient features.

3 Circulation in the Ross Sea Area

3.1 Numerical Simulations

In this section we consider the circulation induced by the wind forcing of Fig. 2a with temporal variation given in Fig. 2c. Figure 3a shows the transport Hu (plotted at every other grid point in each direction) obtained by solving Eq. (1) with vanishing initial conditions and closed free-slip boundary conditions all around the domain after 15 days of integration (a quasi-steady solution is attained already after 1 week). The first thing to be noticed is the formation of the ACC along the northern boundary and of the EWD along the northern coasts of the Antarctic continent as shaped by coasts and topography. Moreover, a large-scale closed cyclonic system is present in the north-eastern part of the domain, representing a schematic Ross Sea gyre. Its shape is determined also by the unrealistic northern and eastern boundaries, but, nonetheless, this dynamic structure bears important similarities with the well-known local climatological circulation. Figure 3b shows a zooming of

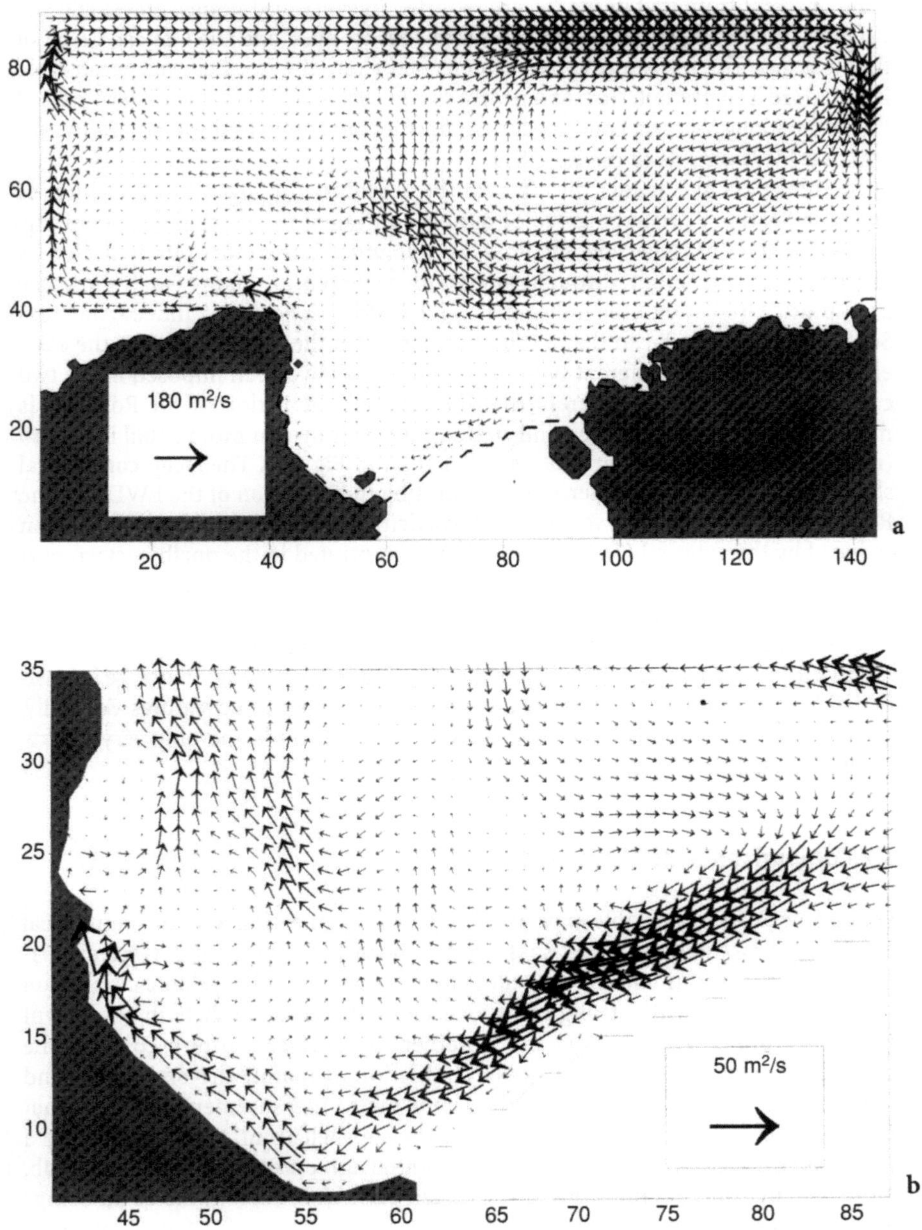

Fig. 3. **a** Vertically integrated transport induced by wind forcing of Fig. 2a. **b** Zoom of the Ross Sea

the wind-driven circulation inside the Ross Sea (here the velocity vectors, with a scale different from that of Fig. 3a, are reported in every grid point). A cyclonic circulation can be observed all along both the Ross Ice shelf and the western Coast of the Ross Sea. The transport appears intensified along the ice shelf; sub-basin scale features are also present, such as a nearly closed cyclonic gyre in the eastern part of the basin. Along Victoria Land the flow intensifies and, more to the north, converges with the EWD outside of the Ross Sea.

We now pass to consider the effect of the boundary forcing of Fig. 2b which, as stated in Section 2, is supposed to be almost equivalent to the wind forcing of Fig. 2a outside the Ross Sea (now the Ross Sea dynamics is therefore purely driven by the external circulation through topographic coupling). Figure 4a,b show the asymptotic circulation thus induced. It is interesting to note that, indeed, the Ross Sea gyre is very similar to that of Fig. 3a, apart from the area adjacent to the eastern boundary where different boundary conditions have been imposed in the two cases. It is immediately apparent that the internal circulation of the Ross Sea is much weaker than that of the wind-driven case, as shown in more detail in Fig. 4b (note that the scaling is ten times smaller than in Fig. 3b). The steep continental shelf acts as a dynamic barrier, strongly limiting the intrusion of the EWD into the Ross Sea so as to conserve potential vorticity. The weak residual circulation induced by the external EWD is seen to be concentrated in the north-eastern part of the basin in the form of a cyclonic gyre, while in the north-western part the circulation is fairly similar to that of Fig. 3b.

In conclusion, two main results can be drawn from our numerical experiments. Firstly, the barotropic transport inside the Ross Sea has a cyclonic structure intensified along the southern boundaries and with sub-basin cyclonic features. Secondly, the internal circulation is basically driven by the local winds and only very weakly by the external circulation.

3.2 Comparison with Observations

Let us now analyze available experimental data in connection with our numerical results. The circulation in the Ross Sea has been the subject of several experimental investigations and the prevailing idea is that a mainly cyclonic circulation is present (e.g. [9]) with strong transports along the Ross Ice shelf, in agreement with our results. Lagrangian data obtained by surface drifters confirm the cyclonic character of the upper ocean circulation as mainly driven by the wind [10]. Pillsbury and Jacobs [11] used current meter measurements to show that between 170 and 180°W, in the vicinity of the ice shelf front, the flow presents a U-shaped cyclonic structure, again in substantial agreement with Figs. 3b and 4b. More recently, in the framework of the Italian PNRA, six current meter moorings were deployed in the Ross Sea [12, 13]. The data (courtesy of P. Picco) averaged over the summer 1994-1995 are presented in Fig. 5: the black arrows represent the velocities measured by intermediate depth current meters (at 150- to 280-m depth) while the gray ones give bottom currents (see [13] for details).

The most notable feature is the strong barotropic character of the circulation, which suggests that the wind is probably the main forcing of the climatological cir-

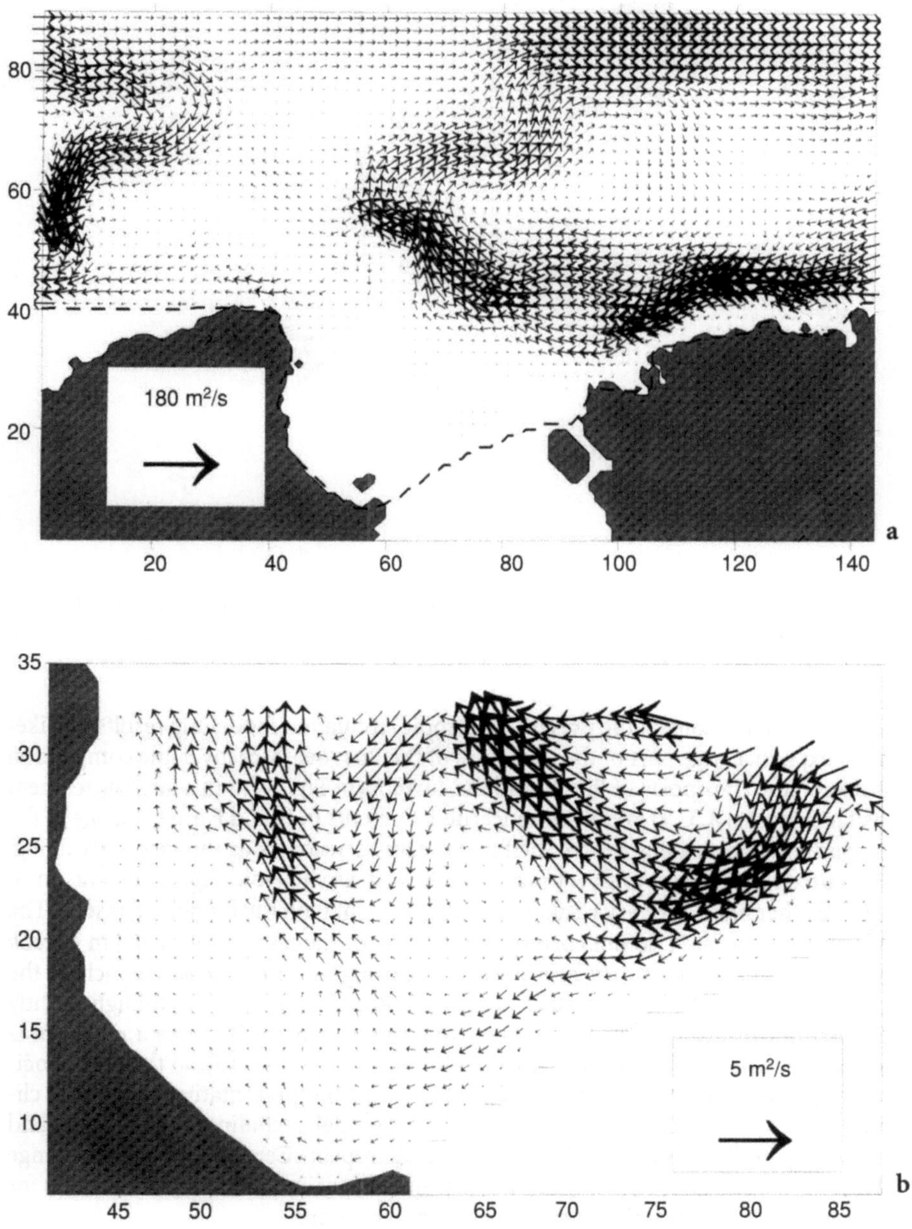

Fig. 4. a Vertically integrated transport induced by boundary forcing of Fig. 2b. **b** Zoom of the Ross Sea

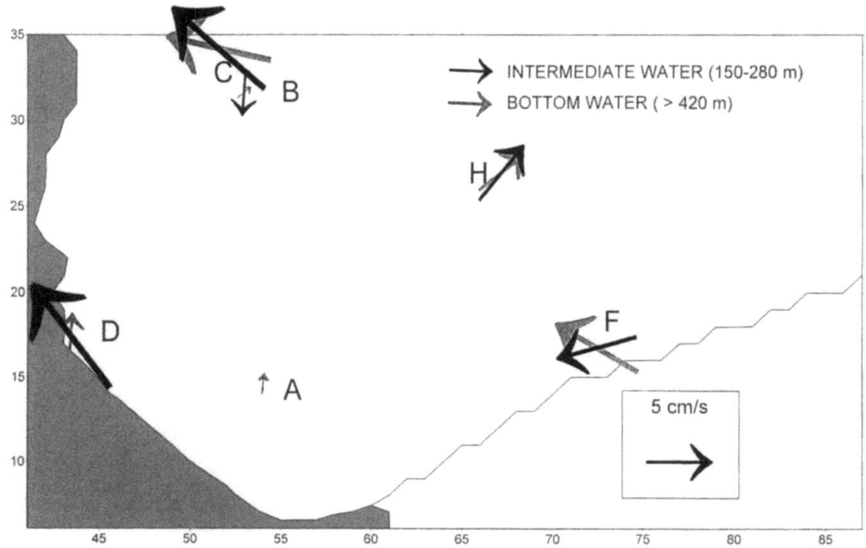

Fig. 5. Intermediate (*black arrows*) and bottom (*gray arrows*) currents measured in the framework of the Italian PNRA averaged over summer 1994-1995. (Courtesy of P. Picco; see [13])

culation. In this context the present barotropic circulation modelling results are likely to explain a large part of the dynamics of the Ross Sea. Moreover, the comparison between the transports of Fig. 3a and the currents of Fig. 5 shows a good agreement as far as points A, C, D and F are concerned. Currents in D and F are evidence of the strong circulation along the ice shelf and the western coasts, as shown clearly in Fig. 3b; currents in C tend to flow toward north-west, as shown in Fig. 3; currents in A, on the other hand, are very weak, again in agreement with the numerical results. The velocities in point H do not correspond to those of the numerical model in exactly the same location, but they can be seen as belonging to the upper branch of the cyclonic gyre in the south-eastern part of the Ross Sea (see Fig. 3), though slightly shifted northward. Finally, the current time series of point B shows a remarkable temporal variability, sensibly larger than in the other points [13], so that the velocities reported in Fig. 5 at point B may be not so much representative of the local climatological circulation. The application of a model including stratification and forced by realistic winds such as ECMWF (European Centre for Medium-Range Weather Forecasts) momentum flux data is under way in order to provide a basis for a more realistic comparison with experimental currents.

4 Circulation in Terra Nova Bay

We analyzed the coastal circulation in the Terra Nova Bay area where the Italian base is located by means of a high resolution model nested with the circulation model of the Ross Sea described above. The domain is defined in the region $\{16 \leq j \leq 23;\ 41 \leq i \leq 48\}$ of the grid of the previous model and the resolution is now

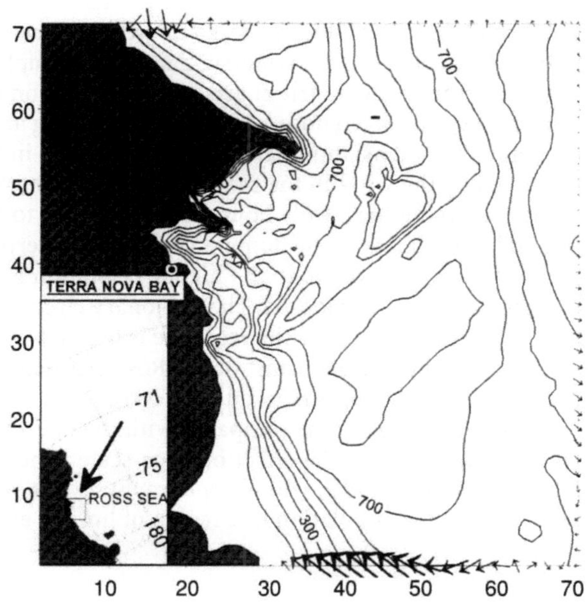

Fig. 6. Domain of integration of the high resolution model of the Terra Nova Bay area. Boundary forcing defined by *lateral arrows* derives from the wind-driven numerical experiment of the Ross Sea

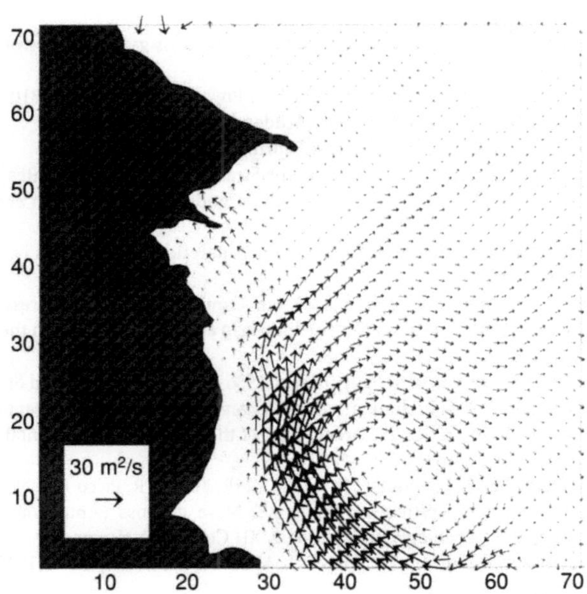

Fig. 7. Vertically integrated transport induced by boundary forcing and by easterly winds

$\Delta x=\Delta y=2$ km (Fig. 6). The bathymetry was obtained by a databank with resolution of only 7'30" in both latitude and longitude, so in order to have a satisfactory resolution in Terra Nova Bay the bathymetric data were complemented by additional data digitized from bathymetric charts with a resolution of 1 mile in both directions. The model was forced both by a boundary forcing and the wind. The nearly steady state obtained asymptotically by resolving the initial-boundary value problem forced by the wind in the run of Fig. 3 (with the coarse resolution of 20 km) was used to define steady boundary conditions to be imposed along the "open" lateral boundaries of the domain (Fig. 6). An interpolation was performed by means of a cubic spline onto the higher resolution grid of 2 km. An easterly wind of 15 m/s was also imposed. The stationary circulation is given in Fig. 7: the cyclonic structure evident in the response is basically a zooming of the strong jet present along the western coast of the Ross Sea (Fig. 3b), while a small-scale anticyclonic structure appears to interest Terra Nova Bay. The effect of the local wind is found to be negligible compared with the boundary forcing, which, however, is itself the result of the effect of large-scale winds. In conclusion, an even higher resolution model, whose topography is not yet available, would be required to resolve the circulation in a region of limited extent such as Terra Nova Bay.

Acknowledgements. This research was funded by the Italian National Programme for Antarctic Research. The authors wish to thank G. Spezie, P. Picco, G. Budillon and A. Bergamasco for valuable discussions.

References

1. Locarnini RA (1994) Water masses and circulation in the Ross gyre and environs. PhD dissertation, Texas A&M University
2. Huthnance JM (1984) Slope currents and "JEBAR". J Phys Oceanogr 14:795-810
3. Gill AE (1982) Atmosphere-ocean dynamics. Academic Press, New York
4. Pedlosky J (1987) Geophysical fluid dynamics. Springer-Verlag, New York
5. Pierini S (1996) Topographic Rossby modes in the Strait of Sicily. J Geophys Res 101:6429-6440
6. Pierini S, Simioli A (1998) A wind-driven circulation model of the Tyrrenian Sea area. J Mar Sys 18:161-178
7. Schwerdtfeger W (1984) Weather and climate in the Antarctic. Developments in atmospheric science. Elsevier, Amsterdam
8. Deacon G (1984) The Antarctic Circumpolar Ocean. Cambridge University Press, Cambridge
9. Ainley DG, Jacobs SS (1981) Sea-bird affinities for ocean and ice boundaries in the Antarctic. Deep Sea Res 28A:1173-1185
10. Moritz RE (1988) The Ross Sea data buoy project, 1986-88. Antarctic J. United States 23:78-80
11. Pillsbury RD, Jacobs SS (1985) Preliminary observations from long-term current meter moorings near the Ross Ice Shelf, Antarctica. In: Oceanology of the Antarctica Continental Shelf. American Geophysical Union, Washington, DC, pp 87-107
12. Bergamasco A, Demicheli L, Langone L, Manzella GMR, Meloni R, Picco P (1996) Analisi di serie temporali di corrente, temperatura e salinità nel Mare di Ross (Antartide) nell'anno 1995. Associazione Italiana di Ocenologia e Limnologia, XII Congresso Nazionale
13. Picco P, Bergamasco A, Demicheli L, Langone L, Manzella GMR, Meloni R (1998) Current, temperature and salinity measurements in the Ross Sea Shelf (Antarctica) during 1995/1996: data presentation. Proceedings of the AIOL Congress, Genova II:105-120

Wintertime Expansion and Contraction of the Terra Nova Bay Polynya

M.L. Van Woert

NOAA/NESDIS/Office of Research and Applications E/RA3, World Weather Building Room 711, 4700 Silver Hill Road Stop 9910, Washington, DC 20233-9910, USA

Abstract

A one-dimensional coastal polynya model, forced by data from a near-by automatic weather station, is used to examine wintertime ice concentration fluctuations in the Terra Nova Bay polynya. It has long been believed that the Terra Nova Bay polynya opens in response to strong offshore katabatic wind forcing. This study shows that not only is the sensible heat flux important in controlling fluctuations in polynya extent, but that the longwave heat fluxes is important also. Moreover, this study shows that wind forcing becomes important to the polynya dynamics as an amplification factor acting on the longwave flux term. Together these two terms can explain up to 40% of the observed variance in open water fraction. Some large-amplitude fluctuations in open water fraction at Terra Nova Bay were not well modeled in this study. These anomalies are believed to be related to large-scale changes in the Ross Sea pack ice, which are driven by strong winds blowing off the Ross Ice Shelf. Ice production rates were found to be comparable with historical estimates. To maintain a conversion of Warm Core Water or Low Salinity Shelf Water (S≅34.5PSU) to High Salinity Shelf Water (S≅34.8 PSU) consistent with the observed ice production rate, a transport of approximately 1 Sv is required at Terra Nova Bay.

1 Introduction

Polynyas are recurring regions of either partially or totally ice-free ocean occurring within the pack ice at times when air temperatures are well below the freezing point of seawater. Polynyas forming in coastal regions where strong and persistent offshore winds drive newly formed sea ice offshore are often referred to as latent heat polynyas [1], because latent heat is produced by the freezing of ice. These polynyas are particularly important because they typically occur on the polar continental shelves where the water is already close to the freezing point. Brine rejected during the ice formation process further densifies the water column leading to cold saline water masses that spill off the continental shelves into the deep ocean basins. Studies in the Arctic (e.g. [2, 3]), Antarctic [4-6], and high latitude marginal seas [7] have attempted to estimate water mass formation rates in coastal polynyas by combining satellite-derived estimates of open water fraction with estimates of the ice production rate based on net atmospheric flux data.

These studies have largely treated the ice production rates in the polynya and the size of the polynya as independent quantities. However, results from simple analytical models that describe polynya dynamics [8-10] suggest that polynya size, and hence open water fraction, is maintained through a delicate balance between ice edge advection at the polynya boundary and ice production within the polynya.

Here we examine the relationship between open water fraction derived from satellite observations and estimates of polynya size simulated with a one-dimensional polynya model driven by time-coincident atmospheric flux data from Terra Nova Bay, Antarctica (75°S, 164°E); (Fig. 1). These estimates of polynya extent are

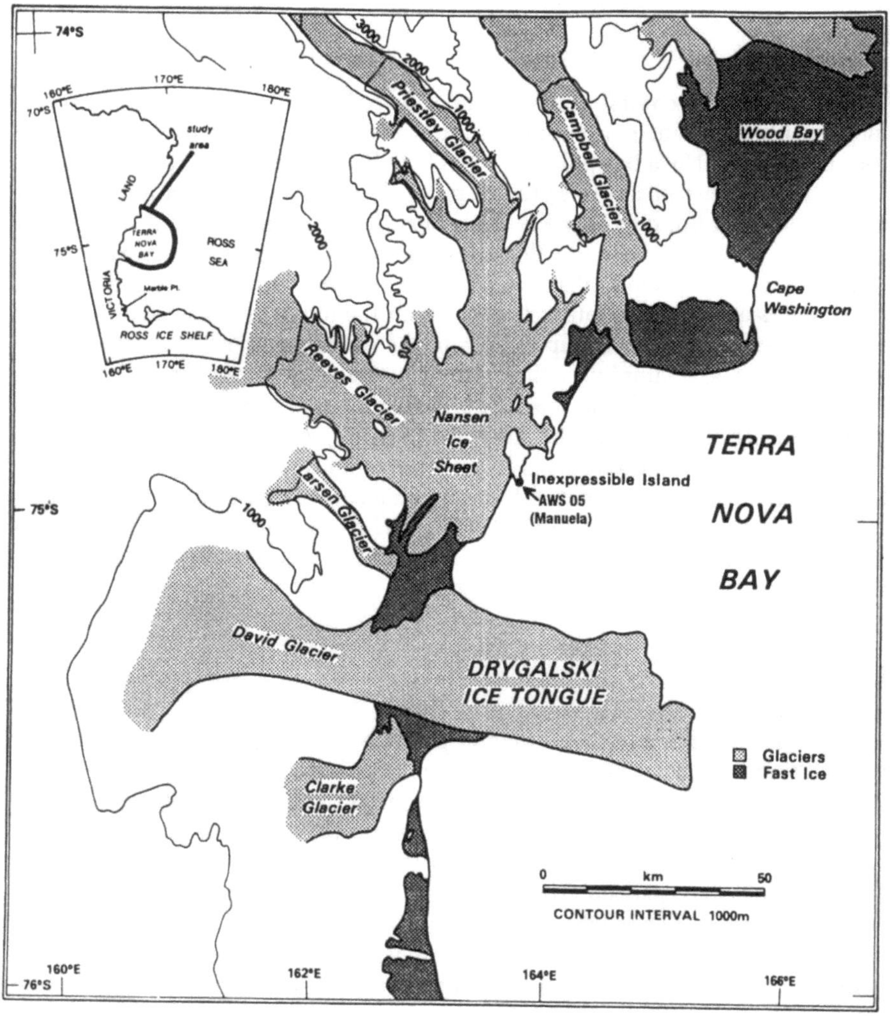

Fig. 1. Location map showing Terra Nova Bay and surrounding features. Location of the Manuela autonomous weather station (AWS) site and the region extracted from the Special Sensor Microwave Imager satellite data are indicated (After [24])

then used to describe the character of wintertime ice production and water mass modification in Terra Nova Bay. Terra Nova Bay was selected as the site for this study because an extensive meteorological data set exists with which to test this relationship, and because the winds are persistently strong and offshore with a high degree of directional constancy, thus making a comparison with a one-dimensional polynya model practical.

2 Environmental Conditions at Terra Nova Bay

2.1 Weather Conditions

Terra Nova Bay occupies an area of approximately 6000 km^2 (65 km north/south by 92 km east/west) [11] in the western Ross Sea near 75°S, 164°E (Fig. 1). It is bordered on the south by the floating Drygalski Ice Tongue, on the north by Cape Washington, and on the west by the Nansen Ice Shelf, which is fed by the Reeves and Priestley Glaciers.

An autonomous weather station (AWS) was installed on February 6, 1984, on Inexpressible Island (74.92°S, 163.60°E) along the western shore of Terra Nova Bay (Fig. 1) [12, 13] and has operated intermittently since that time. Air temperature and vector winds at 1.5 m above the surface and atmospheric pressure at ~1-m height are measured at 10-min intervals and broadcast via the Service ARGOS Data Collection System to the NOAA polar orbiting meteorolgical satellites [14, 15]. In many years, equipment failure during the winter months due to extreme wind events produced significant gaps in the data record. Nonetheless, this station has been the only significant source of detailed wintertime meteorological measurements from the region. When two satellites are on orbit, nearly a continuous sequence of 10-min observations can be collected. Once collected, the 10-min data are scanned to pick out the nearest observation within ±1 h of 00, 03, 06, 09, 12, 15, 18, and 21 h to produce the 3-hourly data set. Occasionally, due to the particular orbit geometry of the satellites, gaps of several hours, duration do occur in the data record. However, this situation occurred generally less than 3% of the time. These small gaps were filled with linearly interpolated values using the adjacent good points.

Bromwich [16] reports on the AWS data for the months of February, March, and April during 1984, 1985, and 1987; there were no data for 1986. Bromwich et al. [17] describe the 1988 and 1989 katabatic wind regime at Terra Nova Bay using data from an extended AWS network. Here we focus on data from Inexpressible Island for the years 1988, 1989, and 1990, which correspond to the years when the Defense Meteorological Satellite Program (DMSP) satellite, F-8, was on orbit. Aside from the period February 17 - April 4, 1990, when wind speed and direction data are not available [18], the AWS data records are complete (Fig. 2).

Table 1 summarizes monthly mean air temperatures and standard deviations, monthly mean wind speed and ±34 percentile limits, and monthly median wind direction and directional consistency (defined as the ratio of the magnitude of the vector-averaged wind speed to the mean wind speed) based on the 3-hourly meteorological observations. Of particular interest to this study is the directional con-

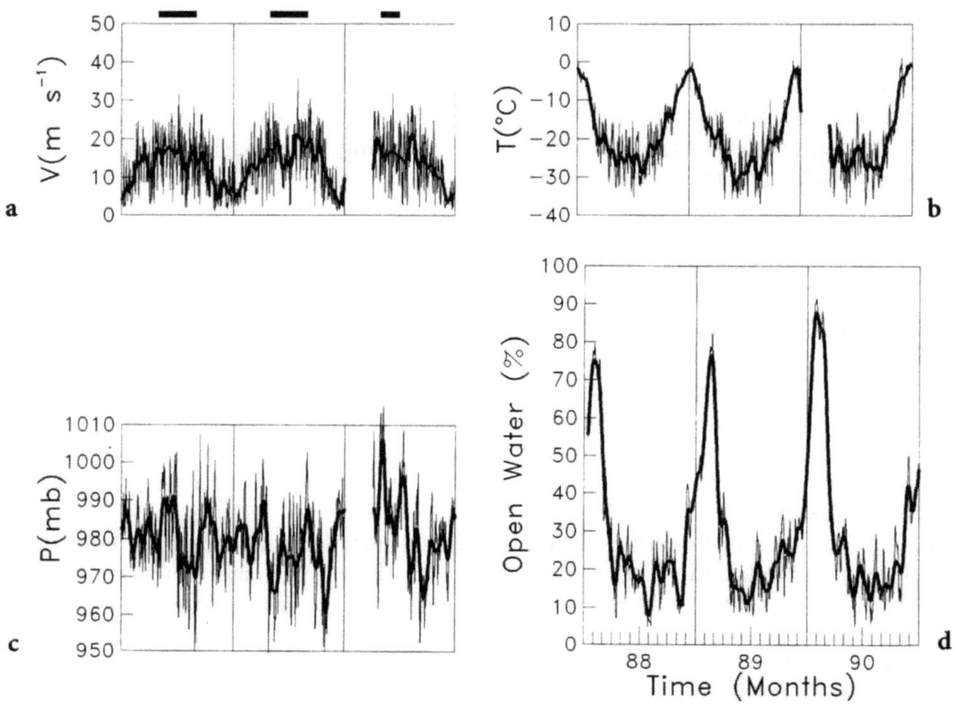

Fig. 2. Wind speed (**a**), air temperature (**b**), and atmospheric pressure data (**c**) from 1.5-m height at the Manuela AWS site. Areal average open water fraction (**d**) was derived from the SSM/I data. *Thin lines* portray daily averaged values and *bold lines* are low-passed filtered versions of the series. *Long tick marks* represent July 1 for each year. *Bars* at top of plot indicate the time periods modeled in this study

sistency, which, for values near 1, indicates that the winds blow predominantly from a single direction – in this case indicating katabatic flow down the Reeves Glacier.

Directional consistency for the winter months in all 3 years was consistently greater than 0.9 except for July and August 1990 when unrealistically low values were observed. Thus, in the analysis that follows, only data from months shaded in Table 1 (May-August 1988, 1989, and May-June 1990) are utilized. The observed high degree of directional consistency is especially important to the design of this study because the polynya model is one dimensional and valid only for offshore winds. The restriction of this study to the winter months, as discussed more fully below, is predicated on the need to simplify the surface energy budget.

2.2 Ice Conditions

It has been known since 1912, when Scott's northern party was left stranded at Terra Nova Bay, that wintertime open water exists at Terra Nova [19-21]. Priestly [19, 20], however, had no way of knowing the spatial extent of open water at Terra

Table 1. Monthly mean air temperature (T, in °C) and standard deviation, monthly median wind speed (V, in m s^{-1}) and ±34% wind speed limits, and monthly mean wind direction (θ) and directional consistency (γ) for 1988, 1989, and 1990. Only months indicated by *shading* are used to drive the polynya models described in this study

	1988.0						
	T	σ	V (16%)	V(50%)	V(84%)	θ	γ
Jan	-4.0	2.5	1.9	4.9	12.8	283	0.84
Feb	-11.3	4.5	4.1	10.9	18.2	288	0.94
Mar	-20.8	3.7	9.2	14.7	20.4	291	0.98
Apr	-21.9	4.6	3.5	11.6	23.4	292	0.97
May	-26.1	4.7	10.8	16.5	21.7	297	0.98
Jun	-25.4	5.0	11.0	16.5	23.0	294	0.97
Jul	-26.6	5.5	4.6	15.6	23.9	292	0.96
Aug	-24.1	5.2	8.7	14.8	21.7	288	0.97
Sep	-22.3	4.3	5.4	15.1	23.7	294	0.97
Oct	-15.5	4.7	2.4	10.4	19.0	292	0.94
Nov	-11.2	4.5	1.9	5.4	11.3	290	0.88
Dec	-4.5	2.7	1.9	4.9	14.0	281	0.86
	1989						
	T	σ	V (16%)	V(50%)	V(84%)	θ	γ
Jan	-4.3	3.5	1.7	5.1	10.6	288	0.83
Feb	-12.5	4.2	5.4	11.0	15.1	292	0.96
Mar	-18.9	3.4	6.3	12.8	20.3	295	0.97
Apr	-21.1	6.7	8.3	15.5	24.0	292	0.97
May	-26.7	4.9	11.3	17.9	25.1	294	0.99
Jun	-30.5	4.3	8.2	13.5	20.5	294	0.97
Jul	-25.8	4.5	5.9	16.7	25.4	290	0.98
Aug	-28.9	5.7	11.1	18.8	25.6	290	0.98
Sep	-22.8	5.8	9.6	17.9	25.6	288	0.97
Oct	-19.6	3.0	5.4	16.2	23.9	292	0.98
Nov	-10.6	4.2	2.6	7.7	11.8	293	0.89
Dec	-2.6	2.4	0.9	2.6	7.4	234	0.59
	1990						
	T	σ	V (16%)	V(50%)	V(84%)	θ	γ
Jan	-4.2	2.9	1.9	5.2	13.6	288	0.91
Feb	-12.4	4.0	2.7	7.7	14.8	271	0.88
Mar	-20.1	4.7	3.7	5.1	13.1	252	0.96
Apr	-25.9	5.0	11.1	18.7	26.4	291	0.99
May	-25.3	5.0	8.3	16.1	22.0	292	0.99
Jun	-26.8	4.4	8.6	16.8	23.6	291	0.98
Jul	-22.1	4.6	3.2	16.0	25.2	278	0.41
Aug	-28.6	5.3	9.6	18.3	27.8	184	0.24
Sep	-26.8	4.7	3.5	14.8	22.7	171	0.11
Oct	-19.4	4.3	5.7	11.9	18.7	170	0.12
Nov	-8.8	6.2	2.4	8.5	15.8	258	0.34
Dec	-2.4	3.1	1.4	3.7	8.3	277	0.71

Nova Bay; it was not until infrared images from the first Earth-observing satellites became available in the mid-1960s that it became possible to examine the temporal and spatial evolution of Antarctic coastal polynyas [22, 23]. These studies indeed confirmed Priestly's earlier finding that open water exists throughout the winter months. Kurtz and Bromwich [6, 24] elaborated on these early satellite studies by examining the extensive archive of visible and infrared images at the National Snow and Ice Data Center in Boulder, Colorado. They concluded on the basis of their analysis that the mean polynya area is 1300 km^2 (65 km north/south by 20 km east/west), but that it varied in size from near 0 to 5000 km^2 (65 km north/south by 75 km east/west). This maximum polynya area corresponds roughly to a polynya with an offshore extent equal to the length of the Drygalski Ice Tongue (~83 km in 1979 [11]). Kurtz and Bromwich [6] also noted that the open water nucleus was surrounded by a roughly 25000-km^2 area of thin or loosely consolidated ice.

Unfortunately, the Antarctic Ocean and ice surfaces are often obscured by clouds in visible and infrared satellite imagery, thus producing a potentially biased account of the actual polynya fluctuations. In this study, daily sea ice concentration estimates from Special Sensor Microwave Imager (SSM/I) are used to examine temporal fluctuations in ice concentration in the vicinity of Terra Nova Bay. The SSM/I is a seven-channel microwave radiometer that senses the Earth's emitted radiation at 19, 37, and 85 GHz vertical and horizontal polarization and at 22 GHz vertical polarization [25]. The first of these sensors was launched June 19, 1987, onboard the Block 5D-2 DMSP satellite F-8 and remained operational through December 1991. The satellite flew in a sun-synchronous, near-polar orbit at a height of ~860 km, providing 4 to 5 passes daily at high polar latitudes.

Antarctic, daily-averaged, 25-km resolution, gridded sea ice concentration data products were obtained from the National Snow and Ice Data Center [26]. These ice products are derived by applying the NASATEAM algorithm [27, 28] to the 19- and 37-GHz vertical polarization data and 19-GHz horizontal polarization data from the SSM/I. Higher resolution methods of estimating open water fraction in coastal regions have been proposed using data from the 85-GHz channels [29]. Unfortunately, 85-GHz data are not available for the entire period of this study and therefore are given no further consideration.

To quantify temporal changes in the local ice concentration, a region representing the apparent maximum extent of polynya expansion (Fig. 1; defined here as a contiguous region of low ice concentration extending northeast from Terra Nova Bay [e.g. 30]) was extracted from the individual daily ice concentration images. Areal ice concentration, C, was computed by summing the total ice concentration within the region and dividing by the total study area (14 000 km^2). Using an areal average damps smaller polynya events, but, given the inherent inaccuracies of the present sea ice algorithms (~15% [31-34]), this level of smoothing seems warranted. Open water fraction in percent was computed as $(1-C) \times 100$. The time sequence of daily-averaged open water fraction for the period July 1987 through December 1990 is shown in Fig. 2.

Figure 2 indicates that during winter 1988, roughly defined here as the period May to October, the ice open water fraction was typically around 20% except for brief periods in August and early November when the open water fraction underwent dramatic reductions. In contrast, the winter of 1989 saw a significant drop in open water fraction in mid April and then a steady increase in open water fraction through December 1989. The winter of 1990 was characterized by generally lower (~15%) open water fraction than in the previous 2 years. Explaining both the low frequency wintertime fluctuations and the superimposed high frequency fluctuations in open water fraction is the primary focus of this study.

3 Polynya Model

In an attempt to explain the observed fluctuations in open water fraction, Pease's one-dimensional, wind-driven coastal polynya model [10] is considered. The model is based on an assumed balance between the offshore advection of the consolidated ice edge by the wind and ice production in the polynya, which tends to close the polynya. This balance can be represented mathematically by:

$$\frac{DL(t)}{Dt} + \frac{P(t)L(t)}{H_i} = U_i(t), \tag{1}$$

where H_i is the thickness of the consolidated ice at the ice edge (taken here to be 0.1 m), L(t) is the width of the polynya, $U_i(t)$ is the offshore advection speed of the consolidated ice (here taken to be 3% of the wind speed measured at 10-m height [35]), and $P(t)$ is the total ice production rate given by:

$$P(t) = \frac{Q_{Net}(t)}{\rho_i L_f}. \tag{2}$$

$Q_{Net}(t)$ is the net heat flux to the atmosphere, ρ_i is the density of ice (0.95×10^3 kg m^{-3}), and L_f is the latent heat of fusion (3.34×10^5 J kg^{-1}) [7].

Equation (1) can be integrated numerically using an Euler integration scheme [36] for specified forcing when initialized for the three individual winter periods using the maximum polynya width:

$$L_m = \frac{H_i U_i(t_0)}{P(t_0)} \tag{3}$$

[10, 37], where t_0 is the initial time for the model run (May 1, 00:00 h for each year).

4 Model Results

4.1 Regression Analysis

To minimize the impact of uncertainty in cloud cover and albedo estimates on the results of this study, attention has been restricted to the winter months of May

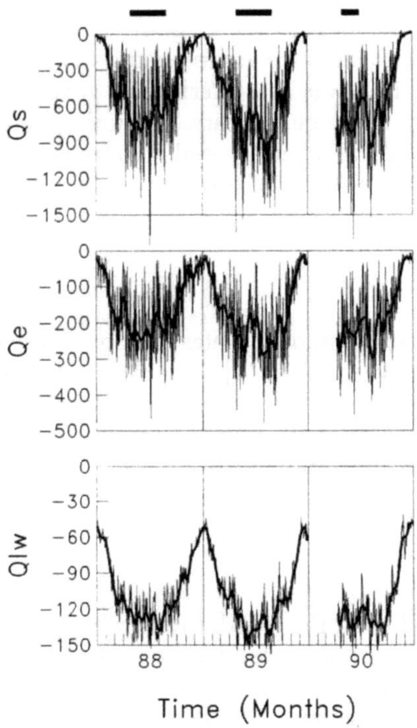

Fig. 3. Sensible (**a**), latent (**b**), and longwave (**c**) heat fluxes assuming a fractional cloud cover of 50% and a relative humidity of 70%. $C_{h10}=C_{e10}=1.1\text{x}10^{-3}$ was used to produce this figure because the formulation of Andreas and Murphy [39] is not appropriate for stable summertime conditions. *Thin lines* are daily-averaged time series and *bold lines* are low-passed filtered versions of the series

Table 2. Wintertime ice and water mass production values for 1988, 1989 and 1990

Parameter	1988	1989	1990
Sensibile heat flux (wm^{-2})	727	836	748
Latent heat flux (wm^{-2})	242	263	249
Longwave heat flux (wm^{-2})	127	134	128
Net heat flux (wm^{-2})	1096	1233	1125
Ice thickness (m)	37.9	42.7	38.9
Ice production (km^3)	46.4	52.2	47.7
Salt production (10^{12} kg)	1.0	1.1	1.0
HSSW production ($10^{13}m^3$)	1.0	1.1	1.0
HSSW trasport (Sv)	0.9	1.0	0.9

HSSW, High Salinity Shelf Water.

through August when the sun is below the horizon at Terra Nova Bay and the surface energy budget is dominated by the sensible heat flux, the latent heat of evaporation, and longwave radiation. Thus, Q_{Net} can be written as:

$$Q_{Net} = -[Q_s + Q_e + Q_d - Q_u], \tag{4}$$

where Q_s and Q_e are the sensible and latent heat fluxes respectively, Q_d is the downward longwave radiation emitted by the atmosphere, and Q_u is the upward longwave radiation emitted by the sea surface.

Priestly [19] and Bromwich and Kurtz [38] assert that it is the strong offshore winds that are responsible for maintaining wintertime open water at Terra Nova Bay. Figure 3 shows the individual heat flux terms based on an assumed 50% cloud fraction, 70% relative humidity, and meteorological data from the Manuela AWS station assuming bulk sensible and latent heat parameterizations developed by Andreas and Murphy [39] and a longwave parameterization developed by Maykut and Church [40]. The dominant term in the heat budget is clearly the sensible heat flux (Fig. 3a), with average wintertime values for the 3 years of –770 W m^{-2} (Table 2). To test the hypothesis that it is the strong katabatic winds that are responsible for the fluctuations in polynya extent, the polynya model was stepped forward in time using only the sensible heat flux as input. The daily averaged estimates of polynya extent were then regressed against the daily averaged open water fraction.

Figure 4a shows the scatter plot and least-squares regression between the modeled polynya width and the open water fraction estimated from the satellite data. The correlation coefficient is 0.39, explaining 15% of the variance (Table 3). Figures 4b-d show the open water fraction derived from the satellite data and the open water fraction estimated by the polynya model and regression analysis. The observed open water fraction shows significantly more variability than the modeled results and the model misses several major fluctuations in the open water fraction. In particular, the model dramatically underestimates the observed open water fraction during three major polynya events during May and June 1988 (Fig. 4b). In addition, the model significantly overestimates the open water fraction during the later half of July and the first half of August. Similarly, a major polynya expansion during July 1989 is missed entirely by the model. During other periods, there is no obvious correspondence between increases or decreases in open water fraction and the modeled estimates.

The mean modeled polynya extent for this case is 21±4 km (Table 3). This is only slightly larger than the 20-km polynya extent derived from AVHRR data [6] assuming a polynya size of 1300 km^2 and a fixed north/south polynya dimension of 65 km. Note, however, one would expect the AVHRR to underestimate polynya extent because the instrument is only capable of estimating polynya size during cloud-free periods. These periods are associated with increased radiative cooling and hence should lead to smaller polynyas. In contrast, the microwave analysis is expected to overestimate the open water fraction because the analysis region for the microwave data includes both the polynya region proper and some fraction of the offshore loosely consolidated ice.

Assuming a constant relative humidity of 70%, the second largest term in the heat budget is the latent heat flux (Fig. 3b) with a 3-year wintertime mean loss from the polynya of –250 W m^{-2} (Table 2). The overall modeled structure (not shown) looks nearly identical to the structure computed from the sensible heat flux alone, and the correlation coefficient remains close to 0.39 (Table 3). It is the close correspondence between the sensible and latent heat fluxes that motivated Pease [10] to increase the sensible heat exchange coefficient and to ignore the latent heat flux term in her heat budget.

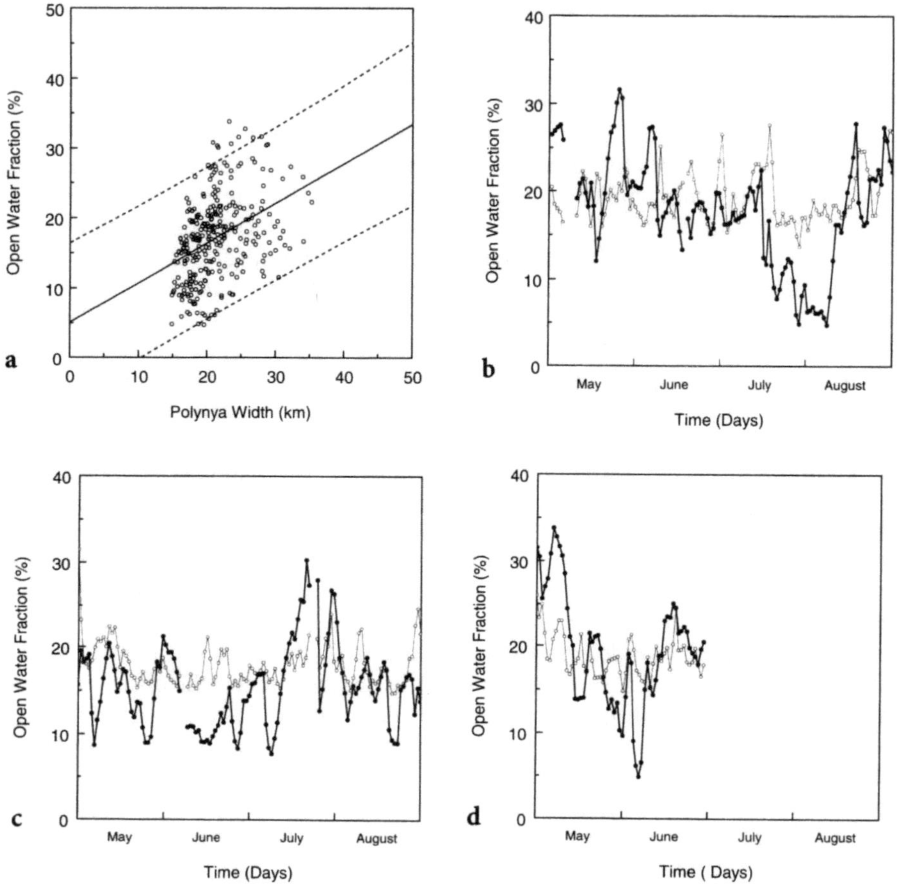

Fig. 4. **(a)** Scatter plot of satellite-derived open water fraction vs. modeled polynya extent under the assumption that net heat flux is equal to sensible heat flux. Satellite-derived open water fraction (*bold line*) and modeled open water fraction (*thin line*) for 1988 **(b)**, 1989 **(c)**, and 1990 **(d)**

The longwave flux (Fig. 3c) is the only remaining wintertime source of heat loss to the atmosphere above the polynya. Assuming a constant cloud fraction of 50%, the 3-year wintertime mean longwave heat flux is -130 W m^{-2} (Table 2). Figure 5a shows the scatter plot and regression line between the satellite-derived open water fraction and polynya extent estimated from the combined sensible, latent, and longwave heat fluxes. The correlation is 0.55 (Table 3), explaining 30% of the variance, and is a significant improvement over the previous results. This correlation is consistent with values from a similar analysis conducted near Halley Station [29]. In contrast to the commonly held belief that the sensible heat flux is solely responsible for controlling polynya width, this result suggests that the longwave heat flux is equally as important as the sensible heat flux in controlling the wintertime open water fraction in Terra Nova Bay.

Table 3. Regression statistics between model estimates of open water fraction and SSM/I-derived open water fraction (= a x L+b) using the formulations for heat and moisture exchange coefficients, C_{h10} and C_{e10}, given by Andreas and Murphy [39]

Q_s	Q_e	Q_{LW}	r	r^2	L	σ_L	a	b
✔			0.39	15%	21.3	4.1	0.6	5.1
✔	✔		0.41	17%	19.7	3.2	0.8	2.0
✔	✔	✔	0.55	30%	17.1	2.7	1.2	-3.5
✔	✔	✔	0.63	40%	17.2	2.7	1.4	-6.3[a]

Q_s, Q_e and $Q_{LW}(=Q_d-Q_u)$, Heat fluxes described in σ indicates that the term was included in the regression.
L, Wintertime mean polynya width.
$_L$, Standard deviation for wintertime polynya width.
r, Regression coefficient.
r^2, Variance explained by the regression.
[a] Statistics based on a U-shaped fractional cloud distribution

Figure 5b-d shows the satellite-derived open water fraction and the model-derived estimates. The two curves generally vary in phase over most of the 3 years studied, suggesting that the model dynamics are at least qualitatively correct. However, significant departures remain. In particular, the several large expansions in open water fraction during May, June, and August 1988 and July 1989 are still not well modeled by Eq. (1).

Including the net longwave heat flux in Eq. (4) extracts additional heat from the polynya and further reduces the mean polynya extent to 17.1 km (Table 3). Thus, although the addition of the longwave heat flux improves the regression statistics between the satellite-derived open water fraction and the modeled polynya extent, the increased heat loss leads to an underestimate of both the mean polynya extent and the amplitude of the fluctuations in polynya extent relative to those estimated from AVHRR data [6].

The largest contributing factor leading to uncertainty in the longwave heat flux is the fractional cloud cover. For example, changing the fractional cloud cover from 0 to 100% gives a RMS uncertainty of 65 W m^{-2} for the longwave flux over the range of temperatures observed. In contrast, for a constant cloud fraction of 50%, the 3-year wintertime RMS deviation of the longwave flux due to temperature fluctuations alone is only 10 W m^{-2}.

The only wintertime in situ cloud data available from Terra Nova Bay are observations recorded in the diaries of Raymond Priestly [41] and Victor Campbell [21]. Priestly was the scientific observer for Scott's northern party and was considered to be a careful and experienced observer. During the nine months that they were confined to the Terra Nova Bay region, Priestly made daily and occasionally twice-daily observations of wind, weather, ice, and drifting snow conditions [38]. A careful examination of the diaries revealed a bimodal, U-shaped, fractional cloud cover distribution with clear conditions occurring ~60% of the time and overcast condi-

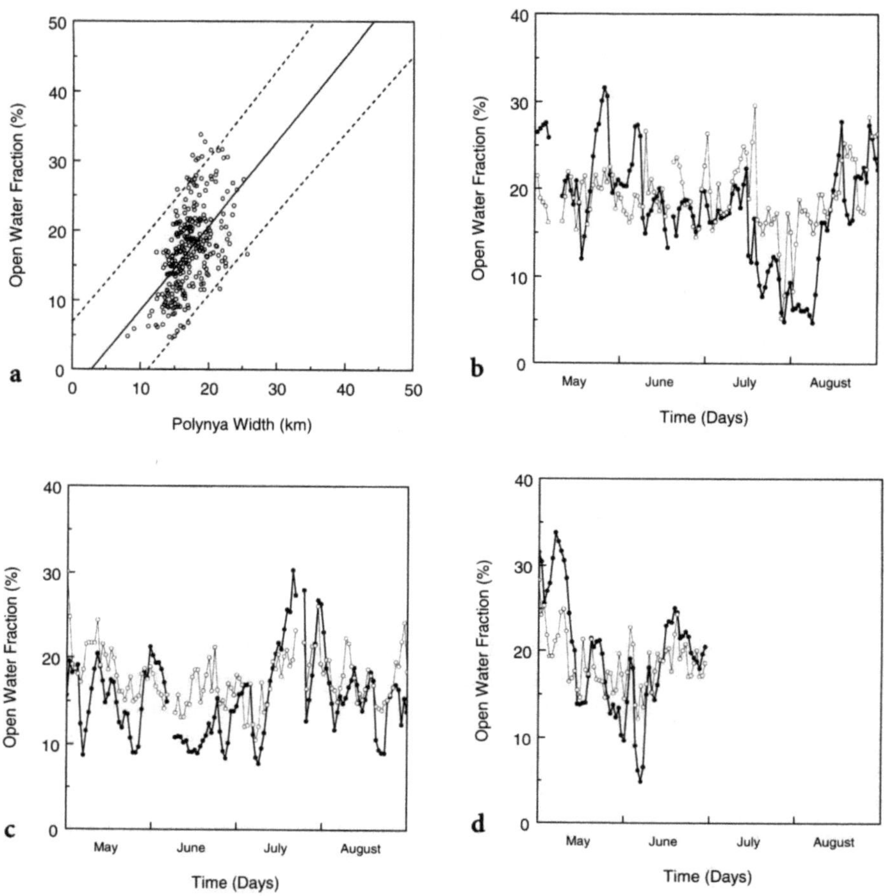

Fig. 5. Same as Fig. 4 except that net heat flux is assumed to be the sum of the sensible, latent, and longwave heat fluxes

tions occurring ~40% of the time for the months May through August; partial cloudy conditions were noted only occasionally in their diaries. The observed U-shaped cloud cover distribution is consistent with cloud cover distributions from recent observations in the Arctic and the Weddell Sea (Andreas, pers. comm.).

Because high-quality, time-coincident, fractional cloud cover data were not available for this study, a baseline of C=0.5 was chosen. To examine the potential impact of cloud cover fluctuations on polynya extent, the polynya model was once again forced with the bulk sensible and latent heat fluxes [39] and the Maykut and Church [40] longwave flux parameterization using 0, 50, and 100% cloud fractions. The time series of open water fraction was first regressed against the modeled polynya extent based on 50% cloud fraction to obtain an initial regression line. A new time series of modeled polynya extent was then generated by select-

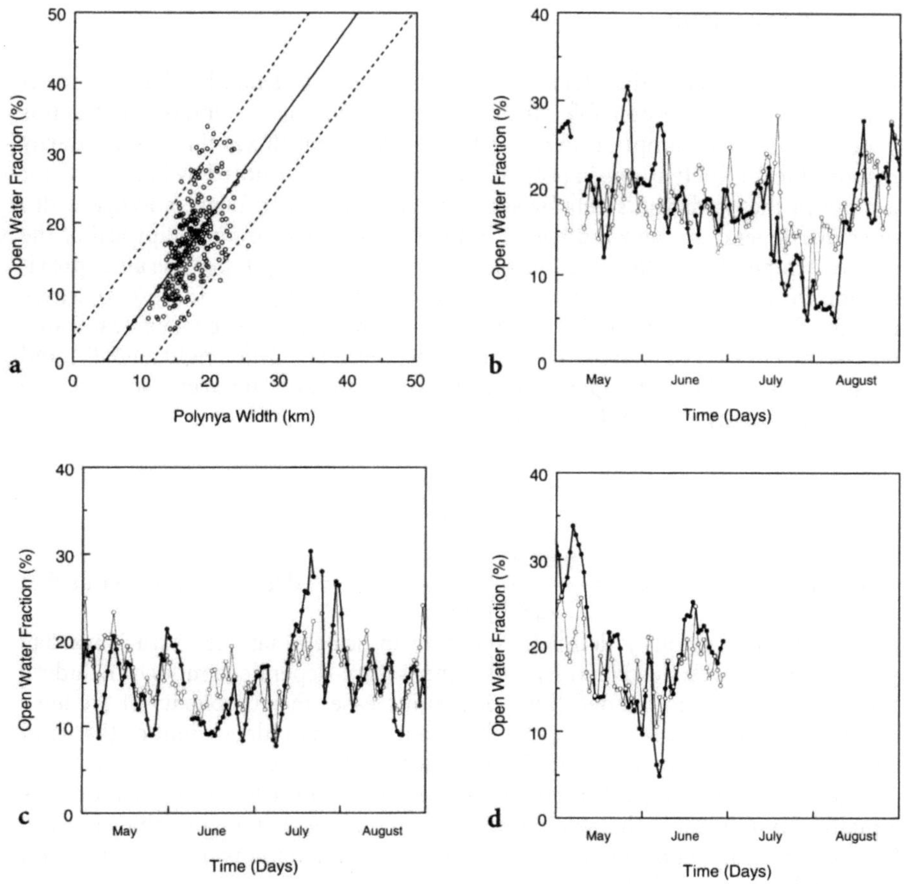

Fig. 6. Same as Fig. 5 except that the fractional cloud cover is allowed to take on either 0 or 100% in accordance with the U-shaped in-situ fractional cloud distribution

ing modeled polynya extents, associated with either 0 or 100% cloud fraction, that were closest to the original regression line. Finally, the regression was recomputed using the revised time series of modeled polynya extent.

The scatter plot and the revised regression line are shown in Fig. 6a. The correlation coefficient is 0.63, explaining 40% of the variance (Table 3). Figures 6b-d show the open water fraction derived from the satellite data and the open water fraction estimated by the polynya model and regression analysis. The mean modeled polynya extent for this case is 17±3 km (Table 3), which is similar to the value derived from the constant cloud cover case. The improvement in the correlation coefficient suggests that the unknown cloud fraction can explain some of the observed fluctuation in polynya extent. However, the model still fails to completely capture the large polynya expansions during May and June 1988 and July 1989. These events are discussed more fully below.

4.2 Model Sensitivity

It is commonly argued that fluctuations in the intense offshore katabatic winds at Terra Nova Bay are responsible for fluctuations in polynya extent. Implicit in this statement is the assumption that it is the sensible heat fluxes that are primarily responsible for fluctuations in polynya extent. However, as noted by Pease [10], for a polynya that is driven solely by sensible and latent heat fluxes, polynya width does not depend on the wind speed, but depends rather on the reciprocal of the air-sea temperature difference. This is easily seen by using Eq. (3) as an approximation for polynya extent and taking $Q_{Net} = -Q_s = r_a C_p C_{h1.5} U_{1.5}(T_0 - T_{1.5})$, where r_a is the density of air, C_p is the specific heat of air, $C_{h1.5}$ is the sensible heat transfer coefficient at 1.5-m height, $U_{1.5}$ is the wind speed at 1.5-m height, and $T_{1.5}$ and T_0 are respectively the temperatures at 1.5-m height and the ocean surface. The maximum polynya extent, L_m, is then given by:

$$L_m = \frac{a}{\Delta T}, \quad (5)$$

where $\Delta T = T_0 - T_{1.5}$ and $a = \dfrac{0.03 \times L_f \times \rho_i \times H_i}{\rho_a \times C_p \times C_{h1.5}}$, which depends on constants that have all been previously defined. Figure 4a indicates that the Terra Nova Bay polynya is not well described by this temperature-dependent model. Including the latent heat flux in the definition of $P(t)$ increases the complexity of the temperature dependence in Eq. (5), but the result remains independent of the wind speed and no additional variance is explained by the inclusion of the term.

A significant finding of this study is that changes in the longwave heat flux are equally as important as the sensible heat flux in forcing changes in polynya width. To assess the impact of the longwave heat flux on fluctuations in polynya extent, Q_{Net} is taken to include both the sensible heat flux and net longwave heat flux. Substitution of these terms into Eq. (3) and (4) and retaining the first two terms in a Taylor expansion yields:

$$L_m = \frac{a}{\Delta T}\left(1 - \frac{b\,(Q_u - Q_d)}{\Delta T U_{1.5}} + \ldots\right), \quad (6)$$

where $Q_u - Q_d$ is the net longwave flux and $b = \dfrac{1}{\rho_a C_p C_{h1.5}}$, which are constants that have all been previously defined. It is clearly seen that the wind contributes to fluctuations in polynya extent by amplifying the net longwave flux. However, for strong winds and cold air temperatures, Eq. (6) asymptotically approaches Eq. (5) and the maximum polynya width becomes independent of the wind speed.

The sensitivity of polynya extent to changes in wind speed and air temperature can be quantified by noting that:

$$\Delta L_m \approx \frac{\partial L_m}{\partial T}\Delta T + \frac{\partial L_m}{\partial U}\Delta U. \quad (7)$$

For a typical wintertime temperature of −25 °C, a wind speed of 16 m s^{-1} (Table 1), and a cloud fraction of 50%, a ±5 °C change (Table 1) in temperature gives rise to a 2.5-km change in polynya width. In contrast, a ±7 m s^{-1} change in wind speed (Table 1) only produces a 1.0-km change in polynya width. Thus, although wind speed variations are important in determining polynya extent, Terra Nova Bay polynya extent is more than twice as sensitive to a 1-s change in temperature as it is to a 1 σ change in wind speed.

4.3 Extreme Polynya Events

A recurring result of this study is the inability for the polynya models to reproduce the large polynya expansions during May, June, and August 1988 and July 1989. Wind speeds and air temperatures were near normal for the time of year (Table 1; Fig. 2) and varying the cloud cover over its observed range could not reproduce the observed fluctuations. Carrasco and Bromwich [42] examined several case studies of katabatic outflow from southern Marie Byrd Land extending out across the Ross Ice Shelf. For the period June 2 through June 5, 1988, surface air temperatures rose from −35.6 to −19.2 °C and wind speed increased by 8.2 m s^{-1} for one site on the Ross Ice Shelf. Associated with this warm air outflow was a major polynya event along the northwest edge of the Ross Ice Shelf. The models in this study examined the response of the polynya to local offshore winds at Terra Nova Bay. However, it appears that the largest unexplained changes in polynya extent are the result of the polynya responding to the regional forcing of the southwestern Ross Sea. In particular, the strong winds coming off the Ross Ice Shelf could set up generally northward movement of the Ross Sea Pack, which is not modeled in this study. Kurtz and Bromwich [24] argued that both katabatic winds and regional geostrophic winds are responsible for the quasi-periodic areal fluctuations in polynya extent. This study suggests rather that it is both the local katabatic winds at Terra Nova Bay and the regional katabatic winds blowing off the Ross Ice Shelf that give rise to the large fluctuations in polynya extent.

The apparent sensitivity of polynya extent to subtle changes in air temperature, wind speed, and cloud fraction, typical of wintertime conditions at Terra Nova Bay, as well as the apparent sensitivity to both local and regional meteorological conditions could explain some of the difficulty that Campbell and his men had in understanding the paradoxical fluctuations in open water fraction at Terra Nova Bay (e.g. [21]).

5 Ice Production and Oceanographic Consequences

The total thickness of ice formed at any open water location in the polynya during winter can be estimated by integrating Eq. (2) over the 4-month period, May through August.

$$H_{Total} = \int_{May}^{Aug} \frac{Q_{Net}(t)}{\rho_i L_f} dt. \tag{8}$$

Using climatological meteorological data, Kurtz and Bromwich [6] estimated that 38 m of ice forms during the period May through August. Using the heat flux data presented in this study, the 1988 and 1990 thicknesses are 37.9 and 38.9 m respectively, which are remarkably similar to the climatological estimates provided by Kurtz and Bromwich [6]. In contrast, the 1989 thickness is 42.7 m, which is 12% higher than the other 2 years (Table 2). This difference is a consequence of a higher net heat loss to the atmosphere during 1989 than in the other 2 years (Table 2). It should be noted, however, that while the general agreement between the historical and present ice thickness estimates is encouraging, serendipity has played a role in this result. Kurtz and Bromwich [6] used $Q_{Net}(t)$ and L_f values that were both 75% lower than the values used in the present study. These factors offset each other in Eq. (8), resulting in the congruence between the two studies.

The total volume of ice produced for any given time period is not derivable from the ice thickness estimates alone; rather, an estimate of the open water fraction participating in the ice production process is also needed. Thus, the total ice production is given by

$$\Sigma_{Ice} = \int_{May}^{Aug} Area(t) \cdot \frac{Q_{Net}(t)}{\rho_i L_f} dt. \tag{9}$$

Kurtz and Bromwich [6] assumed a constant polynya area of 1300 km^2 to arrive at an estimate of 49.4 km^3 of ice formation during the 4-month winter period. It is straightforward to evaluate Eq. (9) using the previously derived heat flux estimates and taking $Area(t)$ to be the product of the north-south dimension of the polynya (65 km) and the modeled polynya extent, L(t). As is the case for the ice thickness, the 1988 and 1990 estimates are close to those observed by Kurtz and Bromwich [6] while the 1989 value is higher than the alternate 2 years (Table 2). Again, serendipity has played a role in achieving the close agreement between the historical and present estimates of total ice produced. As mentioned previously for H_{Total}, the 4-month average ratio of $Q_{Net}(t)/L_f$ is nearly equal for the two studies. In addition, modeled average polynya extent (Table 3) closely matches the historical average estimate. Together, these results produce the close agreement between the historical and present estimates of total ice production in Terra Nova Bay.

Equation (9) also illustrates another important point about the environmental factors controlling total ice production in the polynya. Although a time-dependent polynya model has been used to describe fluctuations in polynya extent, Terra Nova Bay is often well described by the steady state polynya model given by Eq. (2) and (3) [43]. Substitution of Eq. (2) and (3) into Eq. (9) indicates that the total ice production is independent of both $Q_{Net}(t)$ and L_f and is a function only of the surface wind speed, U_i, and the collection depth of the ice, H_i. To the degree that polynya extent is well described by Eq. (2) and (3), total ice production is highest during 1989 because the average wind speed is highest during the winter of 1989 (Table 1). Moreover, because of their dependence on total ice production, these same conclusions hold for annual differences in the total productions of salt and High Salinity Shelf Water (HSSW), which are described below.

From laboratory experiments it has been shown that S_i=0.31S_w [44], where S_i is the salinity of sea ice and S_w is the salinity of seawater, which is taken here to be

34.5 PSU, roughly consistent with the salinity of Warm Core Water (WMCO) or Low Salinity Shelf Water (LSSW) [45]. It follows that the total salt released into the water column is:

$$\Sigma_{Salt} = \rho_i \cdot (1\text{–}0.31) \cdot (34.5 \times 10^{-3}) \cdot \Sigma_{Ice}, \tag{10}$$

where $Ç_i$ is the density of sea ice ($0.95\text{x}10^3$ kg m^{-3}) and the factor of 10^{-3} converts grams per kilogram (PSU) into grams per gram [7]. Roughly $1\text{x}10^{12}$ kg salt was produced during the 4-month period in each of the 3 years with again the results for 1989 being ~10% higher than the other 2 years (Table 2).

The total volume of WMCO or LSSW converted to HSSW, which has a salinity of 34.8 PSU [45], is given by

$$V_{SW} = \frac{\Sigma_{Salt}}{\rho_{HSSW}\,(S_{HSSW} - S_{LSSW}) \times 10^{-3}}, \tag{11}$$

where ρ_{HSSW} is the density of HSSW (1030.45 kg m^{-3}), S_{HSSW} is the salinity of HSSW, and S_{LSSW} is the salinity of LSSW and WMCO. Again the factor of 10^{-3} is used to converts grams per kilogram (PSU) into grams per gram. Roughly $1\text{x}10^{13}$ m^3 HSSW is produced during this 4-month period. To convert WMCO and LSSW to HSSW at a rate consistent with the observed ice production rate, a transport of roughly 1 Sv is required during the winter months (Table 2). This is comparable with estimates from recent current measurements in Terra Nova Bay [46].

6 Summary

The original tenet of this study is that a simple polynya model, driven by in situ heat flux data, could be used to estimate open water fraction at Terra Nova Bay. The major finding of this study is that even for a nearly ideal situation, like Terra Nova Bay, at best the model could explain only about 40% of the variance observed in time-coincident microwave-derived ice concentration data. However, in spite of not fully modeling fluctuations in open water fraction at Terra Nova Bay, some insight into the dynamics of the polynya was obtained. Specifically it was found that:

1. Both the longwave and sensible heat fluxes each explain roughly 15% of the observed variance in satellite-derived open water fraction.
2. Fluctuations in both air temperature and wind speed contribute to changes in the size of the Terra Nova Bay polynya; however, the contribution from the wind speed is introduced as an amplification of the radiative flux terms.
3. The simple polynya models tested here did not adequately explain several anomalously large polynya fluctuations that occurred during the three winters studied. The polynya models studied here assumed that fluctuations in polynya extent take place in response to local one-dimensional meteorological forcing. The large unexplained fluctuations in polynya extent appeared to

occur in conjunction with the rapid expansion of the Ross Sea polynya, which is driven by strong winds blowing off the Ross Ice Shelf. Thus, regional as well as local meteorological forcing both appear to be important in controlling open water fraction at Terra Nova Bay.

4. Annual ice production estimates from the modeled polynya extent are comparable with climatological estimates (~50 km^3). Ironically, these estimates are nearly independent of the net heat flux, depending almost solely on the surface wind speed.
5. To maintain a conversion of WMCO or LSSW to HSSW consistent with the observed ice production rates, the transport of HSSW from Terra Nova Bay must be approximately 1 Sv.

This study hints at possible interactions between the Terra Nova Bay and Ross Sea polynyas. It is hoped that a better understanding of the physical interactions between these two polynyas and their biological consequences will be provided by a recently undertaken international, multi-year field program in the southwestern Ross Sea called "Research on Ocean Atmosphere Variability and Ecosystem Response in the Ross Sea" (ROAVERRS).

Acknowledgements. The SSM/I-derived sea ice data were provided by the EOS Distributed Active Archive Center (DAAC) at the National Snow and Ice Data Center, University of Colorado, Boulder, Colorado. The automatic weather station data were provided by C. Stearn's group at the University of Wisconsin. Robert Headland and William Mills of the Scott Polar Research Institute provided access to Priestly's diaries as well as interesting anecdotal information on life in the snow cave. Their assistance is greatly appreciated. I also thank Ed Andreas for clarifying some subtle points on exchange coefficients and especially Dave Bromwich for many interesting conversations regarding the environs of Terra Nova Bay. This work was supported by NASA Grant W-18821 and the Office of Naval Research Grant N0001495WX30141.

References

1. Smith SD, Muench RD, Pease CH (1990) Polynyas and leads: an overview of physical processes and environment. J Geophys Res 95:9461-9479
2. Schumacher JD, Aagaard K, Pease CH and Tripp RB (1983) Effects of a shelf polynya on flow and water properties in the Northern Bering Sea. J Geophys Res 88:2723-2732
3. Martin S, and Cavalieri DJ (1989) Contributions of the Siberian shelf polynyas to the Arctic Ocean intermediate and deep water. J Geophys Res 94:12725-12738
4. Zwally HJ, Comiso JC, Gordon AL (1985) Antarctic offshore leads and polynyas and oceanographic effects. Oceanology of the Antarctic Continental Shelf. Antarct Res Ser, vol 43, In: Jacobs SS (ed.) 203-226, AGU, Washington, D.C.
5. Cavalieri DJ, and Martin S. (1985) A passive microwave study of polynyas along the Antarctic Wilkes Land coast. Oceanology of the Antarctic Continental Shelf. Antarct Res Ser, vol 43, In: Jacobs ss (ed.) 227-252, AGU, Washington, DC
6. Kurtz DD and Bromwich DH (1985) A recurring, atmospherically forced polynya in Terra Nova Bay. Oceanology of the Antarctic Continental Shelf. Antarct Res Ser, vol 43, In: Jacobs ss (ed.) 177-201, AGU, Washington, DC

7. Alfultis MA and Martin S (1987) Satellite passive microwave studies of the sea of Okhotsk ice cover and its relationship to oceanic processes, 1978-1982. J Geophys Res 92:13013-13028
8. Darby MS, Willmott AJ and Somerville TA (1995) On the influence of coastline orientation on the steady state width of a latent heat polynya. J Geophys Res 100:13625-13633
9. Ou H W, (1988) A time-dependent model of a coastal polynya. J Phys Oceanogr 18:584-590
10. Pease CH, (1987) The size of wind-driven coastal polynyas. J Geophys Res 92:7049-7059
11. Frezzotti M and Mabin MCG (1994) Twentieth century behavior of Drygalski Ice Tongue, Ross Sea, Antarctica. Ann of Glaciol 20:397-400
12. Stearns C (1984) Antarctic automatic weather stations, austral summer 1983-1984. Antarct J Review 19:189-191
13. Stearns C and Weidner GA (1986) Antarctic automatic weather stations, austral summer 1985 -1986. Antarct J Review 21:233-234
14. Stearns CR, and Wendler G (1988) Research results from Antarctic automatic weather stations. Rev Geophys 26:45-61
15. Stearns CR, Keller LMG, Weidner A and Sievers M (1993) Monthly mean climatic data for Antarctic automatic weather stations. Antarctic meteorology and climatology: studies based on automatic weather stations. Antact Res Ser, vol 61. In: Bromich DH, Stearns CR (eds) 47-68, AGU, Washington, DC
16. Bromwich DH (1989) An extraordinary katabatic wind regime at Terra Nova Bay, Antarctica. Mon Weather Rev 117:688-695
17. Bromwich DH, Parish TR, Pellegrini A, Stearns CR and Wendler GA (1993) Spatial and temporal characteristics of the intense katabatic winds at Terra Nova Bay, Antarctica. Antarctic meteorology and climatology: studies based on automatic weather stations. Antarct Res Ser vol 61, In: Bromich DH, Stearns CR (eds.) 47-68, AGU, Washington, DC
18. Keller LM, Weidner GA and Stearns CR (1993) Antarctic automatic weather station data for the calendar year 1991. Dept Atmos Oceanic Sci January 355 pp
19. Priestly RE (1962) Scott's northern party. J Geogr 128:129-140
20. Priestly RE (1974) Antarctic adventure; Scott's northern party. T Fisher Unwin, London, 1914 (reprinted, C Hurst, London) 382 pp
21. King HGR (1988) The wicked mate: the Antarctic diary of Victor Campbell Bluntisham Books, Norfolk, pp 192
22. Knapp WW (1972) Satellite observations of large polynyas in polar waters. Sea ice. In: Karlsson T (ed.) National Research Council, Reykjavik, Iceland, pp 201-212
23. Szekielda KH (1974) The hot spot in the Ross Sea; upwelling during wintertime. Tethys 6:105 -110
24. Kurtz DD and Bromwich DH(1983) Satellite observed behavior of the Terra Nova Bay polynya. J Geophys Res 88:9717-9722
25. Hollinger JP and Lo RC (1983) SSM/I project summary report. Naval Research Laboratory, NRL Memorandum Report 5055, pp 106
26. Weaver RC, Morris, and Barry RG (1987) Passive microwave data for snow and ice research: planned products from the DMSP SSM/I system. Trans Am Geophys Un EOS 68:769, 776-777
27. Gloersen P and Cavalieri DJ (1986) Reduction of weather effects in the calculation of sea ice concentration from microwave radiances. J Geophys Res 91:3913-3919
28. Cavalieri DJ, Gloersen P and Campbell WJ (1984) Determination of sea ice parameters with the Nimbus 7 SMMR. J Geophys Res 89:5355-5369
29. Markus T and Burns BA (1995) A method to estimate subpixel-scale coastal polynyas with satellite passive microwave data. J Geophys Res 100:4473-4487
30. Zibordi G and Van Woert ML (1993) Antarctic sea ice mapping using the AVHRR. Remote Sens Environ 45:155-163
31. Zibordi G, Van Woert ML, Meloni GP, and Canossi I (1995) Intercomparisons of sea ice concentration from SSM/I and AVHRR data of the Ross Sea. Remote Sens Environ 53:145-152
32. Steffen K and AJ Schweiger (1990) A multisensor approach to sea ice classification for the validation of DMSP-SSM/I passive microwave derived sea ice products. Photogrammetric Eng and Remote Sens 56:75-82

33. Steffen K and Schweiger AJ (1991) NASA team algorithm for sea ice concentration retrieval from defense meteorological satellite program special sensor microwave imager: comparison with Landsat satellite imagery. J Geophys Res 96:21971-21987
34. Steffen K and Maslanik JA (1988) Comparisons of Nimbus 7 Scanning Multichannel Microwave Radiometer radiance and derived sea ice concentrations with Landsat imagery for the North Water area of Baffin Bay. J Geophys Res 93:10769-10781
35. Martinson DG and Wamser C (1990) Ice drift and momentum exchange in winter Antarctic pack ice. J Geophys Res, pp 1741-1755
36. Carnahan B, Luther HA and Wilkes JO (1969) Applied numerical methods. John Wiley and Sons, New York, pp 604
37. Lebedev VL (1968) Maximum size of a wind-generated lead during sea freezing Oceanology (English translation) 8:313-315
38. Bromwich DH and Kurtz DD (1982) Experiences of Scott's northern party: evidence for a relationship between winter katabatic winds and the Terra Nova Bay polynya Polar Record. 21:137-146
39. Andreas EL and Murphy B (1986) Bulk transfer coefficients for heat and momentum over leads and polynyas. J Phys Oceanogr 16:1875-1883
40. Maykut GA and Church PE (1973) Radiation climate at Barrow Alaska, 1962-1966. J Appl Met 12:620-628
41. Priestly RE (1912) British National Antarctic Expedition 1910-1913. Copy of part of MS 298/6/2. General diary of RE Priestly 1 Jan to 3 Oct 1912, MS 298/6/3 Scott Polar Research Institute, Cambridge, pp 134
42. Carrasco JF and Bromwich DH (1993) Satellite and automatic weather station analysis of katabatic surges across the Ross Ice Shelf. Antarctic meteorology and climatology: studies based on automatic weather stations. Antarct Res Ser vol 61, In: Bromwich DH, Stearns CR (eds.) 93-108, AGU, Washington, DC
43. Van Woert ML (1998) Wintertime dynamics of the Terra Nova Bay polynya. J Geophys Res submitted
44. Martin S and Kauffman P (1981) A field and laboratory study of wave damping by grease ice. J Glaciol 27:283-313
45. Jacobs SS, Fairbanks RG and Horibe Y (1985) Origin and evolution of water masses near the Antarctic continental margin: evidence from H218O/H216O ratios in seawater. In: Jacobs Oceanology of the Antarctic Continental Shelf. Antarct Res Ser vol 43, 59-85, AGU, Washington, DC
46. Manzella GMR, Meloni R, Picco P (1998) Current, temperature and salinity observations in the Terra Nova Bay polynya area (This Volume)

Current, Temperature and Salinity Observations in the Terra Nova Bay Polynya Area

G.M.R. Manzella[1], R. Meloni[2] and P. Picco[1]

[1]ENEA Marine Environment Research Centre, P.O. Box 316, 19100 La Spezia, Italy
[2]CNR Institute of Physical Oceanography, Forte S. Teresa, 19026 Pozzuolo di Lerici, La Spezia, Italy

Abstract

Long term (February - December 1995) time series of temperature, salinity and current have been analysed for the understanding of the main dynamics in the Terra Nova Bay Polynya area. Two main periods have been identified on the base of the observations: 1) from the ice-free sea conditions to the water column density homogenisation, and 2) the winter polynya period. During the first period the surface fresh water was transformed into High Salinity Shelf Water and sensible heat was delivered to the polynya surface. An analysis of the current-meter data during winter time indicated a transport of HSSW from the polynya area to the Ross Sea of about one Sverdrup.

1 Introduction

The most characteristic water mass of the Ross Sea is the High Salinity Shelf Water (HSSW), which is also formed in the Weddell Sea but with a lower salinity value. This water mass contributes to the formation of Antarctic bottom water and to a significant amount of shelf water circulating around Antarctica. There are strong indications that High Salinity Shelf Water is formed in Terra Nova Bay, where a recurrent polynya has been observed year after year [1].

The two basic mechanisms responsible for polynya formation and maintenance are associated with processes of sensible heat or latent heat [2]. A "sensible heat polynya" is due to vertical convection or upwelling of warm waters driven by winds and/or currents, able to prevent ice formation. The Terra Nova Bay polynya is believed to belong to type called "latent heat polynya", whose maintenance mechanism is due to the ice floes driven offshore by strong katabatic winds [3]. The protection of the Drygalsky Ice Tongue, preventing ice advection from the south, plays a fundamental role in the polynya formation and persistence [4]. Although the dimension of the Terra Nova Bay polynya fluctuates in response to katabatic winds, the presence of an open sea area allows a strong air-sea interaction, and, as a consequence, a continuous ice production and salty shelf water formation.

During February 1995, CTD casts were collected from Terra Nova Bay to Cap Adare (Fig. 1) and one mooring was deployed in the polynya area (named D in

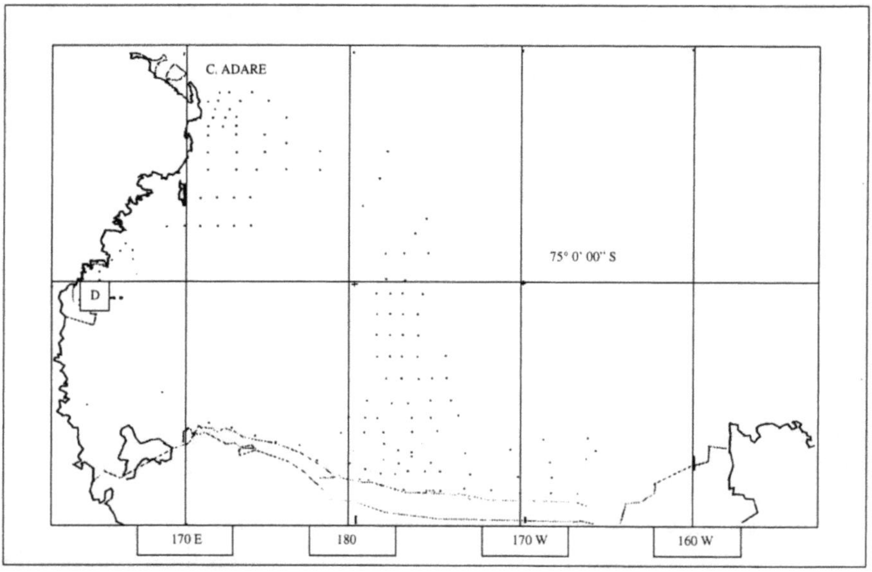

Fig. 1. The Ross Sea and Terra Nova Bay. The mooring in the polynya area is indicated by the *letter D inside a square*

Fig. 1). The CTD casts allowed the characterisation of the vertical water mass structure, but the spatial resolution of the stations was insufficient to resolve the horizontal water pattern.

Previous measurements in Terra Nova Bay during January-February 1987 [5] showed that the surface water masses circulate cyclonically. The coastal zone was characterised by relatively salty water (S>34.7 at the surface), contoured by fresher waters coming from the western opening, supporting the presence of a divergence in Terra Nova Bay. Analysis of the 1987 data suggested the location of the mooring for the study of the polynya processes.

In this paper the main results of the analysis of 1- year time series of temperature, salinity and current collected during 1995 are presented.

2 Materials and Methods

CTD measurements were carried out in the Ross Sea (Fig. 1) during January–February 1995. They were performed using a SeaBird SBE 911Plus, equipped with additional sensors, including one measuring distance from the bottom. The type, range and accuracy of the CTD sensors are reported in Table 1. Temperature and conductivity sensors were calibrated before and after the cruise at the SaclantCen of La Spezia (Italy). In-situ comparison with reversing thermometers (SIS RTM 4002) and salinity samples with Autosal Guildline model 8400B salinometer was performed. On the basis of this inter comparison, offsets of 0.0028 °C for temper-

Table 1. Type, range and accuracy of the CTD sensors

Parameter	Type of sensor	Range	Accuracy/stability
Temperature	SBE3/F	-5 to +35 °C	±0.002 °C/year
Conductivity	SBE4	0.0 to 7 S/m	±0.0003 S/m/month
Pressure	Paroscientific	0 to 10.000 psi	±1.5 psi
Oxygen	SBE 13Y	0 to 15 ml/l	±0.1 ml/l
Light transmission	Sea Tech 25 cm	0 to 100%	±0.5%
Fluorescence	Acquatracka III	0.01 to 100 mg/l	±0.01 mg/l
pH	SBE18	0 to 14	±0.1
Distance from bottom	Datasonic PSA-900	0 to 300 m	±0.75 cm

ature and 0.00095 S/m for conductivity were applied to the raw data. The mooring was composed of five Eulerian current metres (Aanderaa RCM7) (Table 2) with temperature and conductivity sensors, two sedimentary traps and a turbidimeter. The upper current meter was 55 m below the sea surface, the deepest at about 30 m from the bottom. The initial and final records of the temperature and conductivity time series were compared with CTD data collected near the mooring position during February and December 1995, without finding significant differences. Unfortunately some salinity sensors were found broken; as a consequence it was not possible to investigate the salinity changes over the water column from 55 to 800 m.

The current data were rotated to align the v-component with the local isobath. In this way, the major energy was present in this component. Data were low-pass filtered and daily resampled; reported temperatures are in-situ data.

3 Distribution of Water Masses in the Ross Sea During Summer 1995

The Terra Nova Bay bottom topography is made up of a deep pit, reaching more than 1100 m, bounded approximately by the 500-m isobath. The southern end of this pit is open toward the Ross Ice Shelf by a sill deeper than 600 m. In general, the Ross Sea continental shelf is characterised by a high variable bathymetry and by the presence of many banks.

Table 2. Mooring details

Depth	Current	Thermistor		Conductivity	
(m)	meter	Type	Accuracy	Cell	Accuracy
55	RCM7	Seacat SBE3	±0.01 °C/6 months	–	–
140	RCM7	Fenwall GB32JM19	±0.05°C	Inductive 2994	±0.01 S/m/month
402	RCM7	Fenwall GB32JM19	±0.05°C	Inductive 2994	±0.01 S/m/month
748	RCM7	Seacat SBE3	±0.01 °C/6 months	Seacat SBE4	±0.001 S/m/month
882	RCM7	Fenwall GB32JM19	±0.05°C	Inductive 2994	±0.01 S/m/month

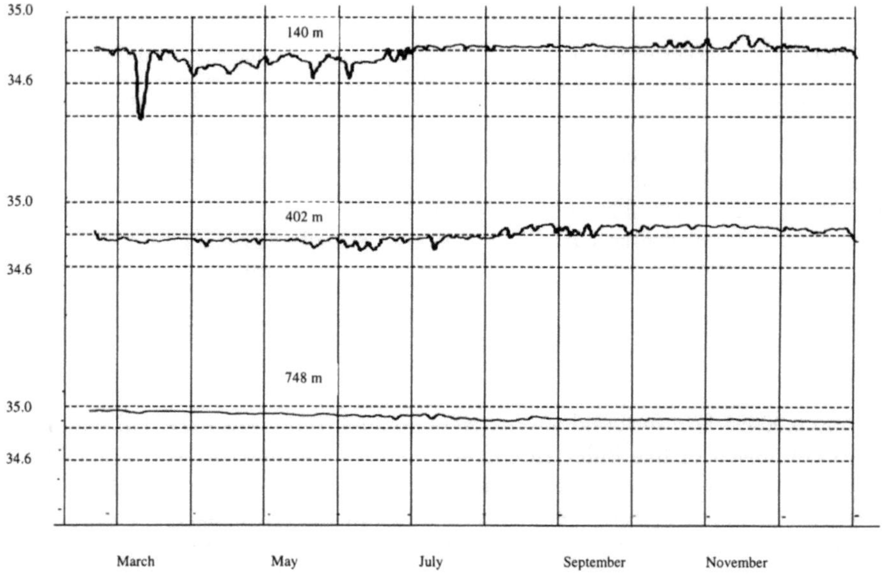

Fig. 2. Salinity time series in Terra Nova Bay

The CTD data collected during 1995 are in agreement with the water mass distribution pattern previously observed by other authors [6] in the Ross Sea and by Stocchino and Manzella [5] in the Terra Nova Bay. In the Ross Sea, the water masses were characterised by a surface layer above the thermocline, about 50 m thick, having a low salinity content from 33.0 to 34.6, and temperature generally close or below 0 °C (0.5 to −1.6 °C). The highest values of salinity were observed in the south-western stations; the higher temperatures were observed in the northern stations.

In Terra Nova Bay, near coastal waters below the thermocline had a relatively high salinity content. Vertical distribution of temperature and salinity during February 1995 was characterised by a surface, shallow, well mixed layer with mean salinity of 34.35 psu and temperature −1.88 °C; a temperature minimum (−1.95 °C) at around 200 m with correspondent salinity 34.77 psu is observed in southern stations, due to the interaction with the Drygalsky Ice Tongue. Water having salinity of 34.87 psu and temperature −1.92 °C fills the deeper layers from about 600–700 m to the bottom.

The northward spreading of HSSW from Terra Nova Bay along Victoria Land coasts has been followed: a well distinct bottom layer, more than 100 m thick, characterised by uniform salinity of 34.83 psu and temperature of −1.91 °C, is observed up to 72° 36'S, extending east near the shelf break. Nevertheless, salinity values of 34.78 psu and temperature −1.91 °C were also detected at depth in the eastern station along the Ross Sea Ice Shelf, indicating a possible southward transport of HSSW (Table 3).

Table 3. Mean values and standard deviations (in cm/s) of the cross-isobath (u) and along-isobath (v) current components

Depth (m)	u-Component	u-SD	v-Component	v-SD
55	-0.76	3.67	7.57	6.30
140	-1.04	3.01	2.97	3.78
402	0.59	2.50	2.08	2.71
748	2.58	2.35	0.88	2.76
882	4.79	2.81	0.35	2.65

4 Terra Nova Bay Time Series

An annual cycle of the water temperature was recorded by the thermistor sensors. At 55 m, the temperature dropped from an average value of −1.50 °C in February to −1.90 °C in April. From July to October the temperature detected from 55 m to the bottom was nearly homogeneous, with values of about −1.91 to −1.92 °C. Starting from late November the temperature increased again to reach maximum daily values of −1.4 °C at 55 m and −1.86 °C at 140 m in December. Temporal variations in deeper layers were of the same order of sensor accuracy.

During the period February–June, the water mass at 140 m had a salinity varying from 34.54 to 34.78 psu (Fig. 2). Starting from July to September the salinity value was less variable, at around 34.78 psu. At 400 m a relatively low salinity (34.6 to 34.75) was detected from February to July. Then the salinity oscillated, reaching values higher than 34.8.

The sensor at 748 m indicated the presence of water having a constant salinity, with characteristics of slightly modified HSSW, i.e. with a salinity of 34.87 higher than those observed in the above layers during the year in the polynya area. A constant decreasing trend to 34.83 could be attributed to the sensor drift or to mixing with the less salty waters detected by the upper sensors. This could signify that this kind of HSSW is older than that formed during the examined austral winter or is coming from a different formation place (e.g. the Ross Sea polynya).

A comparison between the salinity time series and the current data (Fig. 3) shows that the low salinity water observed in the upper layers during February-June in the mooring was mainly directed north-east, i.e. along the isobaths towards Cap Adare (Fig. 1). The upper currents were especially intense during the first 5 months of observation. During the same period, northward movement of the waters was observed over the entire column, with current values decreasing during April in the lower levels. South and south-eastward transport was detected in the upper layers during October, and, with a significant higher intensity, in November and December, below 400 m. This last represents the main contribution to the spread of the HSSW produced in Terra Nova Bay over the Ross Sea shelf. The statistics of the currents for the entire period of measurements show that the currents had a general stability (low standard deviation) along the main direction of the flow (south-north direction). An inversion of the u-components was detected between 140 and 400 m.

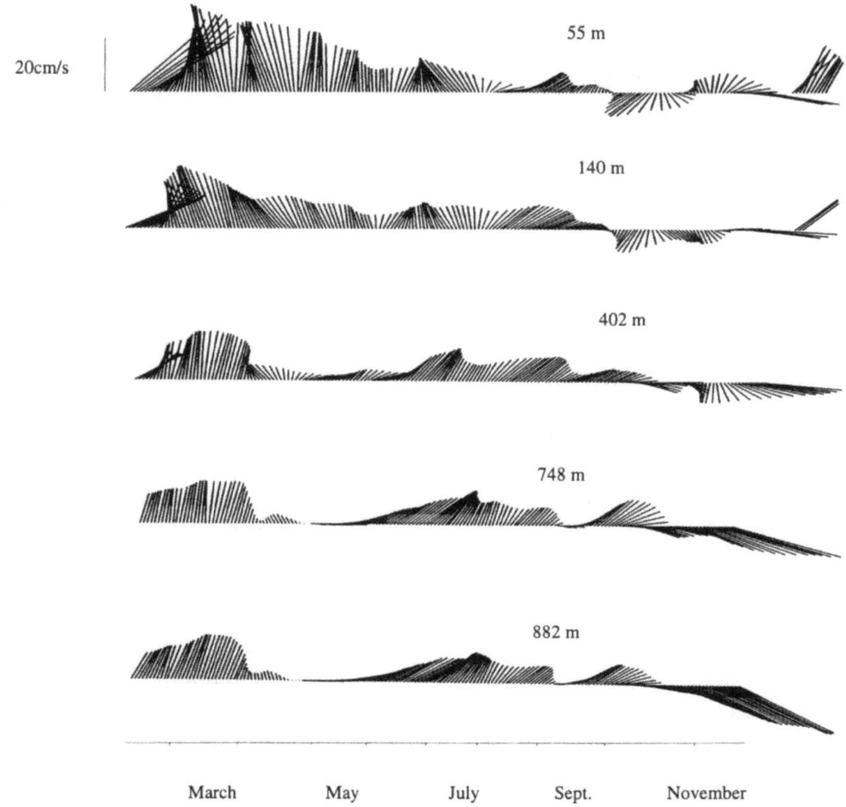

Fig. 3. Stick diagrams of current data in Terra Nova Bay. The *line upper-left* corresponds to 20 cm/s

The main gross classification of the physical characteristics could consider the two periods during which: (1) the Ross Sea is generally ice free and (2) the Terra Nova Bay polynya forms salty water. From time series in D, however, a more detailed classification can be adopted in terms of temperature, salinity and current characteristics (supposing a cyclic behaviour of the parameters): (1) November–April, (2) May–June, (3) July (4) August–October.

Starting from November the sea temperature in the upper layers was increasing. The currents oscillated from south to north. The data indicate the presence of a lower layer capable of transporting HSSW south and offshore (Fig. 3). From mid January to mid February, the entire Terra Nova Bay is mainly ice free [1]; then it is progressively covered by sea ice, until a small open water area remains during the wintertime. From February to April, the currents were almost northward at all depths, and the temperature was decreasing. This period is also characterised by a low salinity from the sea surface to 140 m at least. Short-lived events of about 3–7 days perturbed the salinity character at 140 m. They were probably due to the passage of anti-cyclonic eddies or meanders in the area.

During May–June, the temperature was close to the surface freezing point over the entire water column. The salinity at 140 m was observed to increase to the value of the HSSW, and the currents suggest the presence of two layers: from the surface to about 400 m and from there to the bottom. At 402 m the salinity reached the HSSW value with a delay of 1 month with respect to the upper record (end of July, instead of end of June). In this period the surface haline mixing did not reach this depth, probably due to the advection from the south of water fresher than HSSW or the mixing between the HSSW formed at the surface and the bottom water [7].

The data show that the salinity mixing embraced the entire water column from August to October, making the vertical density gradient minimum.

5 Normal Mode Decomposition

From time series presented in the previous paragraph, it is possible to ascertain the presence of a density-homogeneous water mass from 140 m to the bottom from August to October. The importance of continental shelf processes in the formation of dense water masses has been identified by many authors [2, 8]. Negative buoyancy forcing in polar regions is concentrated in areas of high ice production [1-3], such as the polynyas. In the Terra Nova Bay polynya area, the katabatic winds drive ice offshore, leaving the nearshore waters exposed to the atmosphere. As new ice is formed, brine is rejected and acts as a salt flux [3] producing a negative buoyancy source and, as a consequence, lateral density gradients in the shelf waters. The saltwater formation, in turn, drives a baroclinic circulation involving the locally created dense water, as well as the ambient water.

Although the current time series cannot be directly interpreted in terms of HSSW formation, information on the nature of the water circulation and transport of dense water can be extracted from the empirical decomposition (EOF) of the along-isobath components. The normalised eigenvectors of the first two modes are shown in Fig. 4. The first mode (B) indicates a quasi-homogeneous oscillation of the water masses over the water column. The second mode (C) indicates that the water above and below 300 m oscillates in opposite ways. The interpretation is that this second mode, extracting 12% of the total variance (with respect to the 80% of the first mode), is representative of the transport of HSSW from the Terra Nova Bay polynya area to the Ross Sea. If this is true, the total amount of HSSW drained out from the polynya area can be calculated, using a mean horizontal dimension of about 35 km [1, 3], as equal to 0.85 Sv over the entire column using a velocity of 4 cm/s. An approximate transport of 1 Sv is in agreement with other calculations presented in this book [9].

The modal decomposition makes evident the presence of higher velocities in the lower water column near the bottom. Numerical models applied to polynya areas [10] show that the water mass circulation evolves from a geostrophic adjustment to rapid eddies development and offshore eddy transport. The numerical model shows that maximum along-shelf velocity can be attained at depth.

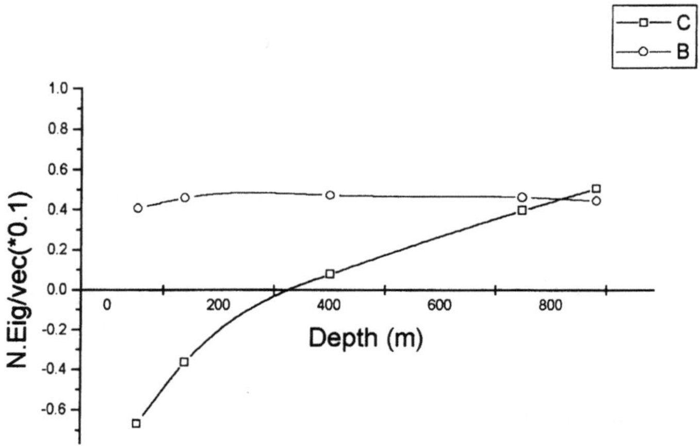

Fig. 4. First (*B*) and second (*C*) empirical modes obtained decomposing the along-isobath current components

6 Discussion

During the austral summer 1995, the coastal area of Terra Nova Bay was characterised by the presence of relatively salty water whose existence was due to an upwelling observed during several campaigns. This phenomenon was related to the winds blowing through the Antarctic glaciers toward the sea. The effects of the katabatic winds were observed in the wind waves and in the presence of strips formed by ice crystals. Close to the coast, temperatures below 0 °C were observed. Furthermore, winds were able to detach from pack-ice some large slabs, which were advected offshore. Over the Ross Sea continental shelf, water temperatures of about -1.5 to 0.5 °C were detected.

One of the effects of the Terra Nova Bay polynya is the formation of HSSW (saltier than its counterpart in the Weddell Sea). During summer 1995, the presence of HSSW was detected in the southern stations close to the Ice Shelf. The continuous removal of sea ice was estimated to be about 60 m per year [1]. The rate of ice formation and the salinity of the bottom water are key parameters for to salty water formation in the polynya area. The salinity of the water was observed to increase slowly from the surface to 600 m, below which depth HSSW was residing during the entire period of measurements.

The empirical decomposition applied to the current time series shows that the main energy was contained in the first vertical sheared mode (about 80%). The second mode contained only a smaller fraction (12%), but it was interpreted as representing an important dynamic aspect of the HSSW transport from the polynya area to the Ross Sea. The estimated transport of 0.85 Sv can contribute to the amount of shelf water entering the Antarctic Circumpolar Current, as well as the bottom water and Ice Shelf Water formation below the Ross Ice Shelf.

Acknowledgements. The authors are grateful to colleagues of the CLIMA Project whose participation in the field campaign allowed collection of the data.

References

1. Kurtz DD, Bromwich DH (1985) A recurring atmospherically forced polynya in Terra Nova Bay. In: Jacobs SS (ed) Oceanology of the Antarctic Continental Shelf. Antarct Res Ser 43:173-202, AGU
2. Smith SD, Muench RD, Pease CH (1990) Polynyas and Leads: an overview of physical processes and environment. J Geophys Res 95:9461-9479
3. Pease CH (1987) The size of wind-driven coastal polynya. J Geophys Res 92:7049-7059
4. Bromwich DH, Kurtz DD (1984) Katabatic wind forcing of the Terra Nova Bay polynya. J Geophys Res 89:3561-3572
5. Stocchino C, Manzella GMR (1991) Analisi delle correnti marine di Baia Terra Nova (Mare di Ross-Antartide). CNR-ARIGE, pp 56
6. Jacobs S, Fairbanks GR, Horibe Y(1985) Origin and evolution of water masses near the Antarctic continental margin: evidence from $H_2\,^{18}O/\,H_2\,^{16}O$ ratio in sea water. In: Jacobs SS (ed) Oceanology of the Antarctic Continental Shelf. Antarct Res Ser 43:59-86, AGU, Washington, DC
7. Topp R, Johnson M (1997) Winter intensification and water mass evolution from year long current meters in the northeast Polynya. J Marine Systems 10:157-198
8. Minnet PJ, Bignami F, Bohm E, Budeus G, Galbraith PS, Gudmansen P, Hopkins TS, Ingram RG, Johmson MA, Niebauer HJ, Ramseier RO, Schneider W (1997) A summary of the formation and seasonal progression of the northeast Polynya. J Marine Systems 10:79-86
9. Van Woert M (1998) Wintertime expansion and contraction of the Terra Nova Bay polynya (This Volume)
10. Gawarkoewicz G, Chapman DC (1995) A numerical study of dense water formation and transport on a shallow, sloping continental shelf. J Geophys Res 100:4489-4507

Particle Fluxes and Organic Matter

Particle Fluxes at the Edge of the Ross Ice Shelf: the Role of Physical Forcing

A. Accornero[1], A. Bergamasco[2], A. Monaco[3] and S. Tucci[4]

[1]Istituto di Meteorologia e Oceanografia, Istituto Universitario Navale, Via A. De Gasperi 5, 80133 Napoli, Italy
[2]Istituto per lo Studio della Dinamica delle Grandi Masse, Consiglio Nazionale delle Ricerche, San Polo 1364, 30125 Venezia, Italy
[3]Centre de Formation et de Recherche sur l'Environnement Marin, Université de Perpignan, Avenue de Villeneuve 52, 66860 Perpignan cedex, France
[4]Dipartimento di Scienze della Terra, Università di Genova, Corso Europa 26, 16132 Genova, Italy

Abstract

In this paper we present the seasonal variation of downward particle flux, as expressed by total mass flux and its main biogenic components (organic matter, biogenic silica and carbonate) in a site located at the edge of the Ross Ice Shelf. The focus of this study is on data from mooring site named F, located at 77°59'.998 S, 177°01'.623 W in the framework of the C.L.I.M.A. Project (Climatic Long-term Interaction for the Mass Balance in Antarctica) during the deployment period January 28, 1995 - January 21, 1996. Total mass flux shows a clear seasonal trend, with the austral summer months (December-March) accounting for about 93% of the total annual flux and the predominance of the biogenic (55-99 % of the total) versus the lithogenic fraction through the whole considered period. In order to understand the role of physical forcing on sedimentation processes (as recorded by the moored trap) we construct a simple Montecarlo model of particle downfall that takes into account both the intrinsic sinking velocity of particles and the actual vertical displacement of the water layer in wich they are contained. On the basis of the potential density anomaly evolution, vertical displacement of the isopycnal surfaces ranging from 27.69 to 27.77 are computed and updated every day. This numerical experiment puts in evidence the great importance of the water column stability and density variability as forcing factors for sedimentation processes. Dynamical constraints are observed to play a crucial role, even in the presence of high production rates.

1 Introduction

Particle fluxes affect a number of important processes in the water column, including nutrient recycling, supply of food to deep-water and benthic communities, sediment composition and stratification. They result from the interaction of all processes that produce, introduce, transform and destroy particulate matter in the marine environment and are consequently influenced by a complex of physical (vertical and lateral transport, aggregation), chemical (adsorption, flocculation, dissolution, oxidation) and biological (primary and secondary production, microbial degradation, egestion, grazing) factors. We have attempted to construct a phenomenological process model to understand the behaviour of particulate fluxes under external physical forcing, considering both the hydrological characteristics of the water masses and the hydrodynamic features in the investigated area. At the same time we have tried to establish a coupling between biological observations and physical data in order to test whether a relation exists between the latter and microscopic observations.

Particulate matter sinking through the water column includes a large variety of particles, with sediment rates ranging from tens to hundreds of meters per day [1]: phytoplankton and zooplankton debris, fecal pellets, molts, various kinds of tests, remains and fecal material, aggregates of small particles, marine snow and mineral matter. The quantitative evaluation of biogenic material and the measure of its vertical transport rates through the water column are essential to describe the functioning of an ecosystem and can significantly help to understand biogeochemical cycles on a wider scale.

The Ross Sea offers very interesting characteristics for the study of biogenic particle fluxes, because it is the site of abundant primary production and significant biogenic accumulation [2-5]. In particular the ice edge zone seems to be of key importance in the framework of the ecological [6] and biogeochemical processes occurring in the Southern Ocean, playing a major role in the annual cycle of biogenic particle production [7-10]. Ice melting is responsible for the creation of local conditions of vertical stability which are essential for the starting of phytoplankton blooms in surface waters [11], which generally result in substantial transport of biogenic particles to the deeper layers and sediments. Other features typical of the Ross Sea are the absence of sediment input by fluvial system, terrigenous contributions being exclusively linked to glacial transport [13], and a relatively simple and confined ocean circulation [13, 14 and bibliography therein].

2 Materials and Methods

In the framework of the Italian National Programme for Antarctic Research, the CLIMA Project (Climatic Long-term Interaction for the Mass balance in Antarctica) planned for the first time in 1995 a sub-project for the study of particle fluxes in the Ross Sea. Three sites were selected for the deployment of mooring arrays equipped with time series sediment traps and oceanographic instrumentations from austral summer 1995 through January 1999. In particular, mooring site F (at the edge of the RIS) and H (in proximity of the continental shelf break) were chosen in order to follow the fate of the Ice Shelf Waters (ISW) after that an ISW core outflowing from below the RIS was localized at CTD stations corresponding with the F mooring location (Fig. 1). The focus of this study is on data from mooring site F (77°59'.998 S, 177°01'.623 W) during the deployment period January 28, 1995 - January 21, 1996.

2.1 Mooring Array

The mooring line carried two sediment traps of identical design (Hydro-Bios Multi Sediment Trap MST 24), equipped with 24 receiving cups, with programmed collection intervals of 15 days. Traps were characterized by a collection area of 0.6 m^2 and fitted with a 5-cm-thick square plastic baffle (3.5-cm cell diameter) mounted to the opening. In order to hinder degradation of settled material and avoid breaking of "swimmers" the receiving cups were filled before deployment with a solution of buffered 5% formaldehyde in 0.45 m pre-filtered seawater. The two traps were placed at 230- and 423-m depth in a 602-m water column, but unfortunately the upper trap did not work, owing to the failure of the stepping motor, and for this reason time series samples are not available at this depth. Particle presence was, however, recorded by a transmissometer fixed to the framework of the 230-m trap, which measured the attenuation of a red monochromatic light beam over a pathlength of 0.25 m, with records resulting from the average of eight measurements performed every 30

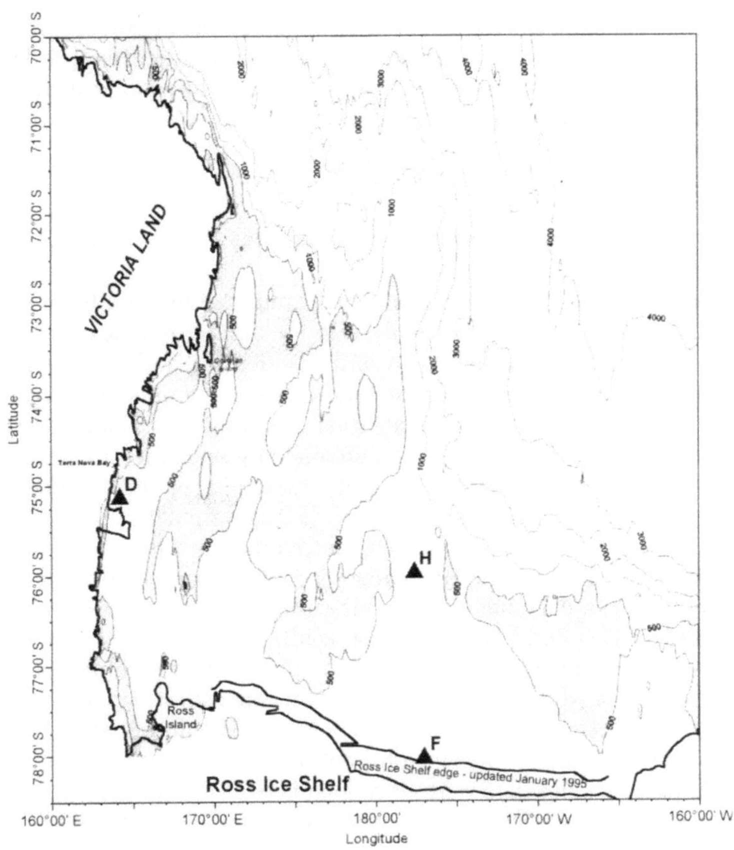

Fig. 1. CLIMA Project mooring sites in the Ross Sea and location of mooring F

min. Physical parameters were monitored by means of two Aanderaa RCM7 current meters with thermistor and conductive cells placed at 244- and 390-m depth, two SBE Seacats with temperature and conductivity recorders positioned at 243 and 398 m, and one Aanderaa RCM7 current meter with only a thermistor sensor located 20 m above the sea floor. An RDI ADCP current meter profiler was fixed at the top of the mooring array to measure the upper layer velocity, but it did not work, due to battery power failure.

2.2 Oceanographic Time Series Data Analysis

Time series of currents, temperature and salinity measurements, acquired every 30 min, were filtered and averaged. The averages were computed on several time intervals (1, 5, 10, 15 and 30 days) to highlight the behaviour of the considered variables on different time scales. Vertical objective analysis [15] was performed with the aim of reconstructing the temporal evolution of the vertical profiles of temperature, salinity, density and velocity. Physical data recording started on January 27, 1995, and stopped on different days. In this study we only consider the time interval in which all the oceanographic instrumentation was working on the whole mooring line, i.e. the first 240 days after the moment of deployment, in order to get a complete hydrological and dynamic setting. Time series data analysis and interpretation are discussed in detail in another work [16].

2.3 Sample Processing and Analytical Techniques

The material collected in each cup was first examined by both light and scanning electron microscopy (a detailed description of microscopic observations is given in [17]), and then divided into a series of identical and homogeneous wet subsamples for the determination of total mass, biogenic silica, organic and inorganic carbon. Sample preliminary treatment was performed according to Heussner et al. [18]. The total mass was obtained by averaging the dry weight of two to four aliquots taken from different subsamples of the original sample. Particulate total (PTC) and organic carbon (POC) were determined by using a LECO CS 125 carbon analyser; for POC determination filters were previously treated with 2N H_3PO_4 and 1N HCl and concentrations were doubled to obtain organic matter [19]. Particulate inorganic carbon (PIC) was calculated by difference (PIC = PTC - POC) and PIC values multiplied by a factor of 8.33 to estimate carbonate concentration, assuming that all inorganic carbon is in the form of calcium carbonate [20]. Determination of biogenic silica was performed by a time-series dissolution experiment in a 0.5 M NaOH solution at 85 °C for 5 h. An aliquot of each sample was taken for analysis after every hour and the relative weight percent silica data were extrapolated back to time zero to correct for the silica eventually originating from coexisting clay minerals [21]. The lithogenic fraction was calculated by subtracting the concentrations of organic matter, biogenic silica and carbonate from total dry weight [20].

Table 1. Current velocity and direction monthly means at 244- and 390-m depth

	244-m Depth		390-m Depth	
	Module (cm s^{-1})	Direction (degrees)	Module (cm s^{-1})	Direction (degrees)
January	3.98	239	11	344
February	9.85	294	5.54	281
March	5.92	301	4.15	325
April	3.60	274	2.61	315
May	4.01	311	5	350
June	0.31	69	3.13	351
July	7.08	351	11.81	335
August	4.65	279	5.27	5
September	5.73	55	9.70	21

3 Results and Discussion

3.1 Oceanographic Setting

In Table 1 monthly averages of current velocity and direction at 244- and 390-m depth are summarized. For the whole period the average was 4.27 cm s^{-1} 306°N at 244 m and 5 cm s^{-1} 337°N at 390 m. All the records show a strong correlation, especially those of current meters at 390 m and 579 m, which exhibit very similar pat-

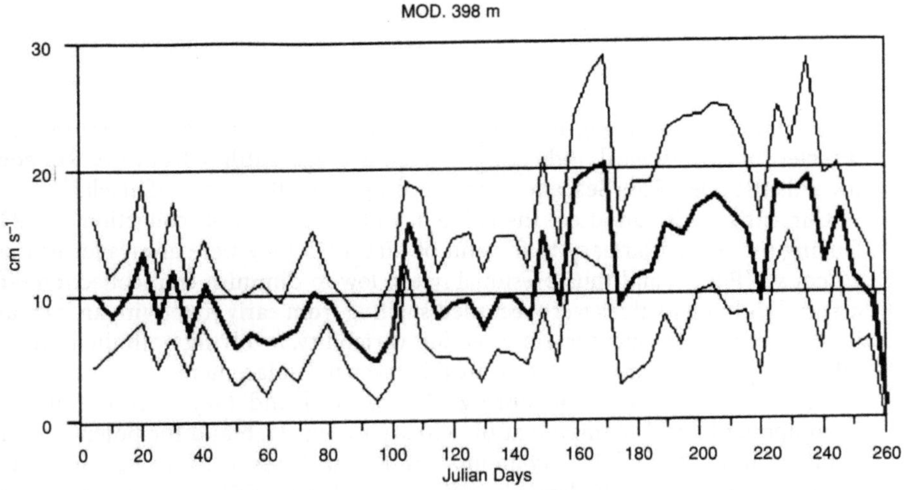

Fig. 2. Current velocity module (5-day means) at 390-m depth. Time in days starting from the beginning of the deployment

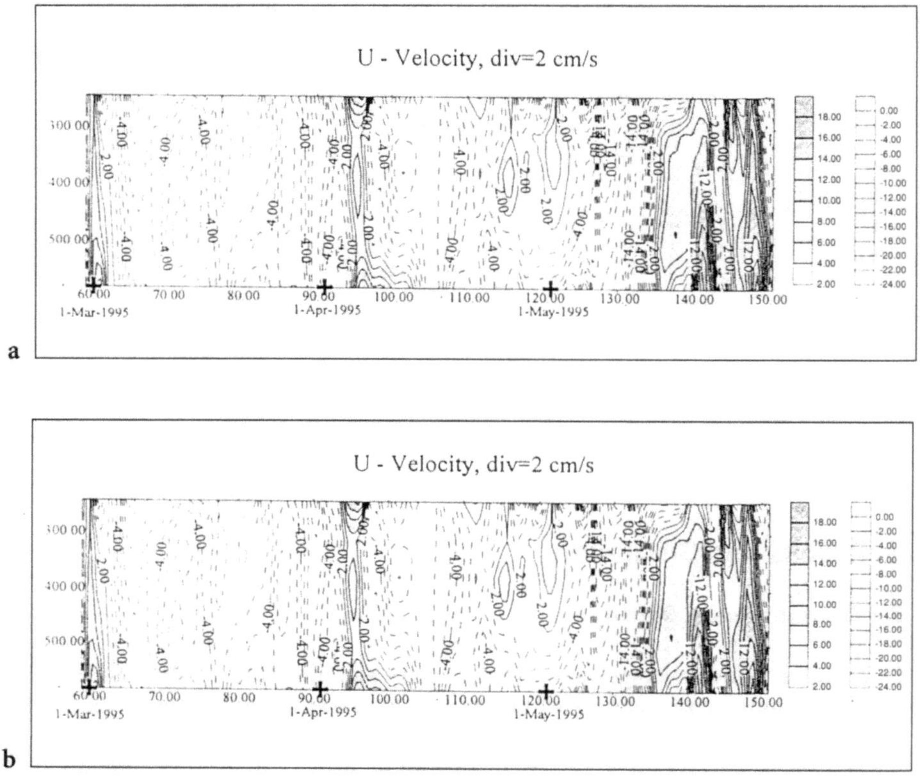

Fig. 3. Vertical profile of the U-component (**a**) and the V-component (**b**) of the current during March, April and May. Time in Julian days (60-150)

terns, indicating a strong barotropic behaviour. The resulting annual mean is directed northwestward, indicating a small average outflow from the RIS zone towards the open Ross Sea, and a strong component along the RIS itself.

Current speeds can affect the efficiency of sediment trap collection [22, 23]. During the period January to the end of May 1995 they were always lower than 20 cm s^{-1} (Fig. 2), and thus presumably too low to diminish trap collection efficiency, but higher values were recorded starting from early June, with an increase in velocity module and current direction variability, involving both the horizontal velocity component and the vertical isopycnal displacement.

Vertical profile evolution during March, April and May is shown for the U-component and V-component of velocity in Fig. 3a, b and for potential temperature and potential density anomaly in Fig. 4a, b, respectively. Looking at the V-velocity trend, the 5.02 days' periodicity found using FFT analysis appears evident and is representative of an inflow-outflow intermittence. This feature characterizes not only the V-velocity but also all the measured variables at every

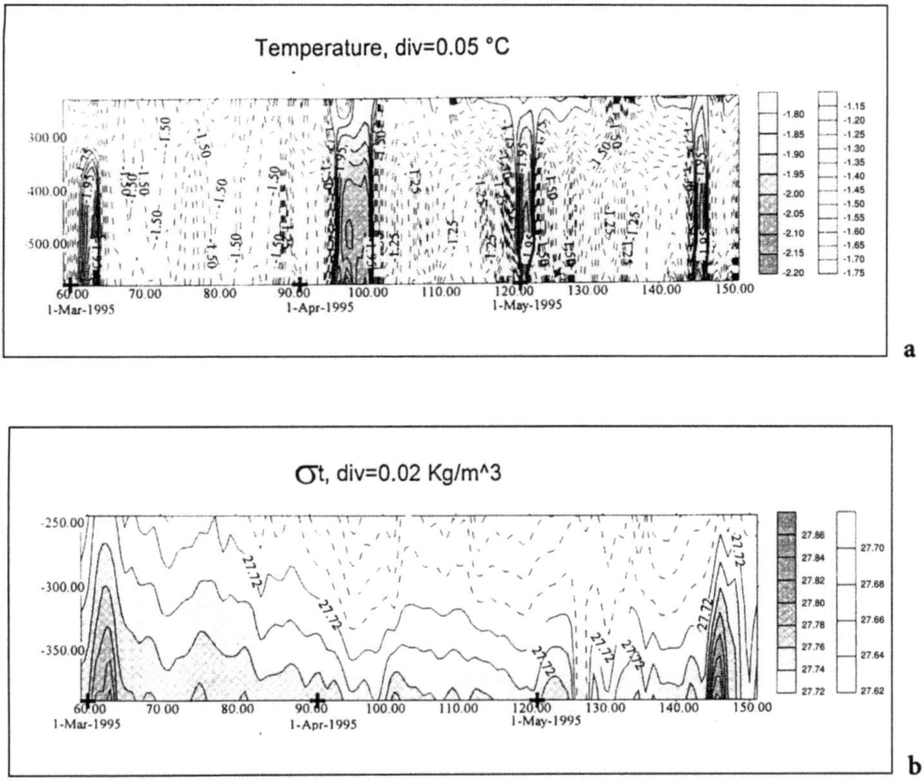

Fig. 4. Vertical profile of potential temperature (**a**) and density anomaly (**b**) during March, April and May. Time in Julian days (60-150)

depth during the whole record period. During the Tenth Italian Antarctic Expedition 1994-1995 a complete vertical section of T and S was acquired along the RIS. This section does not exhibit any other zone where supercold water is outflowing towards the Ross Sea. Other measurements [13] show a substantial agreement with our observations. This suggests the hypothesis of pulses of this water type, localized in space and limited in time, from below the RIS towards the Ross Sea, that occur with strong intensity to balance the amount of incoming water. The U-component evolution shows a modulation of 17.07 days, as revealed by FFT analysis, and a prevailing westward direction along the RIS, that rarely inverts its sign. During the winter period (July-August) the variability increases, coupled with an increase in mean kinetic energy and turbulent kinetic energy. We can identify an intermittent gyre driven by the ISW pulse. Unfortunately the top current meter stopped recording in September and it is thus not possible to complete the analysis for the whole period of the sediment trap observation.

If we compare for the same period the evolution of potential temperature and

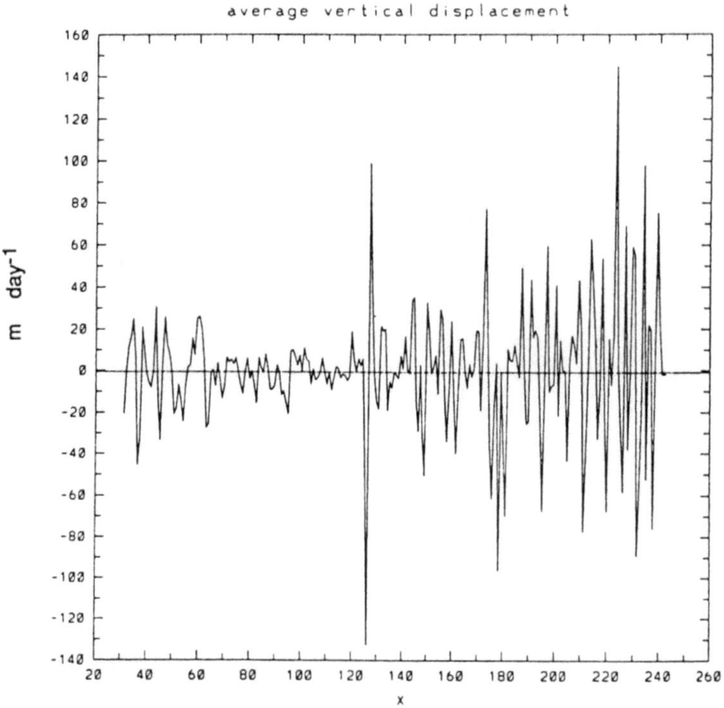

Fig. 5. Average vertical displacement (m day^{-1}) of 27.69–27.77 isopycnal surfaces during the first 240 days of mooring deployment

potential density anomaly vertical profiles (Fig. 4a, b) we can identify the pulses of outcoming supercold water, sometimes associated with recirculation processes, as shown by salinity minimum, probably due to the mixing with the water outside the RIS cavity. During these 3 months, four episodes of intrusion of supercold water are evident, even if a relatively steady circulation system is established. Looking at Figs. 3b, 4a, b we can identify the first episode (Julian day 62-65) as a true intrusion of supercold water flowing from the RIS cavity, owing to the good correlation observed between cold temperature, increased potential density anomaly and V-component of velocity. The second event (Julian day 96-100) is characterized by a much more intense T signal, a decreased density anomaly and a negative V-component. This could be related to a recirculation of the first event or to a new inflow coming from near the mooring location. We are keen to support the hypothesis of a recirculation of the first event, or of previous ones, on the basis of the observed density signal, which we consider to be a good indicator of a mixing event during the recirculation. The third episode (Julian day 121-123) stands again for a recirculation event, owing to the negative V-component of velocity and the decreased density anomaly, while the last one (Julian day

145-147) clearly appears as a strong event of supercold water outflowing from below the RIS. Analysis of the vertical profile evolution of the physical parameters considered in Figs. 3 and 4 shows evidence of the pulsing dynamics existing at the edge of the RIS. Not only is 5.02 days frequency the most characteristic Fourier component but also the pulse events only last a few days. Microscopic observation of the sediment trap samples corresponding to the collection intervals, including the above-mentioned short-term pulse events, does not show any differences that may be specifically correlated to them. This may be due to the discrepancy existing between the duration of the pulse events (2-4 days) and that of the collection interval (15 days): even if the material brought by the short-term pulse exhibits its own specificity, this does not appear evident because it is "diluted" in the bulk of the material collected during the rest of the 15-day collection period.

On the basis of the potential density anomaly evolution, picnoclyne depth variations were calculated and vertical displacement was computed for every isopycnal surface ranging from 27.69 to 27.77. Figure 5 shows the average vertical displacement of these isopycnal surfaces, expressed in metres per day, during the first 240 days of mooring records. We can recognize a seasonal signal, with peaks corresponding to strong movements of the isopycnal surfaces during pulses. During the first 120 Julian days the vertical displacement never exceeds 20 m day^{-1}. Afterwards, during autumn and winter seasons, it rises to a maximum of 100 m day^{-1}: this is a value comparable with the intrinsic settling velocity of the particles collected by the trap and can thus affect the measured flux.

3.2 Particle Fluxes

Particle fluxes (average of the 15-day collecting period), expressed as total mass, organic matter, carbonate and biogenic silica, are shown in Fig. 6. The material collected by each cup was not always sufficient to produce subsamples for the determination of the different biogenic components and thus a complete data set is not available for the period May – early December.

Total mass flux shows a clear seasonal trend, with relatively high values in February (52.7 - 61 mg m^{-2} day^{-1}) that nearly halve in March (27.5-35.9 mg m^{-2} day^{-1}) and drastically diminish in April (9.4 - 4.5 mg m^{-2} day^{-1}). Downward flux appears to be extremely depleted in winter and through mid-spring and is restored again in early December, with values comparable with the previous February, which are maintained and increased through late January 1996, when total mass flux is the highest of the whole deployment period (83.3 mg m^{-2} day^{-1}). The annual particle flux is estimated to be equal to 6.18 g m^{-2}, with an average daily value of 16.9 mg m^{-2}. The seasonal variation of total mass flux well matches with the trend of its biogenic components. Most of the biogenic silica flux occurs during the same period of increased total mass flux, with a maximum percentage value registered in the period mid-February/mid-March. Opal fluxes range from 16 to 39.9 mg m^{-2} day^{-1} in February and March, drop to very low values (down to 0.01 mg m^{-2} day^{-1}) in the period April–November and reach the highest value (59.8 mg m^{-2} day^{-1}) in January 1996. A similar trend can be observed for organic matter and carbonate, which show fluxes ranging respectively from 4 to 13.3 mg m^{-2} d-1 and from 6.4 to 9.9 mg m^{-2} day^{-1} in February and March, that decrease in

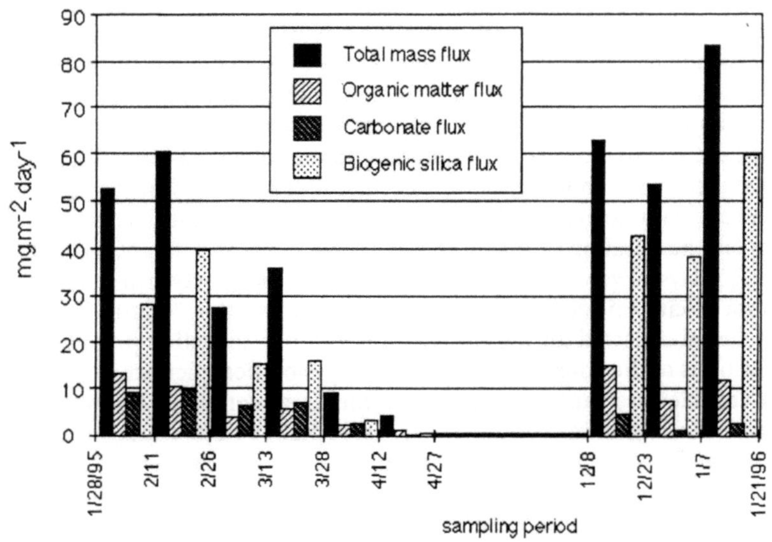

Fig. 6. Particle fluxes as recorded by the 423-m-sediment trap (15-day means)

April and May (2.4 - 0.8 mg m^{-2} day^{-1} for organic matter and 2.6 to 0.3 mg m^{-2} day^{-1} for carbonate). From December, organic matter flux increases again and reaches its maximum annual value (15.2 mg m^{-2} day^{-1}); carbonate flux, on the contrary, exhibits much lower values than in the previous summer.

To summarize the above-mentioned data we can divide the deployment period into three phases: period 1, from the starting of the experiment (January 28, 1995) through April, which shows significant particle flux, with biogenic silica, organic matter and carbonate fractions averaged for this period respectively accounting for 45.5, 21.4 and 18.6% of the total flux; period 2, from May to December, characterized by very low fluxes and during which the determination of the biogenic components was not always possible; period 3, from December to the end of the experiment (January 21, 1996), which includes the highest particle fluxes, with biogenic silica, organic matter and carbonate fractions averaged for this period respectively accounting for 70.5, 17.6 and 4.6% of the total flux.

Through the whole considered period the total biogenic fraction generally represents 55-99% of the total flux. In January 1996 the lithogenic component is higher than in previous February and March. Other authors have reported increased lithogenic flux of terrestrial origin during the same period [24].

The most evident result when looking at the annual trend of the total flux is the huge difference (up to three orders of magnitude) existing between the fluxes of the "productive" (1 and 3) and "non-productive" (2) periods. The annual flux is practically confined to the short period of the austral summer (December-March), with these months accounting for 93% of the total annual flux. The marked decrease of fluxes has already been observed at Bransfield Strait during the period February-October [25], in the Weddell Sea during May-December [26] and under the fast ice in Lützow-Holm Bay during May-October [27]. Enhanced summer biogenic sediment flux has, moreover, been recorded in various sites of the southern Ross Sea [2] and can be considered as a typical feature of Antarctic settings.

Unfortunately flux measurements have been made at only a limited number of locations throughout the Ross Sea and few data are reported on year-round particle fluxes in this area [2, 4]. Our data are comparable with those observed in a site located at 74°S, 175°E (annual averaged flux at 200 m-depth = 11.2 mg m^{-2} day^{-1}; Langone, pers. comm.), and with data from various sites in McMurdo Sound [24] and elsewhere in the Antarctic Ocean [25, 28-30], but appear lower than fluxes recorded by sediment traps in other Antarctic areas [2, 4, 24 in the Ross Sea; 27, 31-33].

Particulate carbon daily flux estimated on an annual basis is in the range of that recorded in the northwestern Ross Sea at 433 m [4] and, on shorter time scales, of those reported for some sites in western McMurdo Sound [24]. Biogenic silica percentage contribution to total mass flux calculated for our samples is lower than in other sites of the Ross Sea, even if sediment trap samples relatively depleted (13-47%) in biogenic silica (and with higher than usual proportions of organic carbon and calcium carbonate) have already been observed in the south-central Ross Sea (76°30'.336 S, 174°59'.128W) [34].

The percentage of total flux represented by opal in our samples agrees with those calculated for sediment trap samples from the Bransfield Strait [25] even if fluxes expressed in milligrams per square metre per day are lower.

3.3 Transmissometer Record

Owing to the malfunctioning of the stepping motor, no samples are available from the 230-m sediment trap, so that transmissometer data (Fig. 7) are the only record we have of particulate material concentrations in the layer above the 423-m trap. From the moment of deployment (January 27, 1995) to the end of August, background values are around 20% of light attenuation, with a significant number of records between 45 and 18% and the highest peaks of particle concentration (up to 84% of light attenuation) occurring in the last week of March. From September through early December the minimum turbidity levels are recorded, with light transmission reaching and maintaining values ≥ 95% through the whole period, while turbidity abruptly increases on June 12, 1995 (64% of light transmission). After this date and until the end of the recording period (January 21, 1996) light transmission measurements show trend and percentage values similar to those observed in the period previous to the maximum light transmission interval, with peaks of up to 50%.

For easier comparison with the sediment trap collection data we can divide the transmissometer record into three periods: period A, from the beginning of the experiment through August, during which significant particle concentrations are recorded at 230 m, with amounts constantly decreasing from the start to the end of the considered time interval; period B, from September to early

December, when the transmissometer measurements practically show the absence of particles; period C, from early December to the end of deployment, during which particle concentration is restored again, with trends and amounts similar to period A.

4 Model Development and Outcomes

It is interesting to note that the transmissometer records the particle presence at 230 m over the whole period from the end of January through the end of August, while the 423-m sediment trap shows dramatic decrease of fluxes starting from May and lasting to early December.

It is known that suspended and settling particles cannot be included in the same category of particles: they have different ecological properties and significance and are detected by means of different sampling techniques. At the same time they are strictly linked in a dynamic equilibrium in which suspended particles can continuously aggregate to form sinking particles and these last undergo, while sinking, disaggregation processes to produce finer particles that remain in suspension. Many authors [35-38] have assumed or demonstrated that sedimentation results from the aggregation of fine particles, which form the standing stock

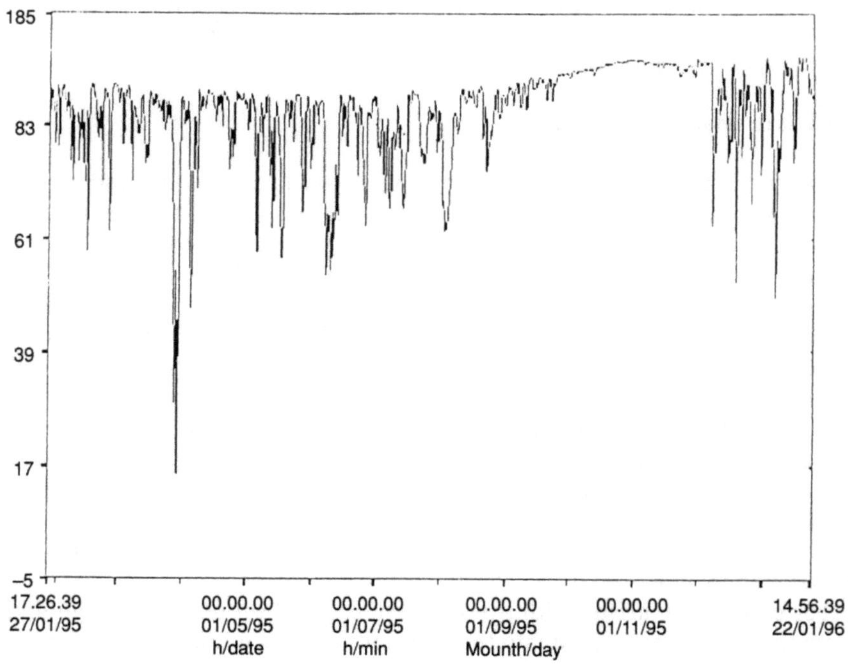

Fig. 7. Transmissometer record at 230-m depth

of particulate matter in the ocean [39-41], by larger particles of biogenic origin. Settling particles resulting from blooms of large dense cells or from aggregation of small particles can constitute short-term pulses rapidly reaching deep layers and sediments and scavenging fine mineral particles of the water column [42-45].

In our study, transmissometer and sediment trap collection records generally show a good correlation, with the exception of the period May-August (i.e. from 120 to 240 Julian days). To try to understand the discrepancy between transmissometer and sediment trap records in this period a simple Montecarlo model of particle downfall was implemented.

Dunbar et al. [46] have found that deep currents are weaker in the investigated area than in other zones of the Ross Sea continental shelf. A one-dimensional model for the settling of biogenic material is thus likely to be successfully applicable there. According to the assumptions of Jaeger et al.[47], uninterrupted downward sinking of biogenic particles (produced by primary and secondary production processes mostly in the 0- to 100 m-layer) supplied from the surface is assumed to originate at the bottom of the surface layer (100 m) and settle to the seabed. The total downfall velocity is computed as the sum of intrinsic downfall velocity of particles plus the actual vertical displacement of the water layer in which they are contained. A set of pseudoparticles is numerically produced in our simulation at 100-m depth, with a Gaussian spectrum of their intrinsic velocity. The value of intrinsic downfall velocity is extracted randomly on the basis of a gaussian shape with an average of 100 m d^{-1} and a variance of 20 m d^{-1} in order to simulate a wider spectrum of real particle size and shape, consequently implying different downfall velocities. Particle intrinsic settling velocities are highly variable. The chosen value of 90 ± 20 m day^{-1} is in the range of the settling velocities of the particles mostly represented in our traps. For fecal pellets we refer to measurements performed by Jaeger and collegues [47] in the study area. As far as diatoms are concerned, it is known that Antarctic species tend to be larger and with more robust frustules than those from silicate-poor waters and can reach, when in assemblages (very common in our trap material), sinking speeds of up to more than 100 m day^{-1} [42, 48].

In our model, vertical displacement is updated every day. Due to the distribution spectrum of particle intrinsic velocities and to total vertical displacement variability, total vertical settling velocity considerably changes after Julian day 120, when the average vertical displacement starts to increase, reaching values comparable with particle intrinsic downfall velocity.

A pseudoparticle daily counter is set numerically in our model at 230-m depth and normalized to simulate the transmissometer measurements. A numerical integrator on a time interval of 15 days is situated at 423 m, to simulate the trap measurements. We ran models with different daily production rates of pseudoparticles at 100 m, considering the hypothetical production of any kind of particles, not just those originated by primary production. Two model outputs seem particularly interesting to observe: the first assuming a constant production, the second simulating a seasonal cyclic production with an amplitude going from 100 to 10%. The cycle is almost a square function going from 100% during December through April and decreasing to 10% during June through October; during May-November a linear interpolation occurs. It is evident that the production rate of settling particle is not constant throughout the year. We chose to explore both the above assumptions (i.e. constant and seasonal cyclic particle production) due to the fact that particles involved in the vertical flux are not only the result of primary and secondary production processes, but also can be produced from transformation of suspended fine particles through mechanisms of aggregation, adsorption, microbial colonization, ingestion and egestion and loss of buoyancy, and in this case the total production rate and the seasonal trend are impossible to estimate. In both cases the response of the numerical transmissometer was similar to the real one. In Fig. 8 the evolution of the numerical transmissometer record for the second run is shown. Even if a strong decrease occurs in particle production, transmissometer signal shows a behaviour that is similar to the real one, shown in Fig. 7. In the run, assuming constant particle production, the evolution of the 423-m-trap simulated behaviour (Fig. 9a) exhibits a very high variable signal after Julian day 120, which means thath the probability of a particle settling in the trap is highly variable. It is evident that the huge probability of quick displacement of the isopycnal surfaces do not foster particle deposition. This effect might, moreover, be enhanced by the increased horizontal velocity that is not considered in this simple simulation. Figure 9b shows the evolution of the numerical sediment trap flux relative to the run with cyclic production of particles. In this case a very good agreement with the real data is achieved. Almost no sedimented particles are shown in the trap, even if the transmissometer signal shows the presence of some particles.

5 Summary and Conclusions

Owing to the limited equipment available and to problems in the functioning of some of the mooring array instruments this experiment must be considered as preliminary. Its results, however, allow us to identify some features that characterize the investigated area and to make some considerations that can constitute a good starting point for future studies:

1. The existence of a characteristic hydrodynamic behaviour along the RIS, modulated by an inflow-outflow intermittence well represented by the 5.02-day periodicity shown by the V-velocity trend (FFT analysis); this 5.02-day periodicity is, moreover, found in all the measured physical variables at every depth.
2. The superimposition of short-term (2-4 days) pulse events resulting in the outflowing of supercold water from below the RIS towards the Ross Sea and sometimes associated with recirculation processes.
3. The low particle downward fluxes and the major role of physical forcing in affecting sedimentation processes at the investigated site.

Fluxes of sedimenting particles are undoubtedly influenced by ecological factors, such as the seasonal cycles of primary production and related secondary production, strictly dependent on environmental conditions such as seasonal patterns of solar irradiance, ice extent, type and thickness, water column stability and meteorological conditions [31, 33, 49-56]. In the first period, from the starting of the experiment through April, moderated current velocity (≤ 20 cm s^{-1}) and limited vertical displacement of the isopycnal surfaces (≤ 20 m day^{-1}) do not seem

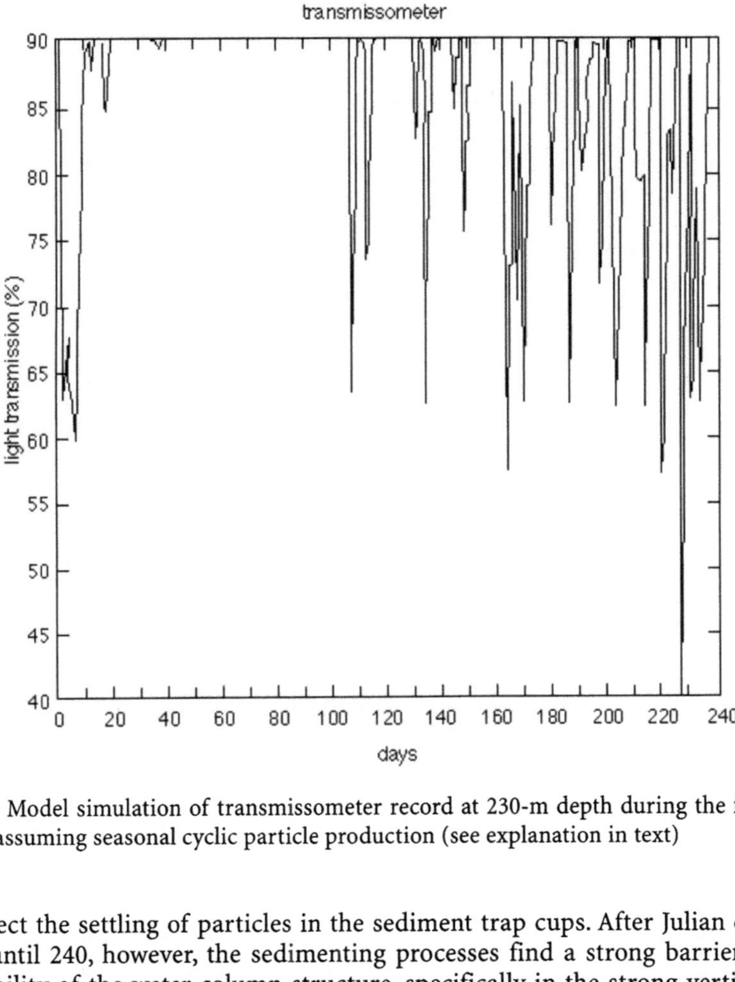

Fig. 8. Model simulation of transmissometer record at 230-m depth during the first 240 days, assuming seasonal cyclic particle production (see explanation in text)

to affect the settling of particles in the sediment trap cups. After Julian day 120 and until 240, however, the sedimenting processes find a strong barrier in the variability of the water column structure, specifically in the strong vertical displacement of the isopycnal surfaces, as shown by the simulated behaviour of the sediment trap in our model (Fig. 9b), even when assuming constant particle production (Fig. 9a). Concluding from this numerical experiments, we can infer that a major controlling factor for sedimentation is represented not only by the actual particle production but also by the water stability and density variability, which play a crucial role even in the presence of high production rates.

Future work will focus on assessing the annual and interannual variability of the downward flux of sinking particles in the same and other sites of the Ross Sea (Fig. 1), also in relation to hydrologic and hydrodynamic features of the water masses. The determination of the origin, transport mechanisms, composition and seasonal variations of the chemical characteristics of sinking particles all year round is included in our future aims. Moreover, we hope that the improvement of the vertical resolution of all measured physical variables and the experimental measurement of the intrinsic settling velocities of the particles collected in our a

b

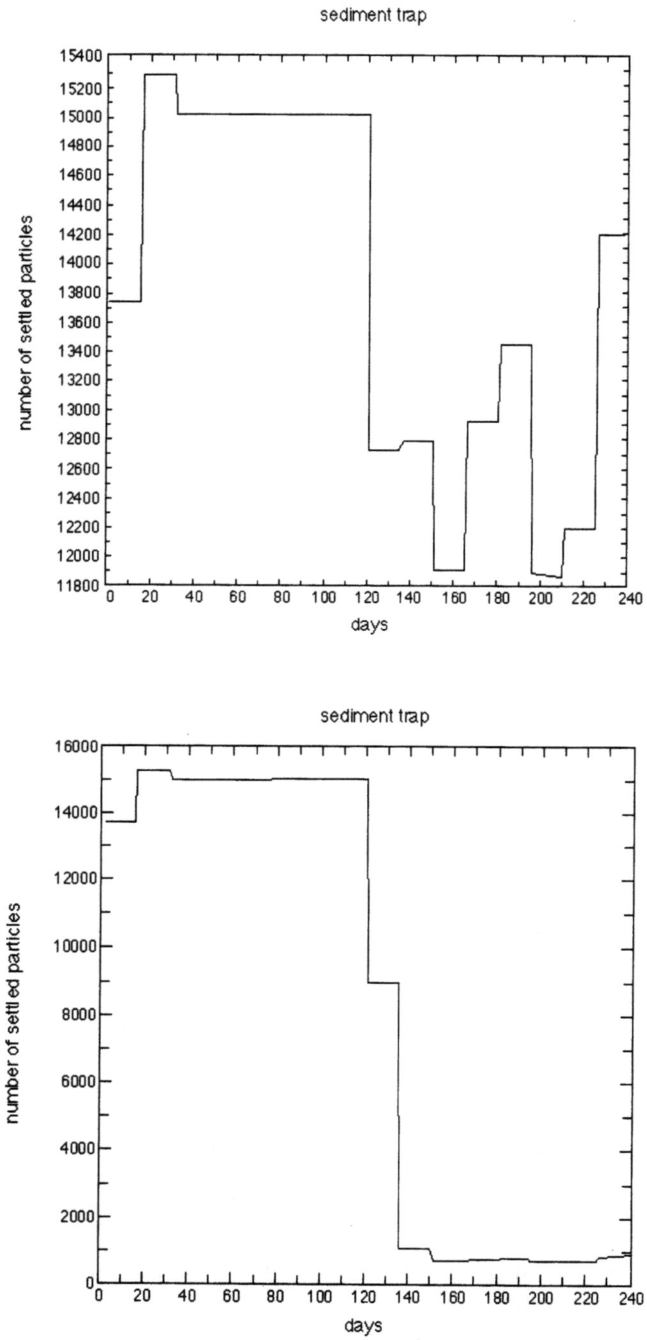

Fig. 9. Model simulation of 423-m sediment trap cup content during the first 240 days of the second numerical run, assuming constant particle production (**a**) and seasonal cyclic production (**b**) (see explanation in text)

trap will enable us to improve our model in order to better simulate the actual total downfall velocity, even in the light of a correct evaluation of our trap efficiency.

Acknowledgements. This research was supported by the Italian National Programme for Antarctic Research (PNRA), funded by of the Italian Government. We are grateful to Prof. Giancarlo Spezie (Coordinator of the CLIMA Project) for valuable discussions and comments. Thanks are also due to Dr. Antonio Artegiani and Ing. Roberto Meloni for the availability of the acquired and pre-elaborated mooring data.

References

1. Fowler SW, Knauer G (1986) Role of large particles in the transport of elements and organic compounds through the oceanic water column. Prog Oceanogr 16:147-194
2. DeMaster DJ, Dunbar RB, Gordon LI, Leventer AR, Morrison JM, Nelson DM, Nittrouer CA, Smith WO Jr (1992) Cycling and accumulation of biogenic silica and organic matter in high latitude environments: the Ross Sea. Oceanography 5(3):146-153
3. Comiso JC, McClain CR, Sullivan CW, Ryan JP, Leonard CL (1993) Coastal zone color scanner pigment concentrations in the Southern Ocean and relationship with geophysical surface features. J Geophy Res 98:2419-2451
4. Nelson DM, DeMaster DJ, Dunbar RB, Smith WO Jr (1996) Cycling of organic carbon and biogenic silica in the Southern Ocean: estimates of water-column and sedimentary fluxes on the Ross Sea continental shelf. J Geophys Res 101(C8):18519-18532
5. Smith WO Jr, Nelson DM, DiTullio GR, Leventer A (1996) Temporal and spatial patterns in the Ross Sea: phytoplankton biomass, elemental composition, productivity and growth rates. J Geophys Res 101(C8):18455-18465
6. Goosse H, Hecq J-H (1994) Modelling the ice–ocean–plankton interactions in the Southern Ocean. J Mar Syst 5:471-484
7. Smith WO Jr, Nelson DM (1985) Phytoplankton bloom produced by a receding ice edge in the Ross Sea: spatial coherence with the density field. Science 227:163-166
8. Smith WO Jr, Nelson DM (1986) Importance of ice edge phytoplankton production in the Southern Ocean. BioScience 36(4):251-257
9. Nelson DM, Smith WO Jr (1986) Phytoplankton bloom dynamics of the western Ross Sea ice-edge - II. Mesoscale cycling of nitrogen and silicon. Deep Sea Res 33(10):1389-1412
10. Wilson DL, Smith WO Jr, Nelson DM (1986) Phytoplankton bloom dynamics of the western Ross Sea ice edge. I. Primary productivity and species-specific production. Deep Sea Res 33 (10):1375–1387
11. Sullivan CW, McClain CR, Comiso JC, Smith WO (1988) Phytoplankton standing crops within an Antarctic ice edge assessed by satellite remote sensing. J Geophys Res 93:12487-12498
12. Tucci S, Budillon G, Capello M, Ferrari M (1996) The suspended matter in water masses near the Ross Ice Shelf. Proceedings International Workshop "Ross Sea Ecology", Taormina, 14-16 May 1996, pp 173-177
13. Carmack EC (1990) Large-scale physical oceanography of Polar Oceans. In: WO Smith Jr (ed) Polar Oceanography. Part A: Physical Science. Academic Press Inc, San Diego, California, pp 171-222
14. Pillsbury RD, Jacobs SS (1985) Preliminary observations from long-term current meter moorings near the Ross Ice Shelf, Antarctica. In: Jacobs SS (ed) Oceanology of the Antarctic Continental Shelf. Antarc Res Ser 43:87-107. AGU, Washington DC
15. Roemmich D (1983) Optimal estimation of hydrographic station data and derived field. J Phys Oceanogr 13:1543-1549
16. Bergamasco A, Meloni R, Carniel S, De Biasio F, Degl'Innocenti F, Arcari G, Curiotto S, Morgigni M (1998) Project CLIMA - mooring F 1995 data analysis. CNR ISDGM Technical Report, in press
17. Accornero A, Marino D, Razouls S, Monaco A (1998) Seasonality and composition of particulate fluxes in the Southern Ross Sea (Antarctica). Paper presented at 30th International Liege

Colloquium on Ocean Hydrodynamics "Hydrodynamical and Ecosystem Processes in Ice Covered Seas of the Southern and Northern Hemispheres", Liege, May 4-8, 1998
18. Heussner S, Ratti C, Carbonne J (1990) The PPS 3 time-series sediment trap and the trap sample processing techniques used during the ECOMARGE experiment. Cont Shelf Res 10(9-11):943-958
19. Gordon DC (1970) Some studies of the distribution and composition of particulate organic carbon in the North Atlantic Ocean. Deep Sea Res 17:233-244
20. Monaco A, Courp T, Heussner S, Carbonne J, Fowler SW, Deniaux B (1990) Seasonality and composition of particulate fluxes during ECOMARGE – I, western Gulf of Lions. Cont Shelf Res 10(9-11):959-987
21. DeMaster DJ (1991) Measuring biogenic silica in marine sediments and suspended matter. In: Marine particles: analysis and characterization. Geophys Mono 63, AGU, pp 363-367
22. Gardner WD (1980) Field assessment of sediment traps. J Mar Res 38:41-52
23. Baker ET, Milburn HB, Tennant DA (1988) Field assessment of sediment trap efficiency under varying flow conditions. J Mar Res 46:573-592
24. Dunbar RB, Leventer AR, Stockton WL (1989) Biogenic sedimentation in McMurdo Sound, Antarctica. Mar Geol 85(2/4):155-179
25. Wefer G, Fischer G, Futterer D, Gersonde R (1988) Seasonal particle flux in the Bransfield Strait, Antarctica. Deep Sea Res 35(6):891-898
26. Fischer G, Futterer D, Gersonde R, Honjo S, Ostermann D, Wefer G (1988) Seasonal variability of particle flux in the Weddel; Sea and its relation to ice cover. Nature 335:426-428
27. Matsuda O, Ishikawa S, Kawakuchi K (1987) Seasonal variation of downward flux of particulate organic matter under the Antarctic fast ice. Proc NIPR Symp on Polar Biol 1:23-34
28. Fujita N, Nishizawa S (1982) Vertical flux of particulate matter in the Antarctic Ocean in summer 1981. Trans Tokyo Univ Fish 5:43-52
29. Dunbar RB (1984) Sediment trap experiments on the Antarctic continental margin. Antarct J United States 19:70-71
30. Biggs DC, Berkowitz SP, Altabet MA, Bidigare RR, DeMaster DJ, Macko SA, Ondrusek ME, Noh II (1989) Cooperative study of upper ocean particle flux. Proc Ocean Drilling Program Part A: Initial Rep 119:109-119
31. von Bodungen B, Smetacek VS, Tilzer MM, Zeitzschel B (1986) Primary production and sedimentation during spring in the Antarctic Peninsula region. Deep Sea Res 33(2):177-194
32. Noriki S, Tsunogai S (1986) Particulate fluxes and major components of settling particles from sediment trap experiments in the Pacific Ocean. Deep Sea Res 33:903-912
33. Karl DM, Tilbrook DB, Tien G (1991) Seasonal coupling of organic matter production and particle flux in the western Bransfield Strait, Antarctica. Deep Sea Res 38(8/9):1097-1126
34. Dunbar RB, Mucciarone DA, Leventer A (1991) Sediment trap experiments in the central and western Ross Sea, January and February 1990. Antarct J United States 26(5):115-117
35. Honjo S, Manganini SJ, Poppe LJ (1982) Sedimentation of lithogenic particles in the deep ocean. Mar Geol 50:199-220
36. Alldredge AL, Youngbluth MJ (1985) The significance of macroscopic aggregates (marine snow) as sites for heterotrophic bacterial production in the mesopelagic zone of the subtropical Atlantic. Deep Sea Res 32:1445-1456
37. Asper VL (1987) Measuring fluxes and sinking speed of marine snow aggregates. Deep Sea Res 34:1-17
38. Alldredge AL, Silver MW (1988) Characteristics, dynamics and significance of marine snow. Prog Oceanogr 20:41-82
39. McCave IN (1975) Vertical flux of particles in the ocean. Deep Sea Res 22:491-502
40. Bishop JKB, Edmond JM, Ketten DR, Bacon MP, Silker WB (1977) The chemistry, biology and vertical flux of particulate matter from the upper 400 m of the equatorial Atlantic Ocean. Deep Sea Res 25:511-548
41. Chesselet R (1980) Modes of settling and organic input to the sediment seawater interface. A review. In: Biogéochimie de la matière organique à l'interface eau-sédiment marin. Colloque International CNRS 293:27-33
42. Billett DSM, Lampitt RS, Rice AL, Mantoura RCF (1983) Seasonal sedimentation of phytoplankton to the deep-sea benthos. Nature 302:520-522

43. Lampitt RS (1985) Evidence for the seasonal deposition of detritus to the deep-sea floor and its subsequent resuspension. Deep Sea Res 32:885-897
44. Biscaye PE, Anderson RF, Deck BL (1988) Fluxes of particles and constituents to the eastern United States continental slope and rise: SEEP I. Cont Shelf Res 8:855-904
45. Alldredge AL, Gotschalk CC (1989) Direct observations of the mass flocculation of diatom blooms: characteristics, settling velocities and formation of diatom aggregates. Deep Sea Res 36:159-171
46. Dunbar RB, Anderson JB, Domack EW (1985) Oceanographic influences on sedimentation along the Antarctic continental shelf. In: Jacobs SS (ed) Oceanology of the Antarctic Continental Shelf. Antarct Res Ser 43:291-312, AGU, Washington DC
47. Jaeger JM, Nittrouer CA, DeMaster DJ, Kelchner C, Dunbar RB (1996) Lateral transport of settling particles in the Ross Sea and implications for the fate of biogenic material. J Geophys Res 101 (C8):18479-18488
48. Takahashi K (1986) Seasonal fluxes of pelagic diatoms in the subarctic Pacific, 1982-1983. Deep Sea Res 33:1225-1251
49. El-Sayed SZ (1984) Productivity of the Antarctic waters - reappraisal. In: Holm-Hansen O, Bolis L, Gilles R (eds) Marine phytoplankton and productivity. Springer-Verlag, New York, pp 19-34
50. Sakshaug E, Holm-Hansen O (1984) Factors governing pelagic production in polar oceans. In: Holm-Hansen O, Bolis L, Gilles R (eds) Marine Phytoplankton and Productivity. Springer-Verlag, New York, pp 1-18
51. Hempel G (1985) Antarctic marine food webs. In: Siegfried WG, Condy PR, Laws RM (eds) Antarctic nutrient cycles and food webs. Springer-Verlag, New York, pp 266-270
52. Palmisano AC, Kottmeier ST, Moe RL, Sullivan CW (1985) Sea ice microbial communities. IV The effect of light perturbation on microalgae at the ice seawater interface in McMurdo Sound, Antarctica. Mar Ecol Prog Ser 21:37-45
53. Palmisano AC, SooHoo JB, Moe RL, Sullivan CW (1987) Sea ice microbial communities. VII Changes in under-ice spectral irradiance during the development of Antarctic sea ice microalgal communities. Mar Ecol Prog Ser 35:165-173
54. Holm-Hansen O, Mitchell BG, Hewes CD, Karl DM (1989) Phytoplankton blooms in the vicinity of Palmer Station, Antarctica. Polar Biol 10:49-57
55. Mitchell BG, Brady EA, Holm-Hansen O, McClain C, Bishop J (1991) Light limitation of phytoplankton biomass and macronutrient utilization in the Southern Ocean. Limnol Oceanogr 36:1662-1667
56. Nelson DM, Smith WO Jr (1991) Sverdrup revisited: critical depths, maximum chlorophyll levels, and the control of Southern Ocean productivity by the irradiance-mixing regime. Limnol Oceanog 36(8):1650-1661

Actual Sedimentation on the Antarctic Continental Shelf (Southern Part of the Ross Sea)

S. Tucci, M. Ferrari and M. Capello

Dipartimento di Scienze della Terra, Università degli Studi di Genova, Corso Europa 26, 16132 Genova, Italy

Abstract

During the austral summer 1994-95 the C.L.I.M.A. Project carried out an oceanographic cruise aimed at improving the caracterization of water masses inside the Ross Sea. A series of 154 CTD casts was performed and water samples were collected to study particulate matter, concentration data and dimensional analyses of particulate matter were correlated with physical characteristics. In the area of the Ross Ice Shelf high concentrations of particulate matter seem to coincide with the caracteristics of "Deep Ice Shelf Waters" and "Warm Core Waters". The relationship between water masses and characteristics of particulate matter have enabled us to use this one as a natural tracer to identify flow pathways; results have also allowed to define a general hypothesis on actual sedimentation during the austral summer in the proximity of the RIS and at the edge of the continental shelf.

1 Introduction

Antarctic areas present a series of ideal conditions for following the progress of suspended matter in the water. The absence of river sediment and the presence of a completely natural organic component make it possible to draw a series of conclusions that would be difficult elsewhere due to variables that are difficult to control and evaluate. The ice, covering almost all the coastline and much of the waters of the internal shelf for between 8 and 12 months a year, limits sediment to that carried by the glaciers. This situation also affects the transport of sediments, which are often redistributed exclusively by oceanographic dynamics [1].

There is a very close relationship between the gravitative dynamics of suspended particles and the physical and dynamic characteristics of the mass of water [2]. The sedimentation of suspended particles follows preferential lines of conveyance where there are well-defined masses of water. It is also heavily affected by thermocline and thermohaline fronts, which act as physical barriers conditioning the gravitative movement of the particles. The relationship between the diffusion of suspended particles and the physical and dynamic characteristics of the mass of water make it possible to use the suspended matter as a natural tracer to identify flow lines and characteristics of different water masses. This provides a very stable record of the conditions of the column of water.

During the 1994-1995 austral summer, the National Antarctic Research

Program organized the X Italian Expedition in the Ross Sea. Within this framework, the C.L.I.M.A. Project (Climatic Long-term Interaction for the Massbalance in Antarctica) carried out its first oceanographic cruise. The aim was to improve the characterization of the most important water masses and to study the general circulation inside the Ross Sea. From 7 January to 20 February 1995 a series of 154 CTD casts were performed, and 3 moorings were deployed in order to monitor the annual variability of water fluxes and suspended sediments. One year later, during the XI Expedition from 12 January to 5 February 1996, the moorings were recovered and repositioned; 28 sampling stations were also recorded during these operations.

This paper discusses some parameters regarding suspended matter in the water off the Ross Ice Shelf (RIS). These results have enabled us to define the origin and dynamics of the suspended matter and outline a general hypothesis on actual sedimentation during the austral summer. Data on the composition of the suspended matter will be assessed together with seasonal flow values in Chapter by Accornero et al., this volume [3].

2 Instrumentation and Analysis

The hydrological stations used a multi-parameter SBE 9/11 plus probe from Sea-Bird Electronics equipped with a Carousel SBE 32 unit with 24 12l-bottles. Samples of suspended matter were taken at 122 of the stations at depths calculated according to the hydrological profile (always one sample on the surface, one under the thermocline, one in the intermediate waters and one close to the bottom); on board, the samples were subdivided and subjected to different filtrations in order to prepare them for the subsequent laboratory investigations to be performed on the suspended load. Determinations were performed according to the methods reported in Strickland and Parsons [4]. In this phase the total concentration data related to dimensional analyses of suspended matter were used [5]. The dimensional analyses were carried out on board using a Coulter® Counter Multisizer with a double sampling cell; a 30μm capillary was used to define the fine fraction and a 140μm capillary to define the larger fraction; this dual analysis provided reliable results in the range 0.6 to 100μm [6].

3 Results and Discussion

The first step was to locate the temperature minimum along the RIS, as these waters are thought, to play an important role in the general dynamics of the Ross Sea. For this purpose a survey was carried out along the RIS front from longitude 170° E to 167° W (Stations 3-21; Fig. 1). A water mass of interest was identified at 178° W, centered on stations 9 to 15; it has temperature values lower than -2.00 °C at depths in the range 300 to 500 m. In this sector, sampling was performed northwards, with stations 85 and 86 up to the limit of the shelf. This confirms the results of Jacobs et al. [7] who identified the presence of "supercold waters" that defined Ross Ice Shelf Waters (RISW). Finding this situation almost in the same sector of the shelf and at

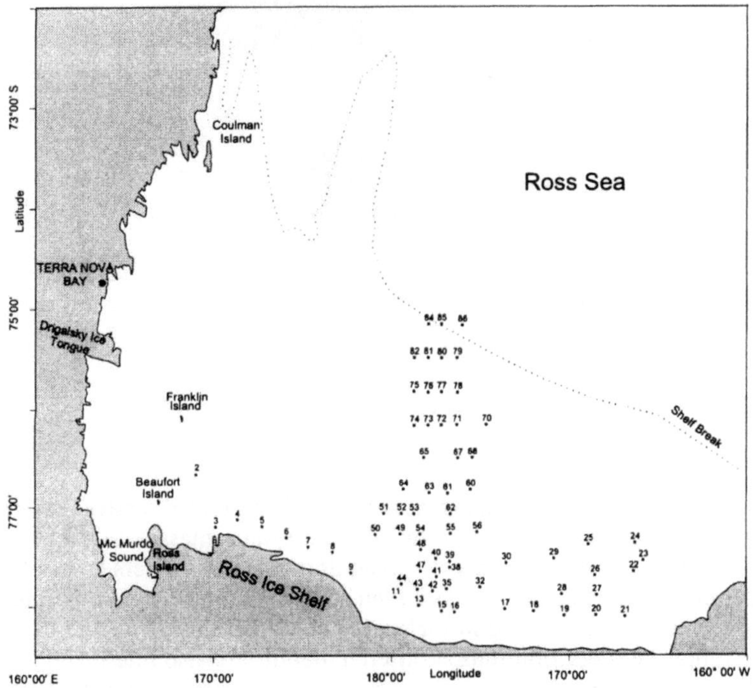

Fig. 1. Suspended particulate matter sampling stations

the same depth both in the 1976-1978 and 1994-1996 periods indicated the constancy of summer conditions and enabled the quantisation of the importance of these waters which extend northward about 150 km and are about 300 m thick [8].

The depth of the continental shelf in the sampled zone is between 500 and 600 m; only laterally, in correspondence with stations 6 and 18, can we find two of "morphological height" with depths lower than 500 m [9, 10]. These topographic constrictions, north-south orientated, would be conditioned by northwards flux of the Cold Waters.

This chapter will not go into detail of the physical characteristics of the water masses, as this subject is examined by other authors (e.g. [8] to which we will make reference). Figure 2 illustrates the concentration of total particulate matter (TPM) data at stations 3 to 21 along an east-west section; the values are in the range 0.17 to 5.81 mg/l, with an average of 1.67 mg/l, and an inorganic fraction greater than 60%.

In cross-section there are high concentrations between stations 10 and 17, i.e. almost coinciding with waters influenced by the Deep Ice Shelf Waters (DISW) and the Warm Core Waters (WMCO) [7, 8]. In almost all the surrounding stations values are less than 2 mg/l. These values are always higher than those found by Carter at al. [11] in the waters under the RIS. Later, a south-north section was drown,

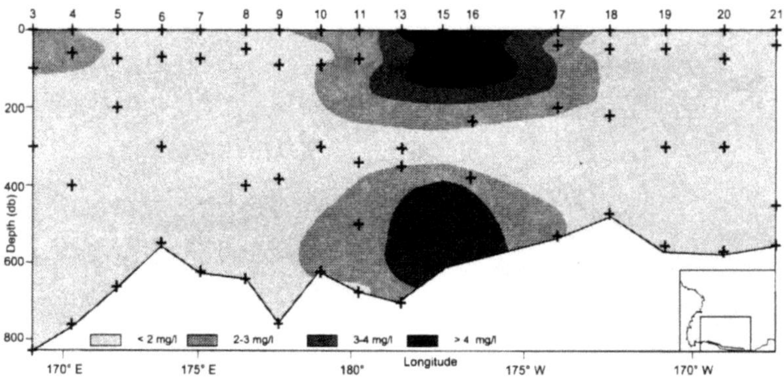

Fig. 2. Transect showing vertical distribution of TPM along the RIS

comprising stations 15 to 85, using values of total suspended matter (Fig. 3).

The higher values are located in two different water masses. In the area of the RIS, the suspended matter seems to spread northwards in waters from 100 to 400 m in depth and near the bottom, and even affects station 67. A second interesting concentration area is centered in surface waters at stations 67, 72 and 81.

Horizontal TPM distributions at different depths are also interesting. The surface distribution (Fig. 4a) indicates two areas of maximum concentration. The first is connected with the RIS and the second with the external part of the continental shelf, at the limit of the shelf-break. In waters close to the bottom (Fig. 4b) the maximum values are relative to the central part of the RIS, with a northwards diffusion along two different paths. There is also a sector of greater concentration at the edge of the shelf.

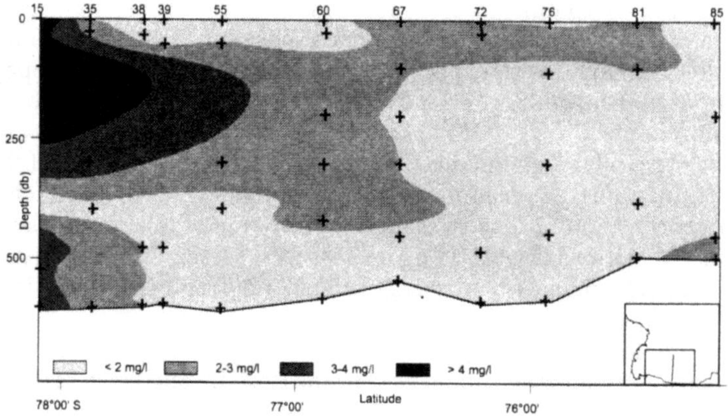

Fig. 3. Transect showing vertical distribution of TPM in a south-north section

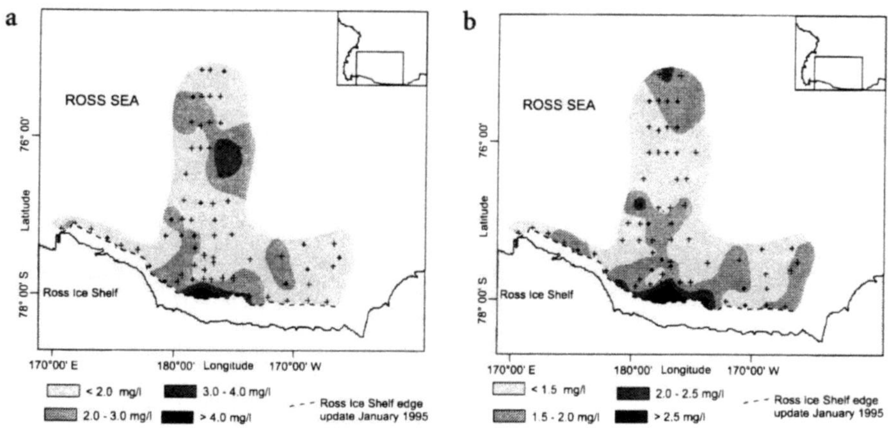

Fig. 4. Horizontal distribution of TPM in surface waters (**a**) and from 400 m to the bottom (**b**)

The "concentration" parameter is generally used to highlight sources of suspended matter and diffusion lines. However, to differentiate the different masses of water a detailed dimensional and compositional analysis is required. For this reason dimensional analyses were performed immediately on board. Modal spectra and statistical parameters were then used to identify a typical mark to differentiate between water masses.

A first interesting indication was provided by studying the distribution of statistical parameter values. If we plot the values of the median along the south-north section previously used for TPM, we can easily see the correlation between "cold waters" and high median values at depths between 400 m and the bottom (Fig. 5).

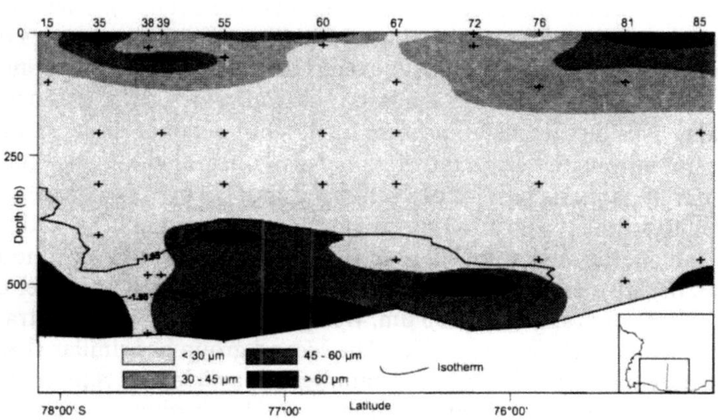

Fig. 5. Transect showing vertical distribution of median in a south-north section (values calculated by dimensional data)

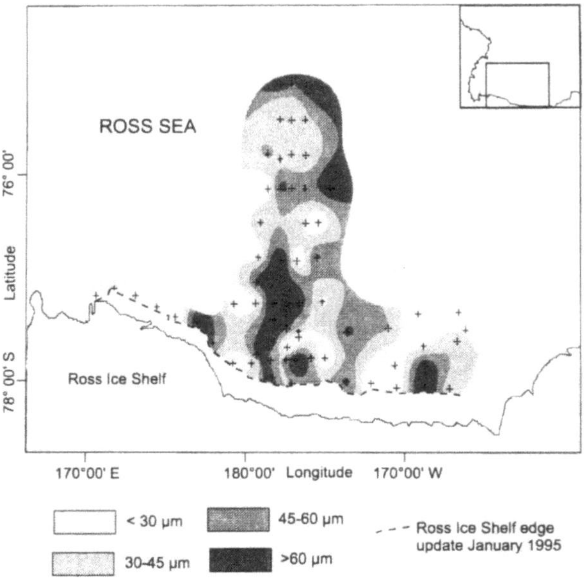

Fig. 6. Horizontal distributon of median from 400 m to the bottom (values calculated by dimensional data)

There is also a surface area, coinciding with the outer stations located in the Antarctic Surface Water (AASW), which goes some way to explaining the relatively high concentration values found in the presence of large particles and/or aggregates.

The horizontal distribution of the median (at the depths 400 m to the bottom previously used for TPM) is reflecting the trend indicated by the concentration in TPM. High median values characterize a northwards central flow originating in the RIS, and an area at the edge of the continental shelf (Fig. 6). These data seem to support the ability of "concentration of TPM" and "size" parameters to define the suspended matter in water masses even if the suspended matter contained in the ISW and in the AASW shows a number of similarities, such as concentration and median. This fact led us to increase the level of detail of dimensional analysis, using not only statistical parameters but also integrated dimensional spectra and number of particles present per volume units.

A summary of the types of 260 spectra found is reported in Fig. 7 and reference values in Table 1; the most frequent spectra refer to a group with plurimodal characteristics, negative asymmetry, high modal values and a high number of particles in the diameter range 0.7 to 90 µm. We refer to this group of spectra as SRP_L (spectrum rich in particles-large size). The second group has similar characteristics to the previous but with a very low number of particles. We refer to this group as SPP_L (spectrum poor in particles-large size). The third group of samples has spectra containing few particles of small size up to 30 µm. We refer to this group as SPP_S (spectrum poor in particles-small size). Using coordinates to locate all the types of spectra, we note that spectra SRP_L are typical of superficial and bottom

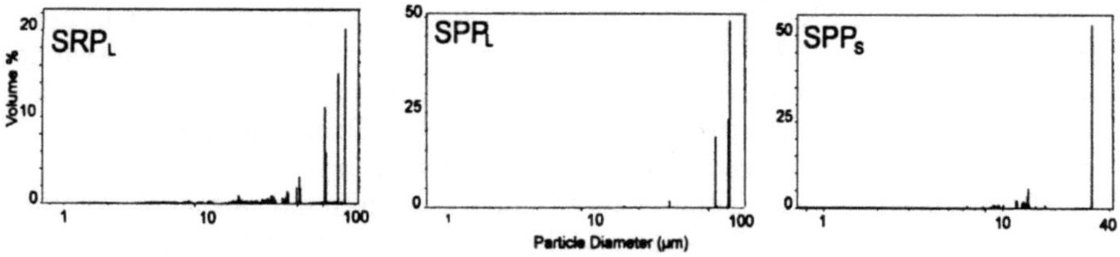

Fig. 7. Dimensional analysis of suspended particulate matter (SPM): spectral typology; *SRP*$_L$ spectrum rich in particles-large size; *SPP*$_L$ spectrum poor in particles-large size; *SPP*$_S$ spectrum poor in particles-small size

Table 1. Statistical data of dimensional analysis with characteristics of particulate matter and reference to the water masses

	Number of Particles	Range (μm)	Mode (μm)	Median (μm)	Samples	Number of Particles	Range (μm)	Mode (μm)	Median (μm)	Water Mass	Particulate Matter	Samples
SRP	25,700 σ=15,300	0.7÷90	60.9 σ=15.8	40.5 σ=20.9	124/260	37,300 σ=14,900	0.7÷85	60.3 σ=15.8	33.6 σ=17.6	SISW	Shallow Ice Shelf Particulate	84/124
						18,000 σ=6,500	30÷85	61.8 σ=17.0	46.0 σ=19.9	AASW	Antarctic Surface Particulate	42/124
						13,000 σ=5,400	50÷90	67.9 σ=24.3	60.9 σ=23.3	HSSW DISW	Hight Salinity Shelf Particulate and Deep Ice Shelf Particulate	18/124
SPP						4,600 σ=2,700	30÷90	58.6 σ=18.9	52.0 σ=21.0	?	Offshore Subsurface Particulate and Intermediate Particulate	78/136
SPP$_S$						3,600 σ=2,800	5÷30	20.5 σ=5.8	14.6 σ=4.9	WMCO	Warm Core Particulate and Warm Core Particulate modified	58/136

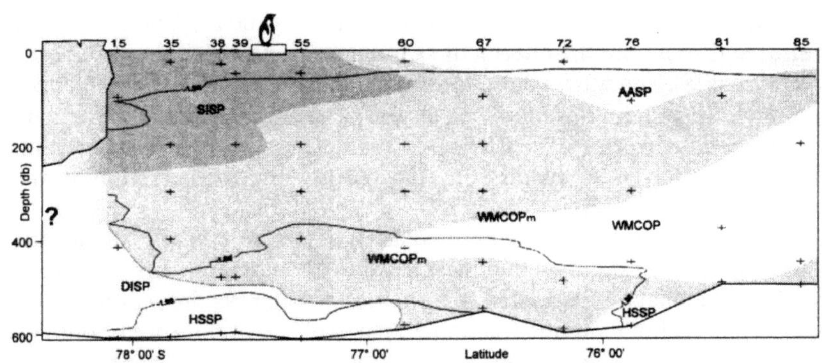

Fig. 8. Vertical section with hypothesis of characterization of SPM

waters; spectra SPP$_L$ and SPP$_S$, with a small number of particles, are generally located in intermediate waters.

To better distinguish them, we have analyzed with multivariate analysis the data for all spectra to assess a number of characteristics, including the number of particles, the range, the main mode and the median. This enabled us to divide the group richest in particles (SRP_L) into three additional subgroups; the two groups of spectra poor in particles (SPP_L and SPP_S) are already distinct. The samples of each of the clusters were then located by coordinates and depth to associate them with the water masses that characterize the area under survey. In Table 1 we report the principal values calculated for each parameter in the five defined clusters.

We have used the representative south-north section to indicate all our results in terms of concentration and sizes, and this has enabled us to hypothesize a characterization of the suspended matter (Fig. 8) that takes account of the water masses present (the thermocline and the isotherm −1.95 °C, limit of DISW, are indicated in red in the figure).

Along the edge of the RIS, in the waters between the surface and the thermocline and up to the stations located at 77 °S, we find samples (Shallow Ice Shelf Particulate, SISP) with higher values in terms of number of particles (25 000–50 000), with spectra with a median of around 30 μm and high values of the main mode (60 μm). In this cluster, 94% of the 64 samples were located in the Shallow Ice Shelf Water (SISW).

In the Antarctic Surface Waters (located in the external part of the shelf), the suspended matter (AASP) is still rich in particles (12 000–25 000), with a higher median, and main mode as above. The samples taken from this water mass account for 79% of the total in the cluster. The third group, 18 samples, is located on the bottom and in the nearby waters, corresponding to the HSSW and the DISW, with a percentage of 77%. This group is distinguished from the other two groups by a lower number of particles (8000-18 000); range, main mode and median values are also higher.

Spectral types SPP_L and SPP_S are rarely present in surface waters (10% of the total), while they characterize the intermediate water mass, which is therefore poor in suspended matter (500-8000 particles). SPP_L spectra with higher modal characteristics and particle numbers in the range 2000–8000 are high in concentration (65%) offshore in the subsuperficial layer (10–100 m). The other samples are diffused throughout the intermediate waters with no specific correlation with the water masses. The waters close to the RIS, that are not directly influenced by the ice shelf, in general have few suspended particles of small size, particularly between 200 and 500 m (65%). The same characteristics are encountered in the intermediate water (70%) further out, the WMCO and the surrounding waters modified by it ($WMCO_m$).

Only suspended material in intermediate waters is not clearly characterized; for this problem we examined the finest fraction and used only the data obtained with a 30μm capillary. We adopted this solution to improve the separation of the samples with an abundant fine fraction. In particular, therefore, we examined the SPP_S class. The results confirmed that this type of spectra is correlated to the WMCO, as 100% of the samples present in this mass of water have the lowest values found in terms of particles, mode and median.

4 Conclusions

From a sedimentological viewpoint, the effect of the barrier created by the strong variations in the thermohaline characteristics of the waters has already been demonstrated. This is very clear in the waters off the AASW with high concentrations of suspended matter above the thermocline and low values immediately beneath them [8].

We have defined a sector of the RIS where the suspended matter concentrations are higher and where the diffusion fluxes are spreading northwards; this spreading is in the surface and in the water close to the bottom, and the suspended matter has different concentration and dimensional characteristics. The concentrations of SPM are lower in the intermediate waters which are also characterized by particles with small dimensions (30 μm); the lowest values in concentration and in dimension are found in the intermediate waters coming from the open sea. There is a general influence of the RIS on the shelf waters and its higher limit seems to be situated around 77 °S. This latitude is the limit for the diffusion of SPM conveyed by AASW and WMCO too.

The SPM of the internal sector (south of 77° of latitude) spreads in surface waters northwards and in north-west direction, while, near the bottom, it seems essentially northwards. In the external sector, the diffusion seems to be southwards in the intermediate waters, while the situation seems to be different in the superficial water with a north-westwards diffusion. These results are reported in a sedimentation model [12] illustrated in Fig. 9 with the percentage values of SPM. We plotted all the results of suspended matter concentration found in two sectors (south and north of the 77°). Superficial and subsuperficial waters (0-100 m) are always the richest in particulate matter and their values are reported to 100% of suspended matter; the other calculated values for the intermediate waters, for the waters between 400 and 500 m and for the bottom water are taken as base 100.

Close to the RIS the difference in suspended matter between the surface and intermediate waters is in the order of 57%, and the suspended matter lost at a depth of 250 m is conveyed laterally and diffuses, probably towards the fourth quadrant. The situation in the low part of the column of water is different, with an increase of 14% at 450 m and of a further 14% at the bottom. This progressive increase in concentration can only be linked to lateral contributions of suspend-

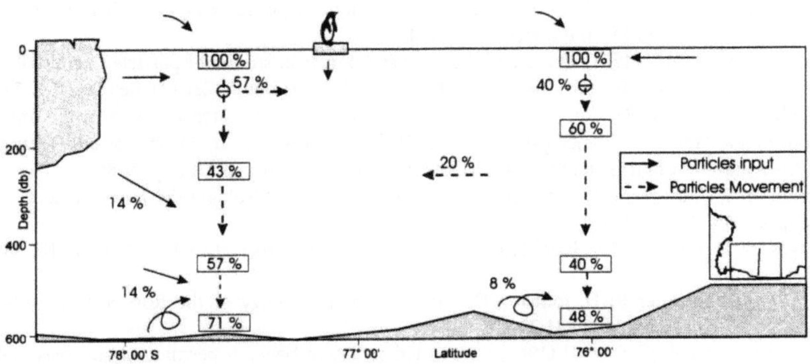

Fig. 9. Hypothesis of sedimentation in proximity of the RIS and at the edge of the continental shelf (% values of TPM)

ed matter inside the DISW and the SISW and only partially to resuspension phenomena. If we consider data on the waters closer to the shelf break throughout the column of water, there is a decrease in suspended matter, reaching 60% at 450 m. There is also an increase of 8% at the bottom which may be due to resuspension contribution [13, 14]. This contribution should be confirmed by higher size values, which are not present in the intermediate waters.

In the external sector of the shelf, our results allowed us to hypothesize that the 40% of suspended matter lost in the subsuperficial waters is due to conveyance westwards, connected with the movement of the Antarctic Circumpolar Water; 20% of the variation verified in the intermediate waters could be linked to the flow of Warm Core waters towards the RIS.

These considerations and others formulated during the discussion place emphasis on subhorizontal movements of the particle which occur in all the mass of water with variations interestingly up to the 60% of the total flux. These movements have been also indicated by a few authors in the Atlantic [15] and in Mediterranean waters [16].

Acknowledgements. This study was performed as part of "National Programme for Antarctic Research" and was financially supported by ENEA through a joint research programme.

References

1. Dunbar RB, Anderson JB, Domack EW, Jacobs SS (1984) Oceanographic influences on sedimentation along the Antarctic continental shelf. In: Jacobs SS (ed.) Oceanology of the Antarctic Continental Shelf. Am Geophys Union, Washington, DC, pp 291-312
2. Anderson JB, Brake FC, Myers NC (1984) Sedimentation on the Ross Sea continental shelf, Antarctica. Mar Geol 57:295-333
3. Accornero A, Bergamasco A, Monaco A, Tucci S (1998) Particle fluxes at the edge of the Ross Ice Shelf: the role of physical forcing (This Vol.)
4. Strickland JDH, Parsons TR (1968) A practical handbook of seawater analysis. Bull Fish Res Board of Canada 167:1-311
5. Pierce JW, Tucci S, Fierro G (1981) Assessing variation in suspensates, Ligurian Sea (northwestern Mediterranean). Geo-Marine Letters 1:149-154
6. Krank K, Milligan T (1978) The use of the Coulter Counter in studies of particle size-distribution. Aquatic Environment, Bedford Institute of Oceanography Dartmouth, Canada, Rep Ser BI-R-79-7, 48
7. Jacobs SS, Fairbanks RJ, Horibe Y (1985) Origin and evolution of water masses near the Antarctic Continental Margin: evidence from $H_2{}^{18}O/H_2{}^{16}O$ ratio in seawater. In: Jacobs SS (ed) Oceanology of the Antarctic Continental Shelf. Am Geophys Union, Washington, DC, pp 59-85
8. Budillon G, Conversano F, Giaquinto G, Morgigni M (1995) Idrologia del Mare di Ross. PNRA Rapp Ant 95(02):428-435
9. Hayes DE, Davey FJ (1975) Geophysical study of the Ross Sea, Antarctica. Initial Rep DSDP, 28:887-907
10. Vanney JR, Falconer RKH, Johnson GL (1981) Geomorphology of the Ross Sea and adjacent oceanic provinces. Mar Geol 41:73-102
11. Carter L, Mitchell JS, Day NJ (1981) Suspended sediment beneath permanent and seasonal ice, Ross Ice Shelf, Antarctica. New Zealand, J Geol Geophy 50, 24:249-262
12. Biscaye PE, Eittreim SL (1977) Suspended particulate loads and transports in the nepheloid layer of the abyssal Atlantic Ocean. Mar Geol 23:155-172

13. Leventer A, Dunbar R (1987) Diatom flux in McMurdo Sound, Antarctica. Mar Micropaleontol 112:49-64
14. Heezen BC, Hollister CD (1971) The face of the deep. Oxford University, Press, New York, pp 659
15. Gardner WD, Southard JB, Hollister CD (1985) Sedimentation, resuspension and chemistry of particles in the northwest Atlantic. Mar Geol 65:199-242
16. Monaco A, Biscaye P, Soyer J, Pocklington R, Heussner S (1990) Particle fluxes and ecosystem response on a continental margin: the 1985-1988 Mediterranean ECOMARGE experiment. Cont Shelf Res 10:809-839

Particle Fluxes and Sediment Characteristics at Three Selected Sites in the Ross Sea (Antarctica)

M. Ravaioli[1], M. Frignani[1], M.C. Gambi[2], L. Labbrozzi[1] and L. Langone[1]

[1]Istituto di Geologia Marina, CNR, Via Gobetti 101, Bologna, Italy
[2]Stazione Zoologica "A. Dohrn", Laboratory of Benthic Ecology, Napoli, Italy

Abstract

We studied three selected sites of the Ross Sea to understand the relationships between their depositional characteristics and benthic fauna. Box cores were collected at each site and sediments were described and analysed for macrobenthos composition and concentrations of biogenic components. The three sites show different lithologies: A and B are subject to present open-sea biogenic sedimentation and differ only slightly, whereas C, being winnowed by strong bottom currents, is characterised by a residual glacial-marine sediment. $^{210}Pb_{ex}$ inventories suggest that both sites A and B are subject to processes of focusing, thus collecting sediments transported from nearby areas. Apparent burial rates were calculated on the basis of ^{210}Pb chronologies and, for core B, corrected using ^{14}C data. A and B show similar apparent mass accumulation rates (89 and 104 mg cm^{-2} yr^{-1}, respectively) as well as burial rates of biogenic species. These rates are overestimated, due to the influence of deep mixing on the $^{210}Pb_{ex}$ profiles: using ^{14}C dates a factor of 2.7 was calculated for site B. Benthic communities differ in composition, density and structure. The presence of infaunal forms, mainly surface and sub-surface deposit feeders, with reworking capability characterise sites A and B, whereas at site C the community is mainly composed by epibenthic filter feeders and is strongly influenced by sediment instability. These differences are mainly due to bottom water dynamics and, when comparing A and B, to water depth and, perhaps, to the slightly different lithologies which reflect the locations of the sites and the different sources of sedimentary materials.

1 Introduction

Sediments of the Southern Ocean, and those of the Ross Sea in particular, accumulate high amounts of biogenic silica. Ledford-Hoffman et al. [1], on the basis of $^{210}Pb_{ex}$ chronologies, estimated that as much as a quarter of the dissolved silica supplied to the world oceans by rivers and hydrothermal vents may be accumulating in the deposits of the Antarctic continental shelf. DeMaster et al. [2, 3] established, using ^{14}C-derived accumulation rates, that the total flux of biogenic silica to the Ross Sea sediments is approximately an order one magnitude lower than the previous estimates. In practice, the extent of biogenic sedimentation, its spatial variability as a function of primary production, bottom morphology and water dynamics, and its relationship with living bottom fauna are aspects of the Ross Sea ecology which are still relatively unknown.

The supply and preservation of organic matter, together with bottom sediment characteristics, may affect macrofaunal abundance and composition. Therefore the water dynamics and burial rates of biogenic materials are key factors in determining the characteristics of the bottom environments. In recent years there has been an increasing interest in the achievement of accurate sediment chronologies and mass accumulation rates for Antarctic continental margin deposits in order to understand environmental processes, calculate fluxes and mass budgets for the various sediment components, and study paleoclimate records. Presently, radiocarbon dating is considered more reliable than $^{210}Pb_{ex}$ data for establishing accurate rates of sediment accumulation in the Ross Sea [2, 3], but they can be better used together to overcome the effect of mixing [4, 5].

This research, carried out within the framework of the Italian Programma Nazionale di Ricerche in Antartide, is aimed at understanding the characteristics of three representative sites of the Ross Sea with respect to biogenic fluxes, sediment composition and macrobenthos populations.

2 Study Area

The study area and the sampling locations are shown in Fig. 1. Frignani et al. [6] described the network of processes that determine the formation, transport and fate of the particulate matter and its role in the biogeochemical cycles in the Ross Sea. The biogenic components formed in the surface layer, which are not dissolved or recycled, sink via fecal pellets, cellular debris and aggregates [7] and accumulate in sediments of areas where water dynamics is weak. Smith and Nelson [8] suggested that algal blooms result from vertical stability imparted by melting ice that begins in the polynya close to the Ross Ice Shelf, and expands northward. The patterns of surficial sediments [9, 10] and satellite imagery [11] suggest phytoplankton biomass and production gradients west to east and south to north. The three study sites, located along the south to north direction of retreat of the marginal ice zone, were chosen to study particle fluxes through the water column using moored instruments (sediment traps, current meters, transmissometers). Site A, located in south-west Ross Sea at 810-m depth, is important to understand particle transport and sediment accumulation in proximity to the Ross Ice Shelf. The area is made interesting by the initial phase of the ice margin retreat and Smith and Gordon [12] found that phytoplankton production was dominated by the prymnesiophyte *Phaeocystis antarctica*, although significant numbers of diatoms co-occurred. Mooring A, operative since January 1994, is the subject of a co-operation with the US program ROAVERRS. Site B, located in the center of the northern part of the Joides Basin at 580-m depth, represents a setting characterised by high biosiliceous sediment accumulation. Mooring B was first deployed in December 1994. Finally, site C, in the northern flank of the Mawson Bank very close to the shelf break, is representative of an environment with high energy and reworked sediment. Mooring C was left in place from December 1994 to January 1996.

3 Materials and Methods

Samples from three representative sites of the Ross Sea (Fig. 1) were collected during the RosSMIZE cruise (1994-1995). A gravity core was taken at site B. To study both sediment features and benthic fauna in the area of mooring points,

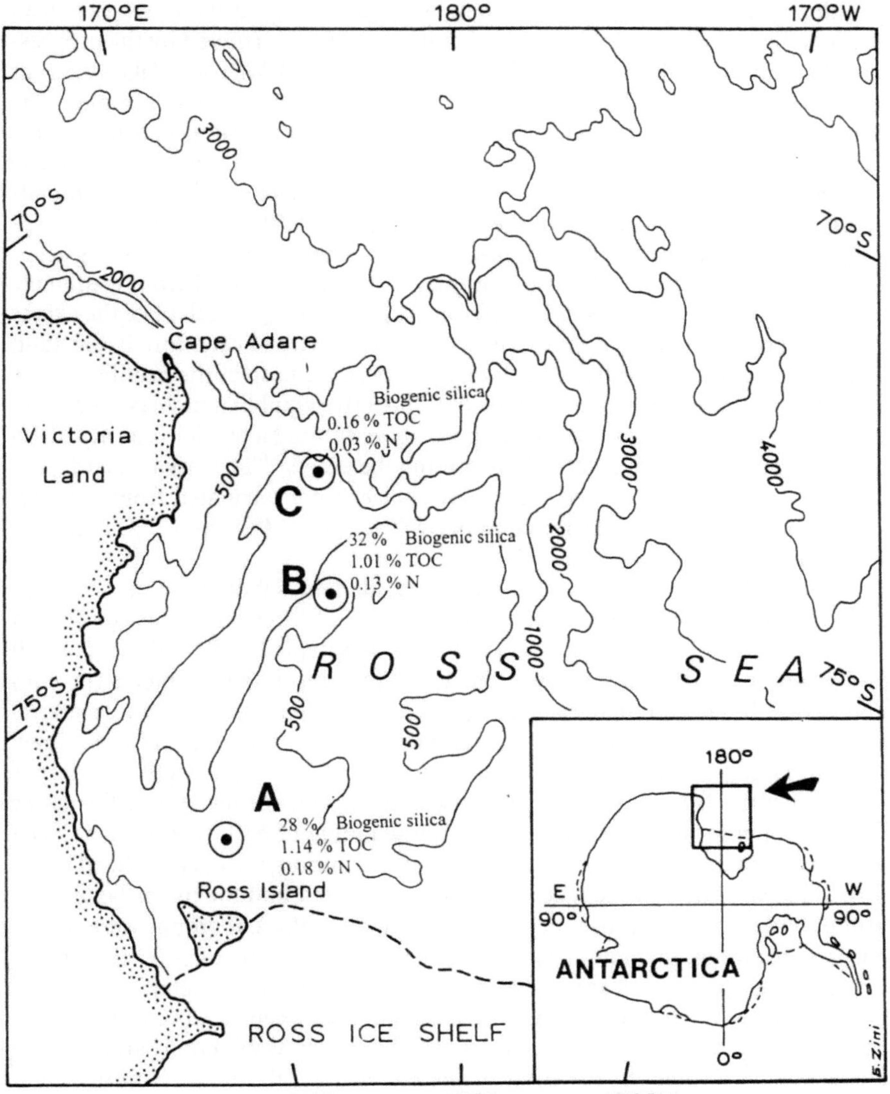

Fig. 1. Map of the Ross Sea with location of study sites. Surficial concentrations of biogenic components are indicated

bottom samples were taken using a box corer (23 cm wide, 42 cm long and 50 cm high), which collects a maximum volume of 38.6 l over a surface of about 0.1 m^2. In order to assess lateral variability of benthic populations, seven replicate cores were collected in the area of sites B and C, whereas at site A only two core replicates were taken, due to logistic constraints. Biogeochemical determinations were carried out on one or two samples per site.

Sediment samples were described and X-radiographed; then grain size composition was obtained using common methods involving the use of sieves and an X-ray sedigraph. Biogenic silica concentrations were determined by the progressive dissolution method suggested by DeMaster [13, 14]. We used NaOH 0.5 N as extractant, in view of large quantities of biogenic silica in our samples. The extraction was carried out on 20 mg of sample at 85 °C, by taking 0.2-ml aliquots for analysis every hour for 4 h. Organic carbon and nitrogen were determined by a Carlo Erba 1108 CHN Element Analyser. The carbonate fraction was eliminated by a preliminary treatment of the sample with 2 N HCl directly in a silver capsule. Alpha counting of ^{210}Po was used for ^{210}Pb determinations, assuming secular equilibrium between the two isotopes. ^{210}Po was extracted from sediments and plated on silver discs as in Frignani and Langone [15]. To calculate the excess ^{210}Pb ($^{210}Pb_{ex}$) we assumed that supported ^{210}Pb activity in each core is constant. The value was estimated from the total activity at the base of the profiles, where ^{210}Pb and ^{226}Ra were considered to be in radioactive equilibrium. Mass depths in cores were calculated from sediment porosity, assuming a particle density of 2.5 g cm^{-3}. Inventories were obtained integrating the excess ^{210}Pb activity-depth profiles vs. mass depth. Expected $^{210}Pb_{ex}$ inventories were calculated from the integrated production in the water column by ^{226}Ra decay (0.2 dpm l^{-1}; Ku and Li, 1976, cited in [4]).

The stratification of the macrobenthos was studied, subdividing the sediment box core into various levels. At site A a maximum of three levels was considered (0-5; 5-10 and 10-20 cm). At site B five levels were sampled (0-5, 5-10, 10-15, 15-20 and below 20 cm). Finally, at site C, due to the features of the sediments (pebbles and coarse biodetritic material), and to the very different quantity of material sampled in each of the replicates, the sediment sections varied from two to three: the interface, generally from the surface to the first 2 to 6 cm depth, the deeper layer, generally from 2 6 to 10 cm depth, and the level below 10 cm, this latter obtained only for two of the replicates. The sediment of each section was sieved with a 0.5 mm mesh size screen. The collected benthic organisms were fixed in 4% formaldehyde. The benthic macrofauna was separated into the major taxonomic groups; for each site, all specimens belonging to the different groups were counted. At each station the dominance of the major taxa and the total mean number of individuals (expressed as number of individuals per square meter) were calculated, as well as the mean abundance in each of the sediment layers considered for the investigated site. A first evaluation of the potential of bioturbation of the fauna was performed on the basis of the eto-ecological literature on the main taxa.

4 Results

4.1 Sediment Features

Figure 2 shows the most evident lithologic features of sediments from visual description and X-radiographs of cores, and Fig. 3 depicts the grain size distribution in selected sediment samples (2bc1, 15bc5, 24bc6). According to Anderson et al. [9] and Dunbar et al. [10] the surficial sediments at sites A and B can be classified as siliceous muds. At site A sediment is brown and contains pebbles and gravels (ice rafted debris) below 10-cm depth. The figures show that there are inhomogeneities among samples, in that surficial sediments are always a biogenic mud, whereas below 8-10 cm sediments of 2bc1 are coarser than that of 2bc2. At site B the sediment is fine, olive-gray siliceous mud and ooze without gravels. Here the sediment is oxic at the surface and becomes sub-oxic below 5-7 cm. The average bulk dry density at A is lower than at B: 0.56 and 0.65 g cm^{-3}, respectively. At site C residual glacial–marine sediment is present, mainly as sand with a conspicuous amount of pebbles, gravels and biogenic carbonate debris.

4.2 Biogenic Silica, Organic Carbon and Nitrogen

Figure 3 shows the concentration-depth profiles of biogenic components at the three sites. The content of biogenic silica decreases with depth in core A from 29 to 13%; organic carbon (OC) and N show a similar pattern varying from 1.14 to 0.5% and from 0.18 to 0.06%, respectively. At site B biogenic silica (21-32%), OC (0.86-1.23%) and N (0.12-0.16%) contents do not decrease significantly with depth in core. At site C the contents of biogenic silica (0.2-1.6%), OC (0.05-0.19%) and N (0.01-0.04%) are very low.

4.3 ^{210}Pb Profiles

$^{210}Pb_{ex}$ activity-depth profiles are shown in Fig. 3. Accumulation rates were calculated only for fine sediments at sites A and B, since there is no present sediment accumulation at site C [5]. Surficial mixed layers (SML) are clearly visible in $^{210}Pb_{ex}$ profiles relative to sites A and B; their thickness is 6 and 4 cm, respectively (Fig. 3). Below SMLs, profiles decrease nearly monotonically with depth. At site A, the change of lithology in 2bc1 (Fig. 2) clearly affects the $^{210}Pb_{ex}$ activity in the surficial layer and cause a discontinuity in the profile. The $^{210}Pb_{ex}$ profile at site C shows two sections with different slopes.

4.4 Macrobenthos

Figures 4, 5 and 6 describe macrobenthos composition abundances and distributions. In fine biogenic sediments (A and B), benthic fauna is mainly composed of infaunal sedentary forms, mostly represented by polychaetes (mean dominance

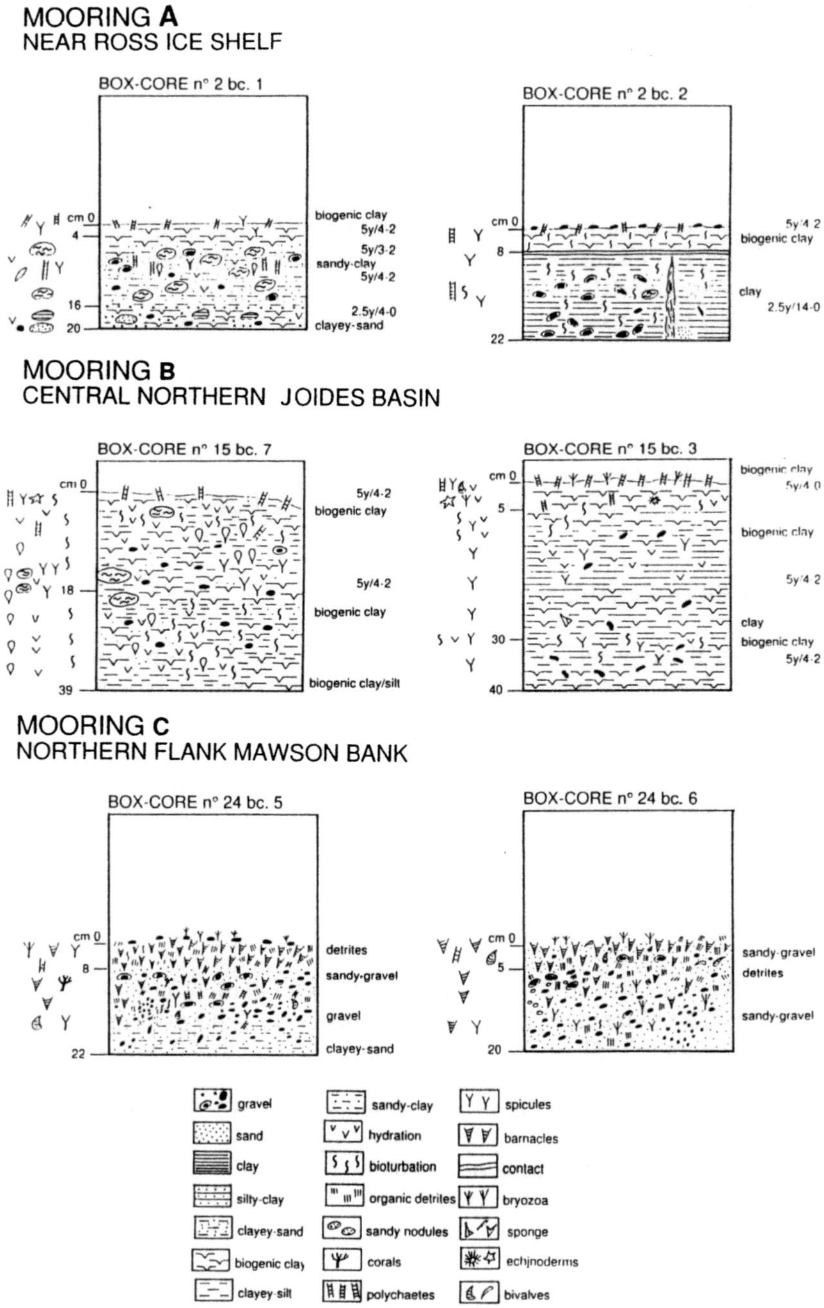

Fig. 2. Sediment features at the three sites (from visual description and X-radiographs). Two replicate box cores taken close to the same site are shown. Codes refer to colors according to the Munsell chart

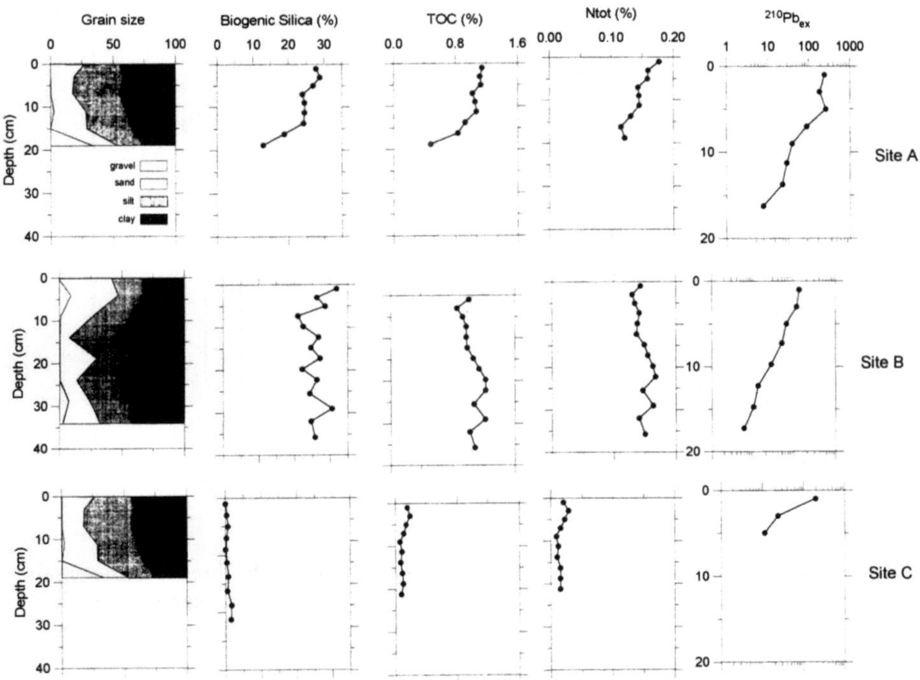

Fig. 3. Grain size composition and concentration-depth profiles of biogenic silica, total organic carbon (TOC), N and $^{210}Pb_{ex}$ at the three sites. Supported ^{210}Pb values are 35.0 and 52.1 Bq kg^{-1} at sites A and B, respectively

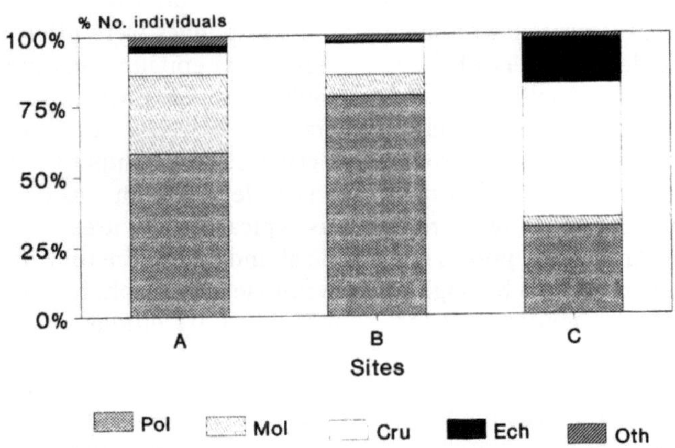

Fig. 4. Quantitative dominance (percentage number of individuals) of the main taxonomic groups of macrozoobenthos collected at sites A, B and C. *Pol*, polychaetes; *Mol*, molluscs; *Cru*, crustaceans (mainly amphipods, isopods and tanaids); *Ech*, echinoderms (ophiuroids and asteroids); *Oth*, others (e.g., sipunculids, bryozoans)

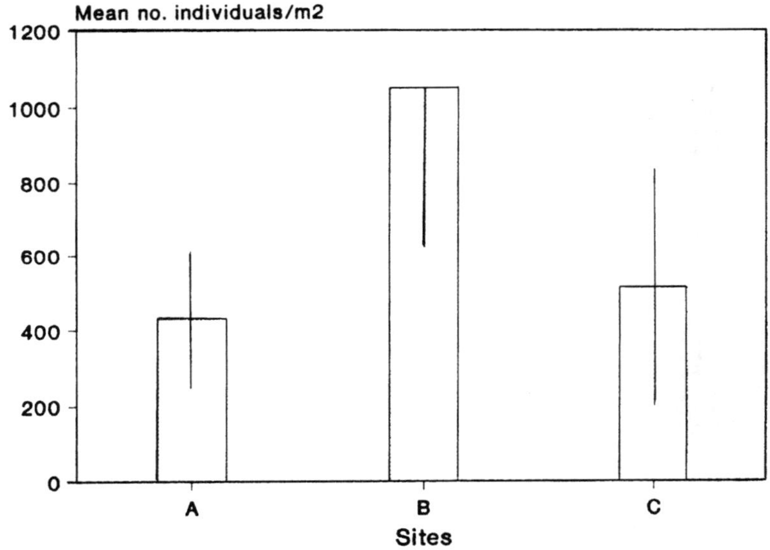

Fig. 5. Abundance of the macrozoobenthic organisms collected at the three sites. Mean values are reported (number of individuals per square meter), and bars represent standard deviations (sites B and C) or intervals between minimum and maximum values (site A)

60 and 77% at A and B, respectively), and crustacean peracarids (28% at A and 11.3% at B). Other groups, such as bivalves mollusc and echinoderms, are less represented (Fig. 4). At site A, the only two samples available contain quite scarce numbers of taxa mainly represented by infaunal polychaetes of the family Fauveliopsidae and bivalves mollusc.

At site B polychaetes represent up to 74% of the quantitative dominance (Fig. 4). Furthermore, aggregates of spicule of silica sponges were often observed, mainly in the deeper layers of the sediment. A few epifaunal organisms (bryozoan colonies, small ophiuroids and asteroids) were also present. The sediment surface was very poor in epifaunal organisms, and only rare bryozoan colonies, few small ophiuroids and asteroids, and crustacean amphipods were found. Arenaceous foraminifera with a stick-shape and length from 1 to 2.5 cm (probably *Hyperammina* sp.), already reported as typical for this area by Asioli [16], are abundant. Among the polychaetes, quite abundant are the tubicolous forms of surface deposit feeders belonging to Oweniidae and Ampharetidae, and also the sub-surface deposit feeders such as Capitellidae, Cirratulidae, Fauveliopsidae, Ophelidae, Paraonidae and Maldanidae. Capitellidae and Maldanidae, in particular, represent the most typical group of polychaetes responsible for bioturbation with a "conveyor-belt" mechanism (*sensu* [17]). As a whole, the faunal composition in the various core replicates seemed quite homogeneous.

At site C the living benthic organisms are composed mainly of epibenthic filter-feeder forms, such as echinoderm ophiuroids (16.3%) and crustacean cir-

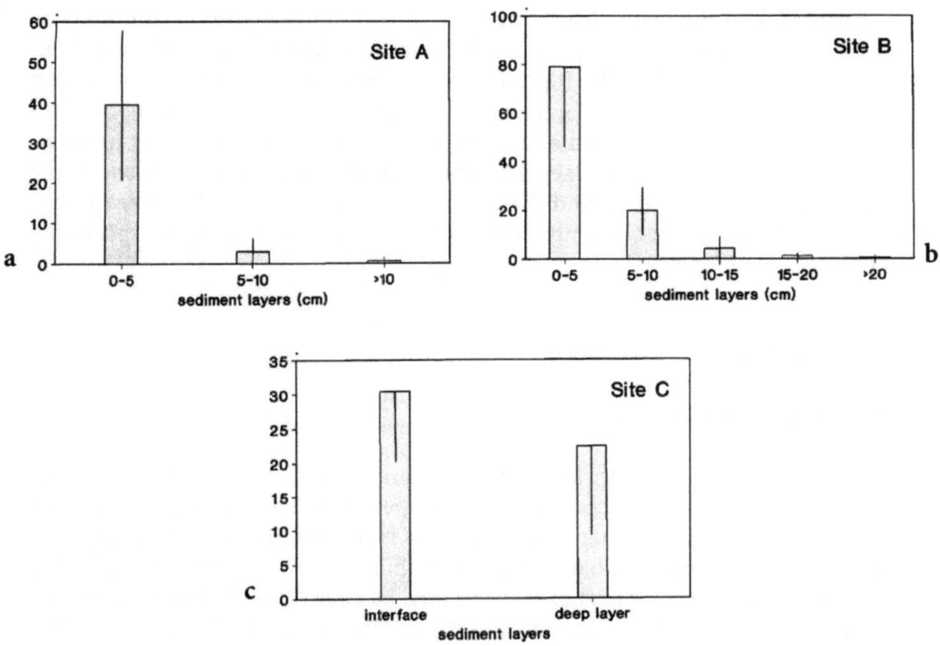

Fig. 6. Abundance of macrozoobenthic organisms in different sediment layers. Mean values are reported (number of individuals of 0.1 m^{-2}), and bars represent standard deviations (sites B and C) or intervals between minimum and maximum values (site A)

ripeds (barnacles), amphipods and tanaids (47.5%). Infaunal polychaetes (31.3%) and bivalves (3.3%) are much less represented with respect to the other sites. The large amount of organogenous *dead* remains present in this sediment is composed of recent and sub-fossil tests of various benthic organisms. The most conspicuous and abundant remains were the tests of the deep-water barnacles of the genus *Bathylasma* (a few of them were also found alive) and mollusc gastropods of the genus *Trophon* (Muricidae). Dead skeletons of solitary and colonial madreporarians, bryozoans and tubes of serpulid polychaetes were also abundant. Such dead remains were also present on pebbles and tests of barnacles. The dimensions of both dead and living organisms were generally relatively small, probably due to the prevalence of young individuals. With respect to the replicates of site B, this site is characterised by higher variability and patchiness in the composition of living fauna.

Faunal density (Fig. 5) showed the highest mean value at site B with more than 1000 individuals m^{-2}, followed by site C (ca. 500 individuals m^{-2}) and site A (430 individuals m^{-2}). These values, which are the first available for the deep sediment of the Ross Sea, are from one to two orders of magnitude lower than those recorded at shallow depth in other soft bottoms of the Ross Sea, such as McMurdo Sound [18] and Terra Nova Bay [19].

Benthic fauna showed a clear stratification within the sediment column (Fig. 6). At site A a mean of 72% and at site B a mean of 76% of the total macrobenthos was concentrated in the topmost 5 cm of the core, whereas below 15-cm depth only occasional living fauna was found (mean 1.1%). The living benthos at depth in core is represented almost exclusively by some polychaete families, mainly Capitellidae, Paraonidae, crustacean isopods and tanaids. At site C the stratification is less evident and the fauna present in the surficial layer (generally 0-2 cm) was almost as abundant (53%) as in the deeper layers (generally 2-10 cm).

5 Discussion

5.1 Accumulation Rates and Fluxes

At all sites the $^{210}Pb_{ex}$ inventories are higher than the expected (Table 1) by factors of 3.5, 3 and 4.5 for sediments A, B and C, respectively. These evidences indicate that fine biogenic material deposited on banks is resuspended by strong currents and settles on the bottom of basins, thus contributing to the increase of the $^{210}Pb_{ex}$ inventory in areas similar to those represented by sites A and B. The same mechanism was observed by Langone et al. [20] in other areas of the Ross Sea. This observation also confirms the role that Anderson et al. [9] and Dunbar et al. [10] have given to the ocean circulation as an important controller of the sediment distribution in the Ross Sea. Sediment at site C is a residual glacial-marine sand. Due to the action of strong bottom currents (velocities up to 90 cm s^{-1} [21]) the sediment is not accumulating because it is subject to periodic winnowing. This type of setting would exclude the presence of $^{210}Pb_{ex}$ and the high inventory seems rather unnatural. The presence of a profile is probably due to the downward transport of fine particles, rich in $^{210}Pb_{ex}$, caused by physical mixing. We used this profile to calculate a mixing coefficient of 0.061 cm^2 $year^{-1}$, applying the solution of the one-dimensional advection-diffusion equation for the case of no sedimentation [22, 23].

To calculate the sediment accumulation rates we used two different models: the CRS model [24], which assumes $^{210}Pb_{ex}$ constant rate of supply, and the CF-CS model [25] based on the assumption of constant flux-constant sedimentation. The effect of mixing has been considered in that we used, applying the CF-CS model, the profile below the surficial mixed layer (SML). To calculate the mean accumulation rate using the CRS model, we can consider the 90% equilibrium

Table 1. Water depth, expected and measured inventories at the three sites

Site	Water depth (m)	Expected inventory (Bq m^{-2})	Measured inventory (Bq m^{-2})
A	810	2700	9505
B	586	1953	5833
C	456	1520	6938

Table 2. Apparent mass accumulation rate and burial rates of biogenic components at sites A and B. The values B*, calculated using both $^{210}Pb_{ex}$ and ^{14}C data, are the most reliable

Site	Mass accumulation rate (mg cm^{-2} year^{-1})	Biogenic silica burial rate (mg cm^{-2} year^{-1})	OC burial rate (mg cm^{-2} year^{-1})	N burial rate (mg cm^{-2} year^{-1})
A	89	22	0.82	0.12
B	104	27	1.13	0.16
B*	38*	9.8*	0.41*	0.06*

depth, i.e. the depth at which the $^{210}Pb_{ex}$ inventory reaches 10% of its value. Sediment accumulation rates obtained with the two models are the same for site B and comparable for site A. According to the CRS model apparent mass accumulation rates are 89 mg cm^{-2} year^{-1} at site A and 104 mg cm^{-2} year^{-1} at site B, whereas apparent sediment accumulation rates are 0.16 cm year^{-1} for both site A and site B.

Multiplying the sediment accumulation rates by the concentrations of biogenic silica, OC and N at the interface or at depth in the sediment we obtained apparent accumulation rates and burial rates for these biogenic components (Table 2). The highest values characterise site B in the Joides Basin. Labbrozzi et al. [5] discussed the influence of deep mixing at site B by resolving the advection diffusion equation [4, 23] both for $^{210}Pb_{ex}$ and ^{14}C data. They calculated a sediment accumulation rate of 0.059 cm year^{-1} (38.4 mg cm^{-2} year^{-1}, B* in Table 2) and a mixing rate of 0.53 cm^2 year^{-1}. This means that fluxes of biogenic species reported in Table 2 are overestimated by a factor of 2.7 for site B and that the value relative to site A may be affected by an error of the same order.

A comparison between rain rates from sediment traps is not yet possible, because the only data presently available are relative to mooring A [6, 26]. Apparent fluxes and burial rates from sediment data suggest that sites A and B receive similar food supply, though fluxes at B appear to be a little higher.

5.2 Benthos

The analysis revealed very strong differences in the composition, density and structure of the benthic communities at the three sites [27].

The above data evidentiate a high degree of consistency between sediment features and benthic community composition and structure, and confirm that both are strictly related to the different dynamics of bottom water and sedimentary regimes. Sites A and B represent areas with high accumulation rates of biogenic fine sediments, stable and dominated by infaunal forms, mainly surface and subsurface deposit feeders, that exert a not negligible reworking of the sediment at the interface and within the first 5-10 cm of the sediment column. The organic input to the sediment is relatively high with a significant component of vertical flux. The differences between sites A and B, which show quite similar sediment features, can be due to the effect of water depth: it is known that faunal abundance decreases strongly with depth, especially in soft sediments.

The stratification pattern at these sites is consistent with the results of the $^{210}Pb_{ex}$ profiles, which indicate the presence of a surface mixed layer in the upper 4-6 cm.

The living benthic community found at site C, mainly composed of epibenthic filter feeders, confirms a high bottom dynamics, high level of sediment instability and flux of particles and food mainly advected laterally. These organisms do not have a high potential of bioturbation, and they live mainly in the interstices of the coarse organogenous component of the sediment. Mixing at this site, confirmed by the presence of an excess $^{210}Pb_{ex}$ profile, is mainly physical in origin. The sessile forms settled on pebbles and other hard substrates probably are not able to grow to the adult size because of the unpredictable and probably frequent catastrophic events (e.g., resuspension and winnowing). In view of this hypothesis, the individuals probably never reach reproduction and the observed community may be composed of pseudo-populations depending on recolonization from larvae produced by adult populations of neighboring areas ("supply-side ecology" [28]).

6 Conclusions

From the discussion above the following conclusions can be drawn:

1. The three sites show different lithologies. A and B differ slightly, in that sediment at A is a little coarser and contains more gravels, due to the proximity to the ice shelf. B is a typical site for present seasonally open marine sedimentation. Due to the action of strong bottom currents, C is not subject to present accumulation and is a residual glacial-marine sediment with a high biodetritic calcareous component.
2. Both sites A and B are subject to processes of focusing, whereas sediment at C is resuspended and winnowed by bottom currents and shows $^{210}Pb_{ex}$ profiles due to the entrapment of fine particles
3. Apparent sediment accumulation rates are similar at sites A and B (89 and 104 mg cm^{-2} $year^{-1}$, respectively), as well as apparent burial rates of biogenic species. It was possible to establish that mass accumulation rates and burial rates for site B are overestimated by a factor of 2.7, due to the influence of deep mixing on the $^{210}Pb_{ex}$ profile. The same situation occurs at site A.
4. There are differences in composition, density and structure of benthic communities: the presence of infaunal forms, mainly surface and sub-surface deposit feeders, with reworking capability characterises sites A and B, whereas at site C the community is mainly composed of epibenthic filter feeders and is strongly influenced by sediment instability. These differences are mainly due to bottom water dynamics. Fluxes of biogenic particles do not seem to account for the differences in faunal composition at sites A and B, which should be ascribed to the influence of water depth. The slightly different lithologies, which reflect the locations of the sites, and the different sources of coarse sedimentary materials may also play a role.

Acknowledgements. This research was carried out within the framework of the Ecology and Biogeochemistry of the Southern Ocean Project (subproject Biogenic Sedimentation) of the Italian National Programme for Antarctic Research (PNRA). The authors are indebted to E. Lipparini, A. Magagnoli and G. Marozzi for gravity core and box core sampling. We are also grateful to E. Lipparini for most of the ^{210}Po extractions, G. Marozzi for X-radiography, and L. Casoni and G. Zini for drawings. This is contribution No. 1155 from the Istituto di Geologia Marina del CNR, Bologna.

References

1. Ledford-Hoffman PA, DeMaster DJ, Nittrouer CA (1986) Biogenic silica in the Ross Sea and the importance of Antarctic continental-shelf deposits in the marine silica budget. Geochim Cosmochim Acta 50:2099-2110
2. DeMaster DJ, Nelson TM, Harden SL, Nittrouer CA (1991) The cycling and accumulation of biogenic silica, and organic carbon in Antarctic deep-sea and continental margin environments. Mar Chem 35:489-502
3. DeMaster DJ, Ragueneau O, Nittrouer CA (1996) Preservation efficiencies and accumulation rates for biogenic silica and organic C, N and P in high-latitude sediments: the Ross Sea. J Geophys Res 101:18501-18518
4. Harden SL, DeMaster DJ, Nittrouer CA (1992) Developing sediment geochronologies for high-latitude continental shelf deposits: a radiochemical approach. Mar Geol 103:69-97
5. Labbrozzi L, Langone L, Frignani M, Ravaioli M (1998) Burial rates for biogenic silica, organic C and N at three sites of the Ross Sea (Antarctica). Proc. XII AIOL Congress, Isola di Vulcano, Italy, 18-21 September 1996, II, 121-130
6. Frignani M, Langone L, Labbrozzi L, Ravaioli M (1998) Biogeochemical processes in the Ross Sea (Antarctica): present knowledge and perspectives. In: Faranda F, Guglielmo L, Ianora A (eds) Ross Sea ecology. Springer, Berlin Heidelberg New York, in press
7. Smith WO Jr, Nelson DM, Di Tullio GR, Leventer AR (1996) Temporal and spatial patterns in the Ross Sea. Phytoplankton biomass, elemental composition, productivity and grow rates. J Geophys Res 101:18445-18465
8. Smith WO Jr, Nelson DM (1985) Phytoplankton bloom produced by a receding ice edge in the Ross Sea: spatial coherence with the density field. Science 227:163-166
9. Anderson JB, Brake CF, Myers NC (1984) Sedimentation on the Ross Sea continental shelf, Antarctica. Mar Geol 57:295-333
10. Dunbar RB, Anderson JB, Domack EW, Jacobs SS (1985) Oceanographic influences on sedimentation along the Antarctic continental shelf. In: Jacobs SS (ed) Oceanography of the Antarctic Continental Shelf. Ant Res Ser AGU 43:291-312
11. Comiso JC, McClain CR, Sullivan CW, Ryan JP, Leonard C (1993) Coastal zone color scanner pigment concentrations in the Southern Ocean and relationship to geophysical surface features. J Geophys Res 98:2419-2451
12. Smith WO Jr, Gordon LI (1997) Hyperproductivity of the Ross Sea (Antarctica) polynya during austral spring. Geophys Res Lett, 24, 233-236
13. DeMaster DJ (1979) The marine budgets of silica and Si-32. PhD thesis, Yale University, New Haven, Connecticut, 308 pp
14. DeMaster DJ (1981) The supply and accumulation of silica in the marine environment. Geochim Cosmochim Acta 45:1715-1732
15. Frignani M, Langone L (1991) Accumulation rates and ^{137}Cs distribution in sediments off the Po River delta and the Emilia-Romagna coast (northwestern Adriatic Sea, Italy). Cont Shelf Res 11:525-542
16. Asioli A (1995) Living (stained) benthic Foraminifera distribution in the western Ross Sea (Antarctica). Palaeopelagos 5:201-214
17. Rhoads D (1974) Organism-sediment relations on the muddy sea floor. Oceanogr Mar Biol Ann Rev 12:263-300

18. Dayton PK, Oliver JS (1977) Antarctic soft bottom benthos on oligotrophic and eutrophic environments. Science 197:55-58
19. Gambi MC, Castelli A, Guizzardi M (1997) Polychaete populations of the shallow soft bottoms off Terra Nova Bay (Ross Sea, Antarctica): distribution, diversity and biomass. Pol Biol 17:199-210
20. Langone L, Frignani M, Labbrozzi L, Ravaioli M (1998) Present day biosiliceous sedimentation in the NW Ross Sea (Antarctica). J Mar Systems, 17, 459-470
21. Picco P, Amici L, Meloni R, Langone L, Ravaioli M (1998) Temporal variability of currents in the Ross Sea (This Vol.)
22. Goldberg ED, Koide M (1962) Geochronological studies of deep sea sediments by the ionium/thorium method. Geochim Cosmochim Acta 51:1897-1921
23. Nittrouer CA, DeMaster DJ, McKee BA, Cutshall NH, Larsen IL (1983/1984) The effect of sediment mixing on ^{210}Pb accumulation rates for the Washington continental shelf. Mar Geol 54, 201-221
24. Appleby PG, Oldfield F (1978) The calculation of ^{210}Pb dates assuming a constant rate of supply of unsupported ^{210}Pb to the sediment. Catena 5:1-18
25. Robbins JA (1978) Geochemical and geophysical application of radioactive lead. In: Nriagu JO (ed.) The biogeochemistry of lead in the environment. Elsevier, Amsterdam, pp 85-393
26. DeMaster DJ, Dunbar RB, Gordon LI, Leventer AR, Morrison JM, Nelson DM, Nittrouer CA, Smith WO Jr (1992) Cycling and accumulation of biogenic silica and organic matter in high-latitude environments: the Ross Sea. Oceanography 5:146-153
27. Gambi MC, Bussotti S (1999) Composition, abundance and stratification of soft-bottom macrobenthos from selected areas of the Ross Sea shelf (Antarctica) Polar Biol (in press)
28. Roughgarden J, Gaines SD, Pacala SW (1987) "Supply side ecology": the role of physical transport processes. In: Gee JHR, Giller PS (eds.) Organization of Communities. Blackwell, Oxford, pp 491-518

General Characteristics of Density–Turbidity Currents in the Ross Sea (Antarctica)

S. Gremes Cordero[1] and E. Salusti[2]

[1]Dipartimento di Fisica, Università "La Sapienza", Piazzale A. Moro 2, 00185 Rome, Italy
[2]INFN, Dipartimento di Fisica, Università "La Sapienza", Piazzale A. Moro 2, 00185 Rome, Italy

Abstract

To investigate thermodynamic currents in Antarctica, we here discuss quasi-steady density currents flowing over a regular slope and their hydrodynamic stability, considering also bottom erosion phenomena: in other words, the current is assumed to exchange sediments with the bottom. To simplify this complex problem a model of sediment evolution is assumed. As in recent work the excess mass due to the bottom erosion and deposition is assumed to depend only on the current stress on the bottom. A nonlinear equation considering both the time and space variability of these "density-turbidity" currents for a two-layer model is obtained and a novel criterion to identify the "ignition" point of these density-turbidity currents is discussed. In Polar Oceans these interactions can play a fundamental role in the generation of new water masses, as a result of violent hydrodynamic instability concerning the downslope motion of dense shelf water along submarine canyons.

1 The Problem

The role of dense water in deep polar basins is well established [1]. The paths that the dense shelf water follows offshore from formation regions to the deep basins nevertheless remain an open problem. Variable bottom topography has long been recognized as an important mechanism for steering the flow of bottom water [2, 3]. This interaction was first emphasized by Nansen [4], who described the tendency of marine currents to follow the deepest channels of the sea bottom.

In general, density currents, whether related to a generation mechanism such as quick storm waves or to long-lasting dense-water formation due to strong climatologic air-sea interaction, are interesting physical phenomena. They are known to flow geostrophically along the isobaths over the shelf, entraining bottom sediments, ultimately reaching the deepest layers [5-8]. However, if this rather regular motion is interrupted by a cross-shelf submarine canyon, then the dense water has to interrupt its geostrophic motion and sink along the canyon. The resulting downslope turbulent flow can either be stable or subject to hydrodynamic instability processes [9]. In oceans, violent mixing is to be expected, and must generate an intermediate kind of oceanic water, a phenomenon of interest in interpreting field observations.

Particularly good candidates for steering dense bottom water flows are submarine canyons and deep troughs. Quadfasel et al. [10] used a sequence of hydrographic sections to follow a bottom water plume down the Storfjord Trough into Fram Strait, where the plume entered the deep basin. Another good example is represented by Barrow Canyon, which runs parallel to the northern coast of Alaska [11].

Surprisingly few studies have attempted to model the formation and offshore transport of dense shelf water in polar seas. Moreover, in general, theoretical analyses of density currents over a regular slope are essential for a better understanding of the behaviour of various natural phenomena [12-14]. For such currents bottom erosion is a particular complication: the bottom stress entrains bottom sediment and also suspended sediment is constantly settling out of the current.

For oceans, such flows have been studied by many authors [1, 2, 15-18] to find the equations of motion specific to these density–turbidity currents [16, 20, 21]. Of particular interest is the analysis by Parker et al. [16] referring to a simple model of a continuous steady flow where the exchange of sediment from the bed is allowed. Stacey and Bowen [17, 18] make a particularly careful analysis of the vertical structure of these currents, also in comparison with the few known marine cases [19]. Salusti [3] considers the hydrodynamic instabilities for a two-fluid problem as discussed in Caserta et al. [2], and has been able to take into account both space and time variability of such density–turbidity currents.

On theoretical grounds our analysis consists of two parts: first, the study of the quasi-steady main flow [22] and, second, the instabilities of these currents, which is the main purpose of the present analysis. We find that these nonlinear oscillations can affect the bottom current in a catastrophic way, giving rise to an "ignition" like those of classical turbidity currents. A novel criterion discriminating between regular flows and catastrophic events is discussed below. In particular, in Section 5, we apply these considerations to a cold dense water observed in the shelf break near Cape Adare and to dense salty shelf water on the oceanic bottom off Victoria Land.

2 Deep Ocean Currents in the Ross Sea and Victoria Land Shelf

The study zone in the Ross Sea, as shown in Fig. 1, is located off Cape Adare, around 72° S, 173° E. The bottom topography is rather irregular; the shelf of about 500-m depth is isolated from the shore by a deep canyon; the ocean bottom can reach 2000 m in depth. It is of interest that the glaceologic and meteorologic conditions of Terra Nova Bay and the adjacent basins differ from most of Antarctica. During the warm season, indeed, it is not difficult to find the ocean free of ice, even for long periods. Consequently, in the Ross Sea catabatic winds dramatically affect the overall marine situation, giving rise to vortices, meanders and other two-dimensional turbulent phenomena.

In particular, Jacobs and Comiso [23] discuss the formation of polynyas associated with violent air-sea exchanges, such as those due to catabatic winds. Interestingly, they explain the maintenance of the Pennell and Ross Sea Passage polynyas (Fig. 1) near the continental shelf break (our zone of interest) as being

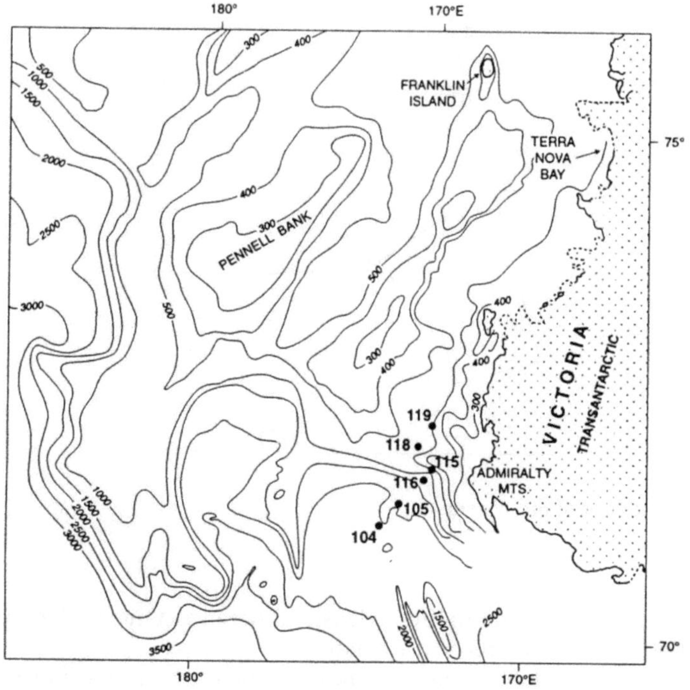

Fig. 1. The Ross Sea continental shelf with hydrological stations of interest (Tenth Expedition, CLIMA Project, February 1995)

due mainly to "upwellings" of warm water near the slope front: temperature (T) and salinity (S) of these "upwellings" correspond to those of the Circumpolar Deep Water, namely $T \cong 0.5$ °C and $S \cong 34.7$ ppt.

The low ice concentrations are a strong indication that upwelling, melting and higher sensible heat flux into the atmosphere occur above the slope front. A description of the water masses in the investigated area is presented elsewhere in this volume [24, 25].

3 Model of Sediment Dynamics

We now analyze the motion of a slightly viscous, dense fluid of density over a regular slope, below a lighter fluid of ρ_0 density at rest (Fig. 2); here, x and z are the along-slope and normal-to-slope coordinates. The boundary between the two fluids is at $z=h$; our treatment follows the classical studies of Yih [22], Parker et al. [16], Stacey and Bowen [17, 18] and Salusti [3]. The current carries uniform non-cohesive sediment with "fall velocity" v_s, which is free to exchange with similar sediment on the canyon bed. Let us stress that $\rho > \rho_0$ both for the entrainment of

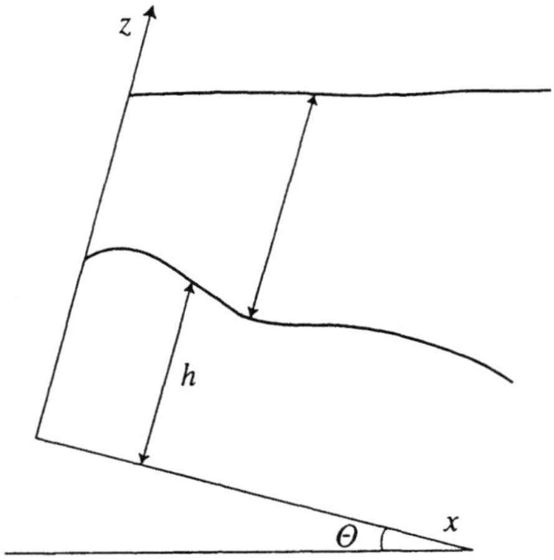

Fig. 2. General view of the system

bottom particulate, and for an initial density difference Δ due to the hydrological characteristics of the oceanic water masses.

To deal with the equations of motion, one needs information on sediment dynamics, namely on a field for which so much experimental data are available that a clear synthesis is difficult to attain today. In particular, in recent interesting studies by Parker [16] and Stacey and Bowen [17, 18] the system time-evolution is described in detail, but the ultimate equations are so complex that either time or one space dimension has to be disregarded. This encourages a more synthetic approach: Caserta et al. [2] recently proposed an heuristic model of a slowly varying density current flowing down a bottom slope and raising some bottom sediment: the density $\rho > \rho_0 + \Delta$ is assumed to vary only as a result of the shear stress on the sea bottom, since it is this stress that raises the bottom sediment and governs its resuspension/deposition. So, following an old idea of Plapp and Mitchell [15, 20, 21], we here assume this effect to be a sensitive function of the bottom stress, i.e. of $t=\nu\rho\ \partial u/\partial z|_{z=0}$ minus a term proportional to the sediment fall velocity v_s.

Following these ideas through, if a quantity of suspended sediment is formed and quickly redistributed uniformly in the whole thin turbulent density current, the resulting density ρ can be represented as

$$\rho - \rho_0 - \Delta = q_0/h_0\ (e^{(\tau-\tau 0)/\tau 0}), \quad (1)$$

heuristically assuming an exponential behaviour for the bottom shears.

It is of interest to compare Eq. (1) with the results of Stacey and Bowen [17] for density evolution, namely

$$\rho - \rho_0 = \rho - \rho_0\,|_{\text{sea bottom}} = \exp\{-v_s \cos\theta \int_0^z (1+B_2R_i)^{1+\gamma}/\nu_0\, dz\}, \quad (2)$$

where γ, B_2 are suitable constants, R_i is the Richardson number and ν_0 is the eddy viscosity for unstratified fluids. This shows that it is not an easy task to fully compare our schematic assumption [Eq. (1)] with those discussed in the literature. Intuitively, our model can be considered valid if the sediment is so fine-grained and light as to be dispersed immediately inside the thin turbulent bottom flow.

On general physical grounds, this situation would correspond to steady thermohaline currents generated by severe climatological conditions, entraining fine-grained sediment as its crossed the shelf or slope and found their way into the deep ocean along submarine canyons crossing the shelf [7]. Note how this situation is truly different from that of a large amount of bottom sediment carried by shelf currents during a severe storm, i.e., a true turbidity current.

4 Bottom Current Evolution

For the current we assume that the longitudinal scale is naturally to be considered large; the lateral confinement is such that the Coriolis effect can be neglected and there is no entrainment of water at the upper surface of the flow [12], that is the Richardson number is $R_i \sim 1$. The equation of motion is thus:

$$\partial V/\partial t + (V\cdot\nabla)V = -\nabla p_*/\rho + \nu\Delta V + \mu\gamma + F$$
$$\gamma \equiv (\rho-\rho_0)g/\rho \qquad \mu = (\sin\theta, 0, -\cos\theta) \sim (\theta, 0, -1), \tag{3}$$

where F is some external forcing for unit of mass, g denotes the gravitational acceleration, V denotes the velocity and p_* denotes the pressure minus the hydrostatic term.

One may now wonder as to the effect of any hydrodynamic instability on these erosion-deposition dynamics. Natural modelling involves computing the bottom stress due to both the quasi-steady current and its short space scale perturbation. After some long computations [3], Eqs. (1) and (3) finally give

$$\partial^2 h'/\partial t^2 = \gamma_0 h_0 \partial^2 h'/\partial x^2 + g\varepsilon h_0/2\partial^3 h'^2/\partial x^3 + \nu(\partial^2/\partial x^2 \partial/\partial t)h' - (\nu\pi\partial h'/\partial t)/4h^2_0 + h_0\partial F/\partial x, \tag{4}$$

with

$$\varepsilon = 1/g[4\gamma^2{}_0\rho_0 h_0(1-e^{-t/4t^*})]/[\pi^2(\tau_0-\gamma_0\rho_0 h_0\theta)] \cong 4/g\pi^2[\gamma^2{}_0\rho_0 h_0]/[\tau_0-\gamma_0\rho_0 h_0\theta]\big|_{t/t^*\sim O(10)},$$

where the fundamental role of the "ignition" time $t^* \equiv h_0^2/(\pi^2\nu)$ will be clear in the following.

The physical meaning of ε as a dimensionless coefficient of non-linear effects now emerges naturally. Equation (4) is a complex non-linear diffusive equation; its treatment in some cases reveals some mathematical pathologies [2], namely an explosive steepening of the signals [26]. A numerical solution, even if it exists, cannot consequently be considered fully satisfactory. So let us first discuss some simpler cases: if $\varepsilon = 0$, the dynamic effect of bottom entrainment vanishes quickly, so the bottom layer dynamics is driven only by the hydrological density difference Δ. After some time, ε can no longer be considered nil. So one is led to investigate perturbative solutions of Eq. (4) in ε and ν such as

$$h' = f(x-Vt, \varepsilon t) \equiv f(y, \varepsilon t) \text{ with } y \equiv x - Vt. \quad (5)$$

Using the following substitutions [2]

$$\xi = e^{-\frac{\pi^2 \nu}{8\varepsilon h^2_0}t}\{f(y,t) - \frac{\nu V_0}{\varepsilon h_0 g}\} = e^{-\frac{\pi^2 \nu}{8\varepsilon h^2_0}t}\,ø(y, e^{-\frac{\pi^2 \nu}{8\varepsilon h^2_0}t}) \equiv \xi \cdot ø(y,\xi), \quad (6)$$

all this finally becomes

$$\frac{\pi^2}{4h^2_0}\frac{\partial}{\partial \xi}ø - \frac{\varepsilon h_0 g}{\nu V_0}\frac{\partial}{\partial y}\{ø\frac{\partial}{\partial y}ø\} + V\frac{\pi^2 \nu}{4h^3_0 \varepsilon}\frac{1}{\xi} + F^* = 0. \quad (7)$$

Solutions to Eq. (7), a non-linear heat type, can be seen intuitively by analyzing the function ø. In fact, if ø > 0 everything is as in the case of diffusion equations, but if ø < 0 everything is as in the case of ø > 0 but with $\theta \to -\theta$, namely the solutions grow exponentially and catastrophic events are to be expected. This corresponds to the quantity $f - \nu V_0/(\varepsilon h_0 g)$ changing sign and explains the classical "ignition point" of the turbidity currents, at least from the hydrodynamic stability viewpoint.

In general, Eq. (7) may be said to allow investigation of the effect of both space and time variation on current evolution, whilst the vertical variability is schematized as a two-layer system. The frictional terms give a general space-time damping as one would expect. Conversely, the non linear term proportional to ε can lead to explosive behaviour: since f is an oscillating perturbation, the quantity

$$|f - f^*| \equiv |f - \frac{\nu V_0}{g \varepsilon h_0}| = |f - \frac{\pi^2 \nu(\tau_0 - \rho_0 V_0{}^2 \theta}{4V_0^3 \rho_0 (1 - e^{\frac{-t}{4t^*}})}| \quad (8)$$

is a critical quantity. This means that if | f | > | f* |, due to the effect of some external forcings, then the system is no longer stable. In such a context the factor $(1-e^{\{t/4t^*\}})$ contained in Eq. (8) gives a kind of "time delay" of considerable interest: for t ~ 0 one has that $f^* \to \infty$ and so any transient forcing gives rise to perturbations that cannot grow. Consequently, the system returns to the original situation of a density-driven current. However, if the perturbation lasts long enough, roughly ~ 10 t*, then the non-linear terms become important, eventually generating ignitions: this relation plays the same role as the "ignition condition" in classical approaches to turbidity currents.

In such a complex situation, for steady density currents flowing downstream along a submarine canyon, an important role is played by $t^* = h_0{}^2/(\pi^2 \nu)$. Indeed, if $\nu \sim 10^{-6}\ m^2\ s^{-1}$ as in laboratory simulations, the ignition time t* necessary for a perturbation to destabilize the current is of the order of 10^7 s. In a more realistic case one can assume $\nu \sim 10^{-2}\ m^2\ s^{-1}$, as for the fully developed turbidity currents; then one obtains the value $t^* \sim 10^3$ s, which constitutes an easier triggering effect since oceanic forcings lasting 10^3 s are rather frequent. So a fundamental difference is found between the kind of diffusivity that plays a role in laboratory simulations and real oceanic eddy diffusion where a much larger value plays an important role. All these considerations finally suggest, as a practical rule, that it is a realistic value of ν that is critical. It is consequently of interest that t^* can be realistically estimated by comparing the two most popular expressions of bottom stress, namely $\tau=\rho_o C_d U^2$ and $\tau=\rho_o \nu \partial U/\partial z$, in this way giving rise to

$$\rho_0 C_d U^2 \sim \rho_o \nu U/h,$$

where a realistic $C_d = 3{,}5 \times 10^{-3}$ (MKS units) can be assumed and, consequently,

$$\nu \sim C_d h U \qquad t^* \equiv h_0^2 / (\pi^2 \nu) \rightarrow h_0/(\pi^2 C_d U),$$

a fairly easily applicable relation.

5 Application to Antarctic Currents

Let us apply these considerations to hydrologic sections along the Antarctic coast off Cape Adare, as shown in Fig. 1. The bottom depth ranges from about 2200 m to about 500 m, the shelf. It is immediately apparent that a very dense salty water is present over the deepest part of the shelf, with $S \cong 34.75$-34.80 ppt and $T \cong -1.9$ to -1.5 °C. Its geostrophic velocity can be estimated as 5-10 cm s^{-1} and the shelf layer is about 100 m thick. This looks like the classical geostrophic situation of any dense water present over the shelf.

It is also of interest that a vein of dense water is evident at the bottom of the shelf break (Fig. 3). Interestingly, this water is very cold and its density is 27.87, slightly denser than the bottom water observed in the adjacent stations. This denser layer has a thickness that can be estimated as $h_0 = 100$-200 m, but the velocities are rather small, namely 1 cm s^{-1}. This bottom water is not particularly salty but is very cold: one is led to assume that its origin is different from that of the salty shelf water, probably due to ice melting, and it arrives at the bottom of the shelf break via a different path, probably along a submarine depression skirting the shelf.

It is of some interest, consequently, to estimate the corresponding values of t^*. For this deep layer we assume that the hydrologic characteristics observed at station 116 are due to dense water reaching this deep layer flowing along a submarine canyon. We thus estimate the friction $\nu \cong C_d h U \sim 5 \times 10^{-3}$ (MKS units), in agreement with classical values. It results that $t^* = h^2_0/(\pi^2 \nu) = 2 \times 10^6$ s and so the observed deep, slow current must essentially be considered stable. This is not surprising since for this deep flow no hydrodynamic instability is evident.

We also apply the same considerations to the salty shelf water discussed earlier. For geostrophic velocity $u \sim 5 \times 10^{-2}$ ms^{-1} and $h_0 \sim 100$-200 m, we obtain an "ignition" time $t^* = 5 \times 10^4$ s. This estimate of an eventual "ignition" time is smaller by a considerable factor than the previous estimate of t^* for the deep layer. Note that the water velocity over the shelf is much smaller than an even-

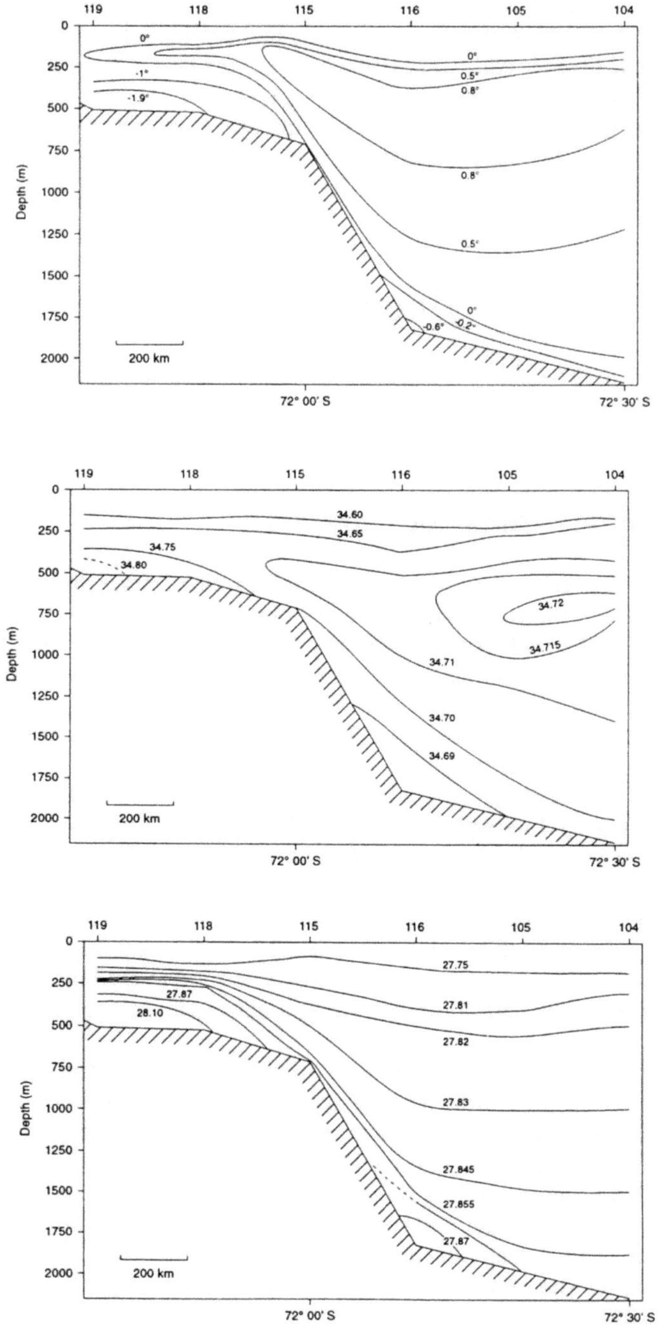

a b c

Fig. 3. Vertical sections of **a** potential temperature, **b** salinity and **c** potential density as observed in the stations shown in Fig. 1

tual along-canyon falling velocity and consequently this "ignition time" of 5 x 10^4 represents an upper limit for t*.

All these considerations underline the fact that no such "ignition" of turbidity currents has effectively been observed off Victoria land, but also stress that the general conditions are such that similar phenomena are probable. Indeed, if the effective pathway of the deep current is questionable (reasonable sources being possible in many points from the Weddell sea to the Ross Sea), if some rough topographic particularity or some dynamic external forcing, such as a violent storm, affects the shelf water then the sinking and ignition of a density-turbidity current appears probable. This is one of the aspects that deserves further analysis on experimental grounds, also taking into account the climatological importance of these phenomena.

Acknowledgements. We warmly thank Prof. G. Spezie and Dr. G. Budillon for providing the hydrologic data shown in Fig. 4. We also thank Profs. Pantin, Parker, Bignami, Levi and Seymour for their help and criticism, as well as Mrs. L. Santonastaso and R. De Gregorio for typing this manuscript and Dr G.M.R. Manzella for the valuable help in redrafting it.

References

1. Cavalieri DJ, Martin S (1994) The contributions of Alaskan, Siberian, and Canadian coastal polynyas to the cold halocline layer of the Arctic Ocean. J Geophys Res 99(18):343-362
2. Caserta A, Mieli E, Salusti E (1990) On a model of bottom erosion by dense water steady veins. Geophys Astrophys Fluid Dyn 55:117-135
3. Salusti E (1996) A new model for marine density-turbidity currents with criteria for ignition. Geophys Astrophys Fluid Dyn 83:233-260
4. Nansen F (1906) Northern waters: Captain Roald Amundsen's oceanographic observations in the Arctic Seas in 1901. Skr Nor Vidensk Akad Kl 1 Mat Naturvidensk Kl 3:145
5. Gardner WD (1989) Periodic resupension in Baltimore canyon by focusing of internal waves. J Geophys Res 94, C12:18185-18194
6. Bignami F, Salusti E, Schiarini S (1990) Observations on a bottom vein of dense water in the Southern Adriatic and Jonian Seas. J Geophys Res 95:7249-7259
7. Chapman DC, Gawarkiewicz G (1995) Offshore transport of dense shelf water in the presence of a submarine canyon. J Geophys Res 100, C7:13373-13387
8. Jiang L, Garwood RW Jr (1996) Three-dimensional simulations of overflows on continental slopes J Phys Ocean 26:1214-1233
9. Sugimoto T, Whitehead JA (1983) Laboratory models for bay-type continental shelves in the winter. J Phys Ocean 13, 1819-1828
10. Quadfasel D, Rudels B, Kurz K (1988) Outflow of dense water from a Svalbard fjord into the Fram strait. Deep Sea Res 35:1143-1150
11. Garrison GR, Becker P (1976) The Barrow submarine canyon: a drain for the Chukchi sea. J Geophys Res 81:4445-4453
12. Ellison TH, Turner JS (1959) Turbulent entrainment in stratified flows. J Fluid Mech 6:432-448
13. Simpson JE (1982) Gravity currents in the laboratory. Atmos Ocean Anny Rev Fluid Mech 14:213-234
14. Emms PW (1997) Streamtube models of gravity currents in the ocean. Deep-Sea Res in press.
15. Drake DE, Cacchione DA (1986) Field observations of bed shear stress and sediment resuspension on continental shelves, Alaska and California. Cont Shelf Res 6:415-429

16. Parker G, Fukushima Y, Pantin HM (1986) Self accelerating turbidity currents. J Fluid Mech 171:145-181
17. Stacey MW, Bowen JA (1988a) The vertical structure of density and turbidity currents: theory and observation. J Geophys Res 93:3528-3542
18. Stacey MW, Bowen JA (1988b) The vertical structure of turbidity currents and a necessary condition for self-maintenance. J Geophys Res 93:3543-3553
19. Seymour R (1986) Near shore autosuspending turbidity flows. Ocean Eng 13, 5:435-447
20. Bagnold RA (1977) Mechanism of marine sedimentation. The Sea. Wiley, Interscience, New York, vol 3, 507-528
21. Plapp JE, Mitchell JP (1960) A hydrodynamic theory of turbidity currents. J Geophys Res 65:983-992
22. Yih CS (1963) Stability of liquid flow down an inclined plane. Phys Fluid 6:321-334
23. Jacobs SS, Comiso JC (1989) Sea ice and oceanic processes on the Ross Sea continental shelf. J Geophys Res 94, C12:18195-18211
24. Jacobs S, Giulivi CF (1998) Thermohaline data and ocean circulation on the Ross Sea continental shelf (This Volume)
25. Gouretski V (1998) The large-scale thermohaline structure of the Ross Sea (This Volume)
26. Whitham GB (1974) Linear and nonlinear waves Wiley, New York, pp 636

Vertical Distribution and Biochemical Composition of Pico- and Microparticulate Organic Matter in the Ross Sea (Antarctica)

M. Fabiano[1], R. Danovaro[2] and P. Povero[1]

[1] Istituto di Scienze Ambientali Marine, Università di Genova, Corso Rainusso 14, 16038, S. Margherita L., Genova, Italy
[2] Cattedra di Ecologia, Facoltà di Scienze, Università di Ancona, Via Brecce Bianche, 60131 Ancona, Italy

Abstract

We studied the vertical distribution and biochemical composition of particulate organic matter in seven coastal stations located in Terra Nova Bay and in five open-sea stations located in the Ross Sea (Antarctica) during summer 1989-90. Protein, carbohydrate, lipid and ATP concentrations were estimated to assess the quantitative role of labile organic fractions in the cycling of organic matter in the water column and the processes related to particle dynamics. Particulate organic matter was filtered through 0.2 and 2.0 m pore-size filters to discriminate two size fractions: picoparticulate matter (particles between 0.2 and 2.0 m) and microparticulate matter (particles between 2.0 and 200 m). Significant quantitative and qualitative differences between the pico- and microparticulate fraction were observed. Total particulate suspended matter was mostly composed of microparticles, which accounted for two thirds of the total concentration. Similar results were obtained for proteins, carbohydrates and lipids. By contrast, ATP concentrations were significantly higher in the picoparticulate fraction. Although most of the picoparticulate matter was detrital (61.3%), this fraction contained the largest proportion of living organisms (i.e. picoplankton), whereas the microparticulate fraction was almost exclusively composed of organic detritus. Significant compositional differences between the two size-classes of particles were observed. The microparticulate matter was dominated by proteins, whereas the picoparticulate matter was mostly composed of carbohydrates. This may indicate a different origin of the two size-groups of particles.

1 Introduction

Particulate matter contains sizable stocks of picoplankton (i.e. 0.2 to 2.0 μm), nanoplankton (2 to 20 μm) and microplankton (20 to 200 μm). During the last decade picoplankton has been identified as one of the major biomass components in the euphotic zone [1, 2] and comprises a major fraction of particulate organic carbon in oligotrophic oceans [2]. Still, little is known about the relative significance of pico-, nano- and microplankton in the polar regions. Although the exact nature and composition of particulate organic matter (POM) is of fundamental importance to our understanding of the food web (including the microbial loop), it cannot be easily obtained. One of the major problems is distinguishing and quantifying different components. Microscopic analysis can yield detailed information, but it is slow, and thus often not suitable for large number of samples taken during the course of an expedition. Accurate and complete characterizations of particulate matter from a chemical point of view are extremely difficult. Simple analyses of elemental composition do not sufficiently characterize the processes associated with the particle dynamics in the water column and their availability to consumers. In contrast, the biochemical analysis of the organic particles, applied to different size classes, might provide new insights into the origin, dynamics and fate of organic particles in the sea [3].

Although a relatively large amount of data on the elemental composition of suspended matter in the Ross Sea is available [4-8], little is known about its biochemical composition [9-12]. Lipid, protein and carbohydrate contents of particles were found to provide useful information about their origin and availability to consumers [3], and ATP content provides information on the living component of the organic carbon [13].

Our principal objectives were to assess, using a biochemical approach, the different particle sizes as a measure of the food potentially available for plankton. The biochemical composition provides indications of the nature and origin of the particles. We measured the ambient concentrations of two different size fractions of the particulate organic matter: picoparticulate organic matter (0.2 to 2.0 μm) and microparticulate organic matter (2.0 to 200 μm) in order to: (1) identify differences in the biochemical composition (carbohydrate, protein, lipid and ATP) among different size fractions; (2) provide new information on the changes in particle composition with water depth and (3) compare the composition and the relative significance of these particles in coastal (Terra Nova Bay) and open sea stations.

2 Materials and Methods

2.1 Sampling Area

Twelve stations were sampled during the Second Italian Oceanographic Expedition in the Ross Sea, on board the R/V "Cariboo", between December 5, 1989, and January 6, 1990, and using the R/V "Malippo", between January 13 and February 2. The sampling schedule included the following stations: 17, 19 and 21 (in the southeastern region of the Ross Sea), 23 and 25 (along 75° S in the southwestern area of the Ross Sea) and 2, 3, 7, 8, 12, 16 and 17 located in the coastal area of Terra Nova Bay (Fig. 1).

2.2 Hydrology

At each station data on temperature and salinity were collected, using a Neil Brown Mark III CTD system [14] during the "Cariboo" cruise and using a Meerestechnik KMS-90 system during the "Malippo" cruise [15]. Vertical attenu-

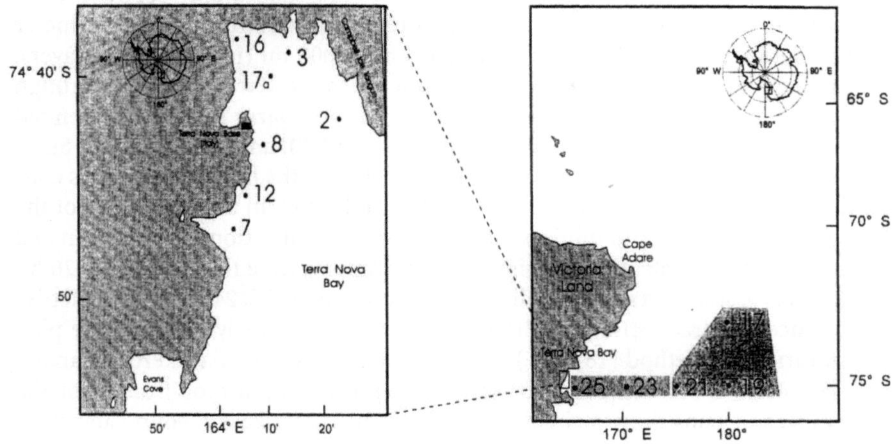

Fig. 1. Study area and sampling stations

ation of irradiance (PAR) was measured with a LI-COR (mod. 801, Idronaut) quantum meter and related data are reported by Innamorati et al. [16]. The base of the photic zone was defined as the depth at which 1% of the surface irradiance was present (Table 1). The depth of the mixed layer (Z_{mix}) was calculated as depth at which the sigma-t gradient is higher than 0.1 unit m^{-1} [14, 15].

2.3 Particulate Matter Analysis

Water samples were collected throughout the water column using Niskin bottles (10 l). Sampling depth ranged between 100 and 1500 m for total POM analysis and

Table 1. Sampling stations for fractionated POM analysis: latitude, longitude, depth, surface temperature, salinity and depth of the photic zone and mixed (Z_{mix}) layer at the sampled stations

Station	Latitude (°S)	Longitude (°E)	Depth (m)	Temp. (°C)	Salinity	Photic zone depth (m)	Z_{mix} (m)
17	73° 13.51	179° 50.28	526	-0.55	34.24	61	72
19	74° 56.88	179° 49.08	440	-0.57	34.37	39	47
21	74° 59.08	174° 52.43	340	-0.25	34.33	55	78
23	74° 59.59	169° 59.84	335	1.30	33.93	15	12
25	74° 56.80	165° 14.79	875	-0.14	33.21	14	14
2	74° 41.45	164° 24.01	541	0.65	32.75	18	19
3	74° 39.06	164° 14.66	180	0.84	33.43	40	43
7	74° 46.59	164° 03.78	195	1.28	33.62	15	16
8	74 °42.70	164° 08.84	160	0.37	33.80	15	13
12	74° 45.26	164° 06.26	100	0.19	33.90	13	15
16	74° 38.27	164° 04.58	232	0.47	33.95	10	9
17a	74° 40.13	164° 11.75	516	0.01	33.51	10	6

between 100 and 875 m for fractionated POM analysis. Bottom depths ranged from 100 to 3500 m (Table 1). The volumes of water filtered for each parameter ranged between 500 ml (in the photic layer) and 2000 ml (in the aphotic layer). Water samples (prefiltered through 200-μm mesh net) were filtered through Nuclepore polycarbonate filters (0.2- and 2.0-μm pore size) for total suspended matter (TSM), carbohydrates (CHO), proteins (PRT) and lipids (LIP). Since Nuclepore filters tend to clog rapidly, it is possible that the larger size class is contaminated with smaller particles. This would also lead to an underestimate of the finer particles. In order to minimize this problem, the filtrations were terminated when the filtration rate dropped significantly. Samples were then stored at -20 °C.

TSM was assessed gravimetrically after desiccation (60 °C, 24 h) using a Mettler M3 balance (accuracy ±1.0 μg [17]). Carbohydrates were analyzed using the phenol-sulfuric acid method [18]. d(+)-glucose was used as a standard. Protein analyses were carried out according to Hartree [19], based on a modification of the Lowry method using the protein reaction with rameic tartrate and Folin-Ciocalteau reagent. Bovine albumin was used as a standard. Lipid extraction was carried out according to Bligh and Dyer [20]. The analyses were performed by carbonization using the method of Marsh and Weinstein [21]. Tripalmitin was used as standard.

Method accuracy was tested against prepared standards. Standard solutions were prepared from dilutions of a 100-μg l^{-1} stock solution. Final concentrations were 0, 2, 5, 10, 20 50 and 100 μg l^{-1} for all standards utilized. For all analyses, blanks were analyzed for 0.2- and 2.0-μm pore size Nuclepore filters. For carbohydrate analysis, which presented the higher blank values, the detection limit was estimated through the analysis of the blank filters with internal standards (final concentrations 0.5, 1.0, 1.5, 2.5, 5.0 μg ml^{-1}). The method was sensitive to addition of 1.5 μg glucose ml^{-1}.

Single samples were normally collected. However, a series of replicate samples was collected in Terra Nova Bay from surface waters, in order to define the coefficient of variation for the different parameters. The inter-sample variability (unpubl. data, based on samples collected using 0.4-μm pore size Nuclepore filters), expressed as the coefficient of variation (CV), was: 13% for TSM, 11% for CHO, 3% for PRT and 11% for LIP. The deviations (CV) from real standard values were 7.8 % for LIP, 6.0% for CHO and 6.6% for PRT.

Samples for the determination of ATP were filtered immediately using 0.2- and 2.0-μm Nuclepore polycarbonate filters. ATP determinations were carried out according to Holm-Hansen and Booth [22] and Bulleid [23]. Standard solutions were prepared with crystalline ATP (Sigma). Analyses were performed on two replicates using an ATP 3000 photometer (Biospherical Instruments).

2.4 Definition of Size Classes

Total particulate matter was defined as the fraction between 0.2 and 200 μm (retained by 0.2-μm pore size Nuclepore), microparticulate matter was defined as the fraction of particles ranging from 2.0 to 200 μm (retained by 2.0-μm pore size Nuclepore) and picoparticulate (0.2 to 2.0 μm) matter was calculated as the difference between total and microparticulate matter.

2.5 Data Analysis

Differences between TSM, lipid, protein, carbohydrate and ATP contents of the pico- and microparticulate fractions of particulate matter were tested using analysis of variance (ANOVA). A Mann-Whitney Test was carried out on non-transformed data, when homogeneity of variance was rejected by an F_{max} test. Pearson's product-moment correlation analysis was used initially to examine relationships between the distributions of different classes of organic compounds and environmental parameters.

3 Results

3.1 Hydrography

The depth of the photic zone and of the mixed layers (Z_{mix}) at the different stations is reported in Table 1. The upper water column was well stratified throughout the study area. Figure 2 shows the typical patterns for temperature and salinity at stations of the different sampling area. The upper mixed layer depth generally ranged between 6 and 72 m. Surface temperatures ranged between -0.57 (Station 19) and 1.30°C (Station 23). Surface salinity ranged between 32.75 (Station 2) and 34.17 (Station 19).

3.2 Total Suspended Matter

The concentration of total suspended particulate matter ranged from 0.15 to 3.9 mg l^{-1}. Highest concentrations were typically found within the mixed layer (average concentrations of 0.7 mg l^{-1} in the southeastern region and 2 mg l^{-1} at coastal stations) and generally showed little change below the mixed layer depth where

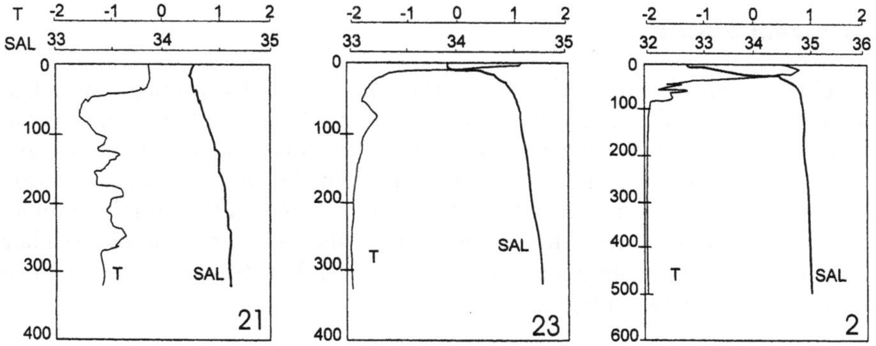

Fig. 2. Vertical profiles of temperature (T, °C) and salinity (*SAL*) in different stations from southeastern (*Station 21*) and southwestern (*Station 23*) regions of the Ross Sea and Terra Nova Bay (*Station 2*)

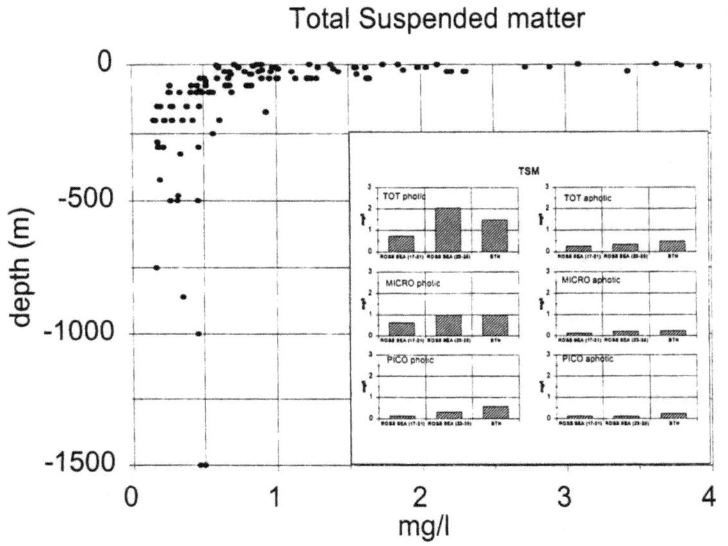

Fig. 3. Vertical profiles and average value of photic and aphotic layers of total, microparticulate and picoparticulate matter (*TSM*) in Terra Nova Bay (*BTN*) and southwestern and southeastern regions of the Ross Sea

we found average concentrations of 0.3 mg l^{-1} in the southwest and southeast Ross Sea and 0.7 mg l^{-1} in Terra Nova Bay. The microparticulate accounted for the majority of the total suspended matter (from 63 to 85 % of TSM in the photic layer and from 51 to 65% in the aphotic layer). Vertical patterns of total particulate matter and pico- and microparticulate matter in Terra Nova Bay and in the open Ross Sea are illustrated in Fig. 3.

3.3 Protein Content

Proteins were the dominant class of organic compounds. The average concentration of total particulate proteins was 109, 286 and 275 μg l^{-1} in the photic layer of the southeastern, southwestern and Terra Nova Bay areas respectively and ranged between 10.7 and 730.3 μg l^{-1}. Protein content in the picoparticulate fraction accounted for from 11 to 40% of the total protein content of the samples. Protein content typically declined with depth both for the pico- and microparticulate fractions. Vertical distributions of total proteins, and protein content of the two size classes, are shown in Fig. 4.

3.4 Carbohydrate Content

Carbohydrates were the second most important class of organic compounds, with

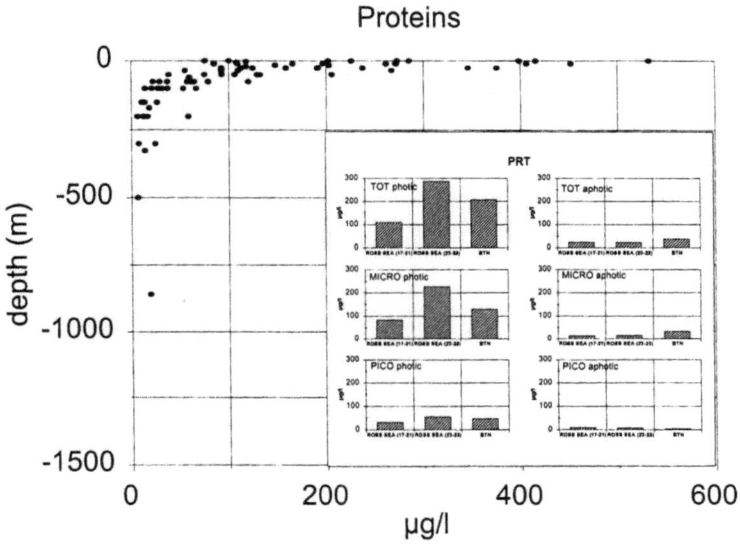

Fig. 4. Vertical profiles and average value of photic and aphotic layers of total, microparticulate and picoparticulate proteins (*PRT*) in Terra Nova Bay (*BTN*) and southwestern and southeastern regions of the Ross Sea

a concentration ranging from 4.8 to 251.1 µg l^{-1} (average surface concentrations of 56 µg l^{-1} in the southeastern region, 108 µg l^{-1} in the southwestern region and 133 µg l^{-1} in Terra Nova Bay). The vertical distribution of carbohydrate content of total, pico- and microparticulate fraction was irregular in the photic zone and vertical gradients were less pronounced in the open Ross Sea than in the coastal waters of Terra Nova Bay (Fig. 5). Carbohydrate concentrations did not decline sharply with depth. The ratio of protein to carbohydrate was, on average, higher in the microparticulate fraction (PRT:CHO = 1.8 and 3.1 in the Ross Sea and Terra Nova Bay respectively) than in picoparticulate matter (PRT:CHO = 0.8 and 1.7 in the Ross Sea and Terra Nova Bay respectively).

3.5 Lipid Content

The average concentration of total particulate lipids was 50 and 15 µg l^{-1} in the photic and aphotic layers and ranged between 4.8 and 75.7 µg l^{-1}. Lipid content of the picoparticulate fraction accounted for a minor fraction of the total (25 38%). The remaining 62 to 75.8% was due to the microparticulate fraction. Vertical distribution of lipid content in the different size fractions is shown in Fig. 6. The pico- and microparticulate fractions clearly declined similarly below the mixed layer depth. Therefore the relative significance of the two size fractions remained constant with depth.

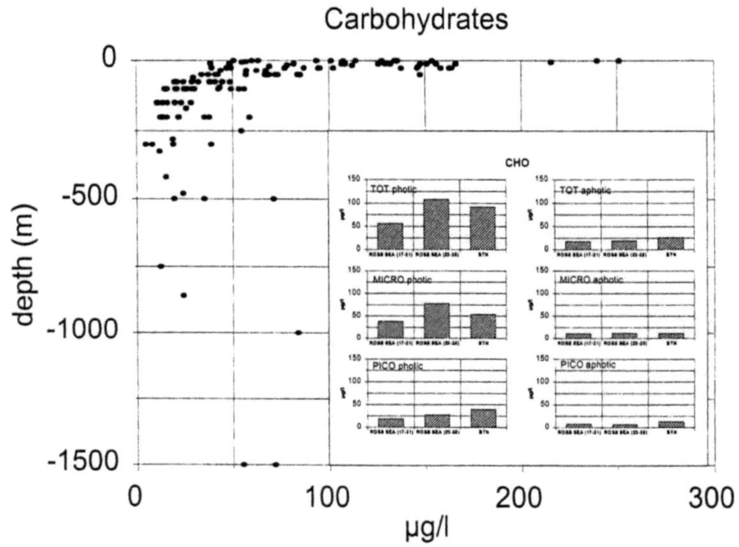

Fig. 5. Vertical profiles and average value of photic and aphotic layers of total, microparticulate and picoparticulate carbohydrates (*CHO*) in Terra Nova Bay (*BTN*) and southwestern and southeastern regions of the Ross Sea

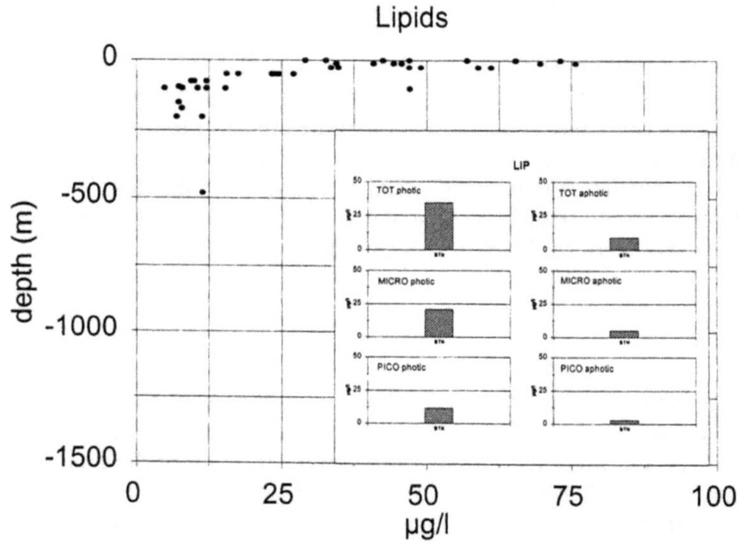

Fig. 6. Vertical profiles and average value of photic and aphotic layers of total, microparticulate and picoparticulate lipids (*LIP*) in Terra Nova Bay (*BTN*)

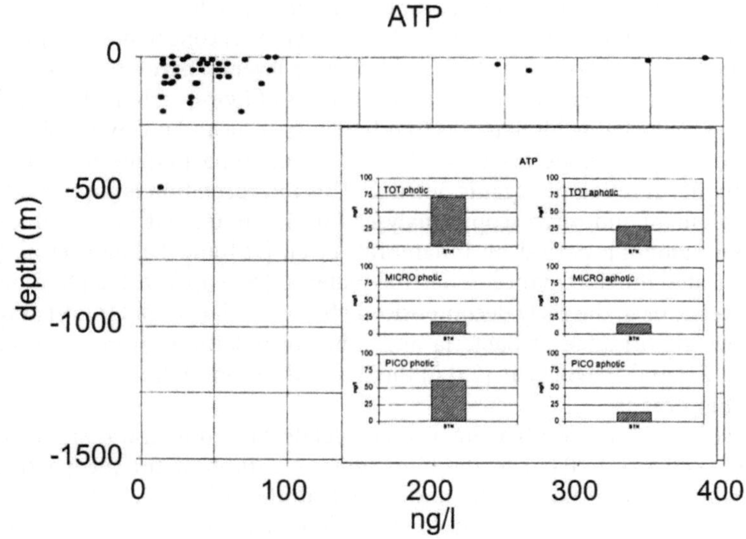

Fig. 7. Vertical profiles and average value of photic and aphotic layers of total, microparticulate and picoparticulate ATP in Terra Nova Bay (*BTN*)

3.6 ATP Content

ATP measurements were carried out only at the seven stations located in Terra Nova Bay. Integrated average ATP concentration was 92 and 47 ng l^{-1} in the photic and aphotic layer. On average, most of the total ATP content (90 and 63.0% in the surperficial and deeper water layers respectively) was due to the picoparticulate fraction. The vertical distribution of ATP concentrations is shown in Fig. 7. ATP concentrations typically declined with depth. However, significant differences were observed comparing the vertical patterns of the picoparticulate fraction of ATP and the microparticulate fraction. In fact, whilst the ATP content of the former fraction decreased strongly with depth (below the mixed layer depth), the concentrations of the larger fraction slightly increased.

4 Discussion

Integrated estimates of the quantitative contribution of the two size classes of particulate matter (picoparticulate fraction: 0.2 to 2.0 μm and microparticulate fraction: 2.0 to 200 μm) in Antarctic waters (the Ross Sea and Terra Nova Bay) showed the quantitative dominance of the microparticulate suspended matter. This fraction (w:w) represented, particularly in the photic layer, about two thirds of total suspended matter, protein, carbohydrate and total lipid concentrations. In contrast, in Terra Nova Bay, most of the ATP content was collected in the picoparticulate fraction (84.5% of total ATP in the surface layer). Therefore, if we assume that ATP represents a measure of the total amounts of living carbon [13], it is likely that most of the living biomass (as free-living bacteria and picophytoplankton) was in the picoparticulate fraction whereas the microparticulate fraction contained a higher fraction of organic detritus. This interpretation is supported by a series of pigment and microscopic analyses (SEM) carried out in the same area during our sampling period by Innamorati et al. [16] and Fabiano et al. [11], which indicated the dominance of detrital material (>10 μm) with a phytoplankton senescence or in the early decompositon phase. Similar results concerning the quantitative significance of bacteria and other picoplankton were reported by Cota et al. [24] and Hewes et al. [25] in the Weddell Sea and Johnson and Sieburth [26] in other oceans.

Moreover, in Terra Nova Bay the picoparticulate ATP content was significantly correlated ($r=0.435$, $p<0.05$) to the biopolymeric carbon of the picoparticulate fraction (the sum of lipid, protein and carbohydrate carbon [3, 9, 27]) while no significant correlations were found for the microparticulate fraction. The intercept value of the y-axis (POM = 24.32 μg C) is equivalent to about 61% of the average picoparticulate carbon in the top 100 m of the water column (picoparticulate matter concentration was on average 39.54 μg l^{-1}). Therefore most of the picoparticulate matter was detrital (i.e. non-living), even though the pico-fraction contains most of the ATP content.

From the data it is possible to conclude that all size classes of organic particles at all depths were mostly detrital, but detritus dominance was more pronounced in the larger particles. Significant differences were found comparing the composition of the two size classes of particles (Fig. 8). The microparticulate matter, in both the photic and aphotic layers, was mostly composed of proteins, followed by carbohydrates and lipids. In contrast, the picoparticulate matter displayed an increase in carbohydrate fraction that dominated in the aphotic layer, followed by proteins and lipids. The different biochemical composition of the two size classes can be referred to the different origin and evolution of the particles. As regards the microparticulate fraction, during summer 1989-1990, high protein concentrations were reported in Terra Nova Bay where the freshly generated organic detritus was extensively colonized by bacteria and other microheterotrophs [10]. In the southwestern region of the Ross Sea (Stations 23 and 25) a typical marginal ice zone (MIZ) area with high phytoplanktonic bloom (prevalently diatoms) was found associated with grazing activity [9, 28]. This situation may have determined a protein enrichment compared to carbohydrates and lipids as reported by Fabiano et al. [9] and Smith et al. [8]. In contrast, the dominance of carbohydrates in the picoparticles, particulary in Terra Nova Bay, would suggest a different origin. One possible explanation was the high picophytoplankton density [10], but also protein depletion from larger particles may have contributed to increase carbohydrate relative significance.

A change in particle composition with water depth was observed (Fig. 9). Total carbohydrate concentrations did not show a strong decrease with depth and, especially in the coastal stations of Terra Nova Bay, their relative significance with respect to proteins and lipids increased in the deeper water layers. These data are in agreement with the results reported by Ittekot et al. [29] and Liebezeit [30] and indicate that the carbohydrates in the sampled POM are more resistant to microbial degradation.

In contrast, total protein concentrations decreased significantly with depth. At stations 23 and 25, total proteins in the aphotic layer were about 10 times lower

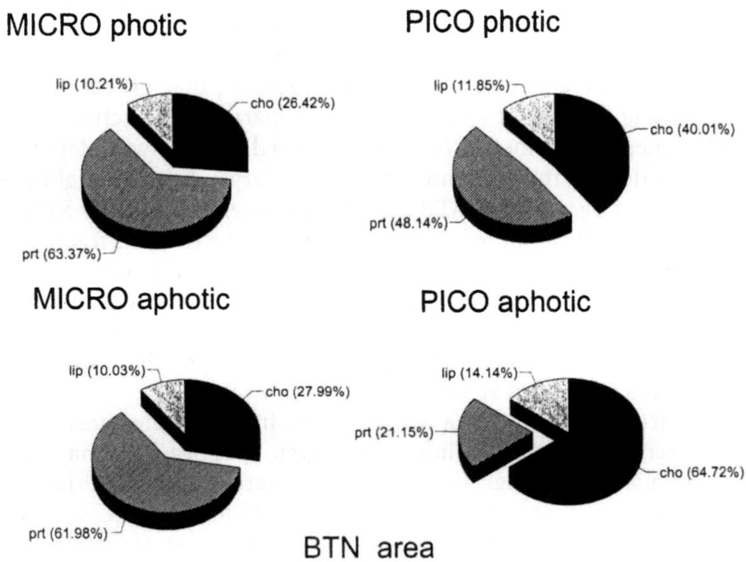

Fig. 8. Differences of microparticulate and picoparticulate biochemical composition in photic and aphotic layers of Terra Nova Bay (*BTN*)

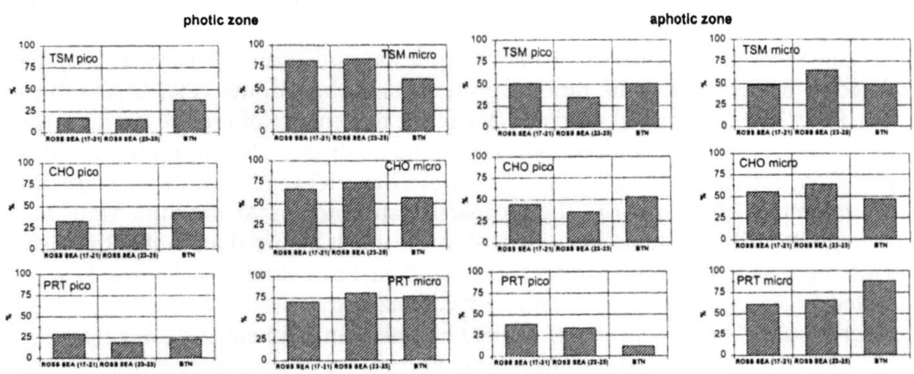

Fig. 9. Contributions of microparticulate and picoparticulate matter (*TSM*), micro and pico carbohydrates (*CHO*) and micro and pico proteins (*PRT*), in photic and aphotic layers in Terra Nova Bay (*BTN*) and southwestern and southeastern regions of the Ross Sea

than surface concentrations. This may indicate that these compounds were readily available for heterotrophic metabolism and/or were converted into dissolved macromolecular compounds [31].

The analysis of the vertical distribution and composition of the organic particles revealed also significant changes in terms of relative significance of the two size fractions. Picoparticulate protein fractions, for instance, in the deeper water layers increased in the Ross Sea area and decreased in Terra Nova Bay (from 20 to 35% and from 22 to 11% for the Ross Sea and Terra Nova Bay respectively). In contrast, the relative significance of picoparticulate carbohydrates and lipids (limited to Terra Nova Bay) increased with depth.

Significant differences were observed comparing the vertical patterns of the picoparticulate fraction of ATP and the microparticulate fraction. In fact, while the ATP content of the former fraction decreased strongly with depth (below the mixed layer depth), the concentrations of the larger fraction slightly increased (Fig. 7). Vertical decreases in ATP content have been reported by Karl et al. [32]. However, our results point to different patterns in the two size fractions, suggesting that changes in ATP content with depth may be the result of the balance between rates of growth (e.g., of nanoflagellates) and consumption (by predation on bacteria or catabolism [33]). Large variations in the cellular ratios of ATP have been observed [13] and this may represent another possible explanation of the inverse pattern of the two size fractions.

TSM concentrations in Terra Nova Bay and in the southwestern region of the Ross Sea were almost double those in the open Ross Sea. This may be related to allochthonous, even though limited, input of terrigenous material [10, 12] for Terra Nova Bay or to a phytoplankton bloom at stations 23 and 25 [9, 28]. Despite this, the characteristics of the particles in coastal and open sea stations were quite similar. In both areas the picoparticles accounted for $<38\%$ of total seston. Similar patterns were shown by proteins. Protein concentrations in Terra Nova Bay were significantly higher than in the open Ross Sea, but again in both areas microparticulate proteins accounted for 70 and 80% of the total protein concentration. Therefore, the two areas are quite different in terms of absolute concentrations of particulate matter but are not significantly different in terms of relative significance of the two size classes.

Other qualitative differences between coastal and open sea stations may be detected by comparing the ratio of proteins to carbohydrates (PRT:CHO ratio). The PRT:CHO ratio varied spatially. Due to the dominance of proteins in the large particles extensively colonized by bacteria and other microheterotrophs and fitoplankton, both in the Ross Sea and Terra Nova Bay, PRT:CHO ratios were higher in the microparticulate than in the picoparticulate fraction. At the same time, the average PRT:CHO ratio in Terra Nova Bay was much higher than in the Ross Sea. This suggests that, during summer 1989-1990, in Terra Nova Bay the freshly generated organic detritus was extensively colonized by bacteria and other microheterotrophs. This may have determined a protein enrichment and an increase of the PRT:CHO ratio. On the contrary, the particulate organic matter of the open Ross Sea was extremely poor in proteins (especially in the picoparticulate fraction) and appeared to be composed of older detritus scarcely colonized by bacteria.

Further investigations of particulate matter are required for a complete understanding of the processes involving the dynamics of the particulate matter in the Ross Sea. This first attempt indicates that an approach based on the biochemical composition of different size classes may provide new insight into the origin, composition and fate of POM in Antarctic waters.

Acknowledgements. We are grateful to Prof. Walker O. Smith for valuable discussions. This work was supported by the Italian National Programme for Antarctic Research (PNRA).

References

1. Fuhrman JA, Sleeter TD, Carlson CA, Proctor LM (1989) Dominance of bacterial biomass in the Sargasso Sea and its ecological implications. Mar Ecol Progr Ser 57:207-217
2. Cho BC, Azam F (1990) Biogeochemical significance of bacterial biomass in the ocean's euphotic zone. Mar Ecol Progr Ser 63:287-298
3. Navarro JM, Thompson RJ (1995) Seasonal fluctuations in the size spectra, biochemical composition and nutritive value of the seston available to a suspension-feeding bivalve in a subarctic environment. Mar Ecol Progr Ser 125:95-106
4. Dunbar RB, MacPerson AJ, Wefer G (1985) Water-column particulate flux and sea floor deposit in the Bransfield Strait and Southern Ross Sea, Antarctica. Antarct J, pp 98-102
5. Nelson DM, Smith WO Jr (1986) Phytoplankton bloom dynamics of the western Ross Sea ice-edge - II. Mesoscale cycling of nitrogen and silicon. Deep Sea Res 33:1389-1412
6. Smith WO, Nelson DM (1985) Phytoplankton bloom produced by a receding ice-edge in the Ross Sea: spatial coherence with the density field. Science 227:163-166
7. Nelson DM, DeMaster DJ, Dunbar RB, Smith WO Jr (1996) Cycling of organic carbon and biogenic silica in the Southern Ocean: estimates of water column and sedimentary fluxes on the Ross Sea continental shelf. J Geophys Res 101(C8):18519-18532
8. Smith WO Jr, Nelson DM, Di Tullio GR, Leventer R (1996) Temporal and spatial patterns in the Ross Sea: phytoplankton biomass, elemental composition, productivity and growth rates. J Geophy Res 101(C8):18455-18465
9. Fabiano M, Povero P, Danovaro R (1993) Distribution and composition of particulate organic matter in the Ross Sea (Antarctica). Pol Biol 13:525-533
10. Fabiano M, Danovaro R, Crisafi E, La Ferla R, Povero P, Acosta-Pomar L (1995) Particulate matter composition and bacterial distribution in Terra Nova Bay (Antarctica) during summer 1989-1990. Polar Biol 15:393-400
11. Fabiano M, Povero P, Danovaro R (1996) Particulate organic matter composition in Terra Nova Bay (Antarctica) during summer 1990. Ant Sc 8(1):7-13
12. Fabiano M, Chiantore M, Povero P, Cattaneo-Vietti R, Pusceddu A, Misic C, Albertelli G (1997) Short-term variations in particulate matter flux in Terra Nova Bay, Ross Sea. Ant Sc 9(2):143-149
13. Karl DM (1980) Cellular nucleotide measurements and applications in microbial ecology. Microb Rev 44:739-796
14. Artegiani A, Azzolini R, Paschini E, Creazzo S (1991) Physical oceanographic conditions in the Southern Pacific Ocean and in the Western Ross Sea. National Scientific Commission for Antarctica, Ocean Cruise 1989-1990. Data Report II:5-62
15. Fabiano M, Povero P, Catalano G, Benedetti F (1991) Hydrological data collected during the biological chemical and geological sampling in Terra Nova Bay. National Scientific Commission for Antarctica, Ocean Cruise 1989-1990. Data Report I:35-71
16. Innamorati M, Lazzara L, Mori G, Nuccio C, Saggiomo V (1991) Phytoplankton ecology. National Scientific Commission for Antarctica, Ocean Cruise 1989-1990. Data Report I:141-252
17. Strickland JDH, Parsons TR (1972) A practical handbook of seawater analysis. Bull Fish Res Board Can 167:1-158
18. Dubois M, Gilles KA, Hamilton JK, Rebers PA, Smith F (1956) Colorimetric method for determi-

nation of sugars and related substances. Anal Chem 28:350-356
19. Hartree EF (1972) Determination of proteins: a modification of the Lowry method that gives linear photometric response. Anal Biochem 48:422-427
20. Bligh EG, Dyer W (1959) A rapid method for total lipid extraction and purification. Can J Biochem Physiol 37:911-917
21. Marsh JB, Weinstein WJ (1959) A simple charring method for determination of lipids. J Lipid Res 7:574-576
22. Holm-Hansen O, Booth CR (1966) The measurement of adenosine triphosphate in the ocean and its ecological significance. Limnol Oceanogr 11:510-519
23. Bulleid NC (1978) An improved method for extraction of adenosine triphosphate from marine sediments and seawater. Limnol Oceanogr 23:174-178
24. Cota GF, Kottmeier ST, Robinson DH, Smith WO Jr, Sullivan CW (1990) Bacterioplankton in the marginal ice zone of the Weddel, Sea: biomass, production and metabolic activities during austral autumn. Deep Sea Res 37:1145-1167
25. Hewes CD, Sakshaug E, Reid FMH, Holm-Hansen O (1990) Microbial autotrophic and heterotrophic eucaryotes in Antarctic waters: relationships between biomass and chlorophyll, adenosine triphosphate and particulate organic carbon. Mar Ecol Progr Ser 63:27-35
26. Johnson PW, Sieburth J (1982) In situ morphology and occurrence of eucaryotic phototrophs of bacterial size in the picoplankton of estuarine and oceanic waters. J Phycol 18:318-327
Fichez R (1991) Suspended particulate organic matter in submarine cave. Mar Biol 108:167-174
27. Hecq JH, Brasseur P, Goffart A, Lacroix G, Guglielmo L (1993) Modelling approach of the planktonic vertical structure in deep austral ocean. The example of Ross Sea ecosystem. Proceedings of Annual Workshop on Belgian Oceanographic Research. Royal Academy of Belgium, pp 235- 250
28. Ittekot V, Degens ET, Brockmann U (1982) Monosaccharide composition of acid-hydrolizable carbohydrates in particulate matter during a plankton bloom. Limnol Oceanogr 27:770-776
29. Liebezeit G (1984) Particulate carbohydrates in relation to phytoplankton in the euphotic zone of the Bransfield Strait. Polar Biol 2:225-228
30. Taylor GT, Karl DM, Pace ML (1986) Impact of bacteria and zooflagellates on the composition of the sinking particles: an in situ experiment. Mar Ecol Progr Ser 29:141-155
31. Karl DM, Knauer GA, Martin JH (1988) Downward flux of particulate organic matter in the ocean: a particle decomposition paradox. Nature 332:438-441
32. Banse K (1990) New views on the degradation and deposition of organic particles as collected by sediment traps in the open sea. Deep-Sea Res 37:1177-1195

Meteorological Processes

On the Heat Energy Fluxes in the Non-stationary Surface Boundary Layer at Hells Gate, Terra Nova Bay (Antarctica)

S. Ferrarese[1], C. Cassardo[2], A. Longhetto[1], D. Bertoni[1], R. Forza[1], G. Ficca[3], M. Pangia[3] and R. Purini[3+4]

[1]Dipartimento di Fisica Generale, Università, Torino, Italy
[2]Dipartimento di Scienze e Tecnologie Avanzate, Università, Alessandria, Italy
[3]Istituto di Fisica dell'Atmosfera, Consiglio Nazionale delle Ricerche, Roma, Italy
[4]Istituto Talassografico, Consiglio Nazionale delle Ricerche, Trieste, Italy

Abstract

Within the framework of scientific activities of the Italian Antarctic Expedition at Terra Nova Bay (74° 41' 42"S, 164° 7' 23"E) during the southern-summer of 1994-95, an uninterrupted three week programme of measurement of sensible and latent turbulent heat fluxes has been run at two locations, 5 km apart from each other, of Hells Gate ice shelf (inner mast: 74° 50'8.3"S, 163° 44' 2.8E, 19.85 m a.s.l.; coastal mast: 74°52' 20.2"S, 163° 49' 6.4"E, 12.47 m a.s.l.). The aim of the programme was to assess the transfer of heat energy (in particular, its latent component) at the air/ice interface as a function of different atmospheric forcings (wind, temperature, solar radiation) in order to describe the state and evolution of the thermal properties of the non stationary atmospheric boundary layer near the surface. To this end, a prognostic enthalpy balance model, whose input data was the information collected during the experimental programme, has been used to establish the relative role played by the different terms concurring to the heat energy budget under different meteorological conditions, ranging from moderate breeze to strong katabatic wind episodes. The experiment involved the collaboration between teams from the Department of Physics, University of Torino and the Institute of Atmospheric Physics of National Research Council in Roma. This paper reviews the experimental and analytic methodologies and discusses the most meaningful results.

1 Introduction

The Antarctic continent is characterized by the presence of katabatic winds which are induced by strong radiative cooling of air masses on the internal plateau. These winds, channelled by the glacier valleys, finally affect some coastal oceanic areas, since they are the main factors responsible for the formation and persistence of the polynyas, that is, an opening in the pack ice. Therefore, the katabatic winds are important features both for Antarctic meteorology and for their impact on the marine environment [1,2]. In this context, during the Antarctic expedition 1994-1995 organized by the Italian National Programme for Antarctic Research (PNRA), the heat energy fluxes in the surface boundary layer in the terminal region of continental ice (in our case Hells Gate, an ice shelf about 20 km from the Italian Summer Station of Terra Nova Bay) were evaluated with two-fold aims: firstly, to investigate the transfer of sensible and latent heat energy at the air-ice surface boundary and the role it plays in determining the evolution of the atmospheric surface layer, so as to better understand the relationship between the dynamics of this last and katabatic winds; secondly, to provide useful information on the existing links between these events, offshore mesoscale cyclogenesis [3] and the polynya of Terra Nova Bay [4]. The measurements of sensible and latent heat fluxes over the Hells Gate ice sheet were obtained by using two masts 10 m high and 5 km apart, both equipped with fast and slow response sensors. Each mast was operated by two teams, namely the Atmospheric Group of the Department of General Physics, Turin University, Italy (GEOFIT), and the Institute of Atmospheric Physics of the Italian National Research Council, Rome, Italy (IFA-CNR). The experiments, the instrument characteristics and data validation are described in Section 2, while in Section 3 the direct evaluation method of turbulent heat fluxes is introduced. In Section 4 the heat energy balance is presented and, finally, in Section 5, the main results of the present study are summarized and discussed.

2 Experimental Set-up, Data Validation and Analysis

The instruments used for the Antarctic expeditions were installed on two masts, positioned at a distance of about 5 km from each other (Fig. 1) and aligned nearly perpendicularly to the continental ice-sea front. On the first mast, located near the continental ice front (coastal mast), the following instruments were placed at different heights: an ultrasonic Kaijo Denki anemometer-thermometer at 10 m, three Sitep thermometers, igrometers, and anemometers at 10, 5 and 2.5 m respectively, a Sitep solarimeter at 2 m, a Sitep barometer at 2 m and three Sitep thermistors buried in the ice at -0.5, -0.25 and 0 (surface) m. On the second mast (inner mast), the following instruments were placed at two heights: two ultrasonic Gill Solent anemometers at 10 and 5 m respectively and a fast Campbell kripton igrometer at 10 m.

The ultrasonic Kaijo Denki anemometer-thermometer recorded the three components of the wind velocity and sonic temperature (T_S) at 20 Hz, allowing the evaluation of turbulent sensible heat flux. The Sitep anemometers, based on solid-state sensors to guarantee the operativeness even in extreme meteorological conditions, sampled at 4 Hz, while the other instruments on the first mast sampled at 1 cycle/min. The two ultrasonic Gill anemometers sampled the three components of the wind speed and the velocity of sound at 21 Hz. The fast kripton igrometer sampled also at 21 Hz and was syncronized with the nearby Gill anemometer, installed at 10 m on the same mast. Table 1 summarizes the list of deployed instruments, their main characteristics and location.

For these instruments, the conversion of the velocity in an orthogonal coordinate system, the derivation of the sonic temperature from the velocity of sound and the correction of the tilt error have been made by a suitable software that uses a three-dimensional rotation of coordinates [5-7]. From these data, it is possible to calculate the turbulent sensible heat fluxes at 10 m on the first mast and at 10 and 5 m on the second mast. Using the fast igrometer of the second mast at 10 m, the turbulent latent heat flux for this site can also be evaluated. During the whole measurement period, from January 20 to February 8, 1995, the micrometeorological conditions of the two sites (the inner and the coastal ones) where our masts have been located looked homogeneous and coherent. Figures 2a and b show the air temperatures at 10 m at the locations of the coastal and inner masts respectively. Figures 3a and b show the horizontal wind speed at 10 m on the inner and coastal masts respectively, while Fig. 4a and b show the corresponding directions. From these last records, the air flow at Hells Gate turned out to be prevailingly offshore during the whole observation period. From the behaviour of these quantities, it has been possible to note two different meteorological conditions: the first, (January 20- February 3) characterized by moderate wind and air temperature of around -4°C, and the second (February 4-8) marked by the presence of katabatic winds combined with a decrease in both air and ice temperatures.

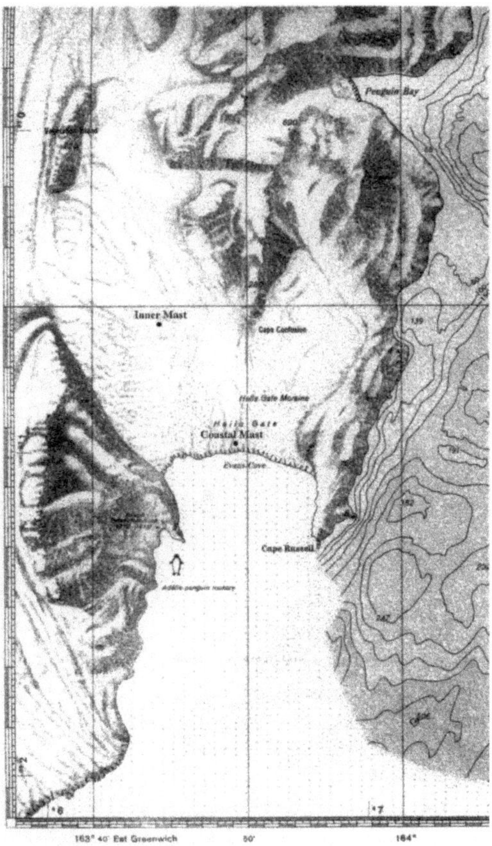

Fig. 1. Map of Hells Gate, showing locations of the two masts

Table 1. Deployed instruments, their main characteristics and location

Instrument	Manufacturer	Sampling frequency (Hz)	Sampling range	Location
Sonic anemometer	Gill Instruments	21	0.26-60 m s^{-1}	Inner mast, 10 m a.i.l.
Sonic anemometer	Gill Instruments	21	0.26-60 m s^{-1}	Inner mast, 5 m a.i.l.
Fast response krypton lamp igrometer	Campbell	21	2.00-19.50 g m^{-3}	Inner mast, 10 m a.i.l.
Sonic anemometer	Kaijo Denki	20	0.2-30 m s^{-1}	Coastal mast, 10 m a.i.l.
Slow response anemometer	Sitep	4	0.26-60 m s^{-1} 0-360 °C	Coastal mast, 10, 5 and 2.5 m a.i.l.
Slow response sensors of air temperature	Sitep	10^{-2}	- 30-70 °C	Coastal mast, 10, 5 and 2.5 m a.i.l.
Slow response sensors of ice temperature	Sitep	10^{-2}	- 50-50 °C	Coastal mast, –50 cm and –25 cm a.i.l., and on the surface
Slow response sensors of humidity	Sitep	10^{-2}	0-100%	Coastal mast, 10, 5 and 2.5 m a.i.l.
Solarimeter barometer	Sitep	10^{-2}	0-1500 W m^{-2} 800-1100 hPa	Coastal mast, 2 m a.i.l.

a.i.l., Above ice level

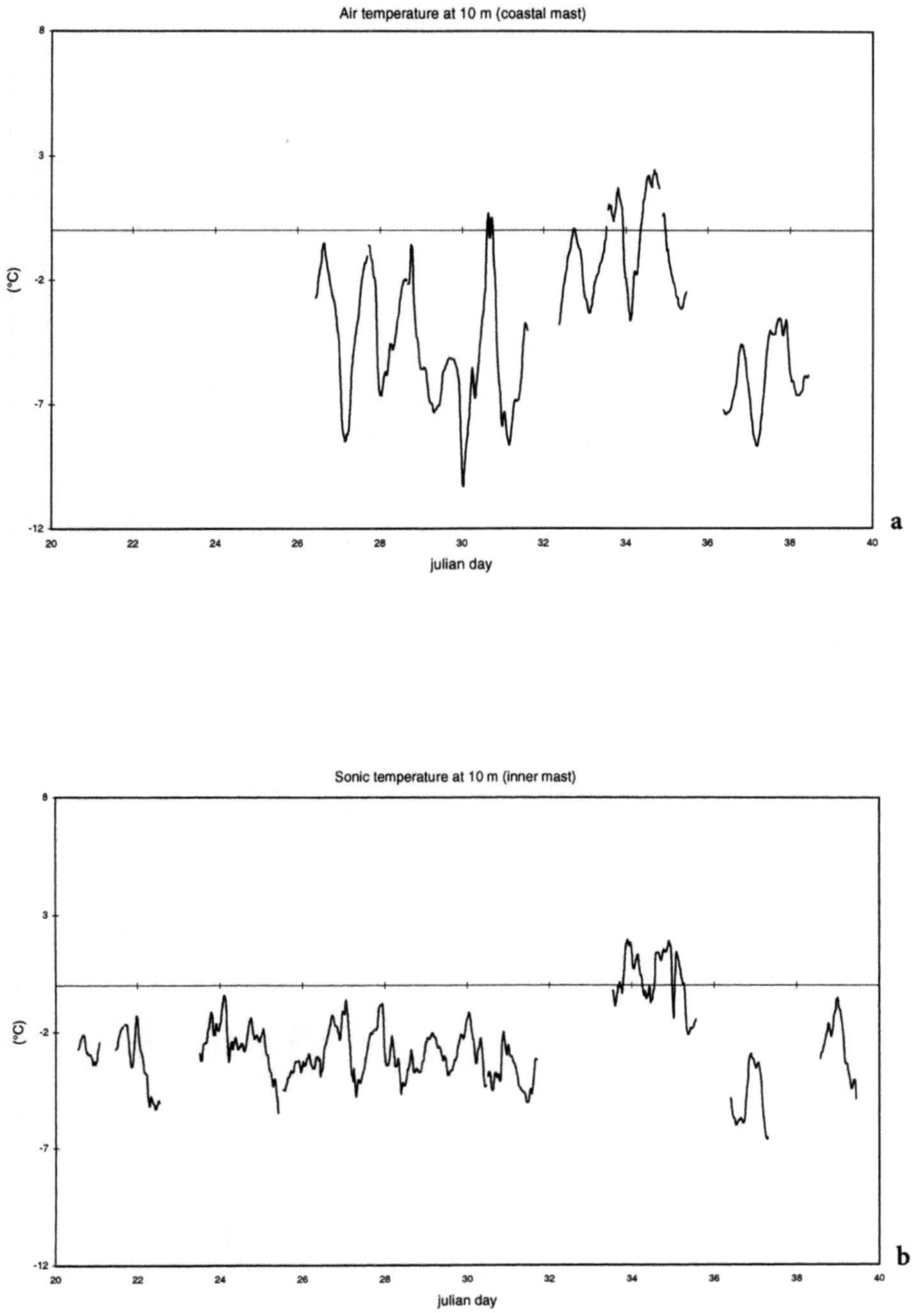

Fig. 2a,b. Air temperature measured at 10 m on the coastal mast (**a**) and on the inner mast (**b**)

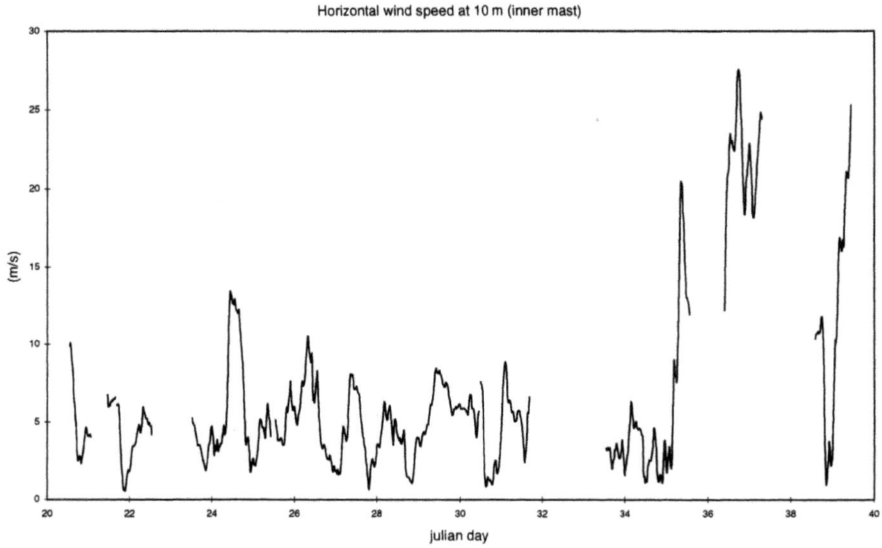

a

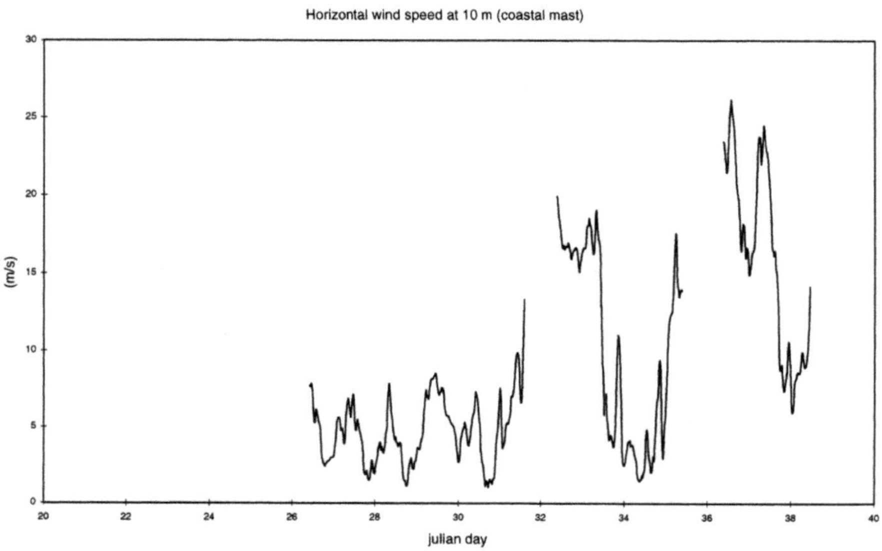

b

Fig. 3a,b. Horizontal wind speed measured at 10 m on the inner mast (**a**) and on the coastal mast (**b**)

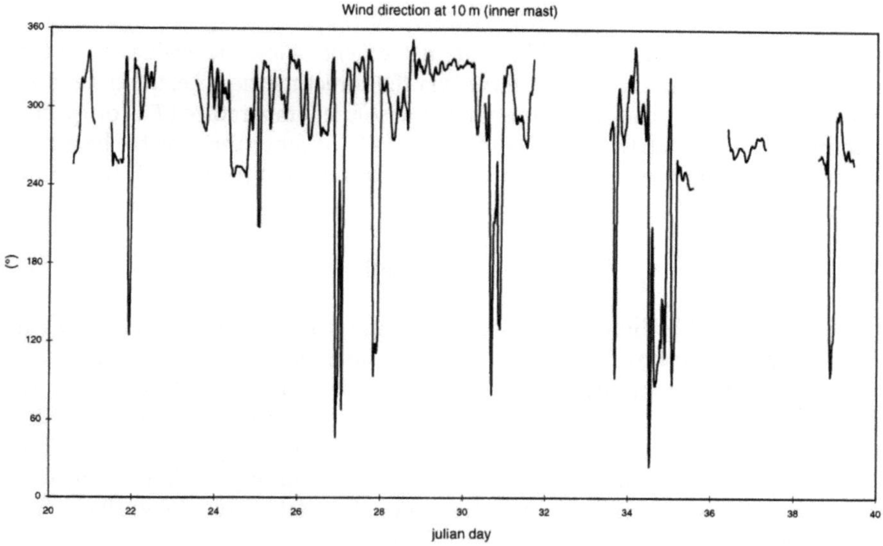

a

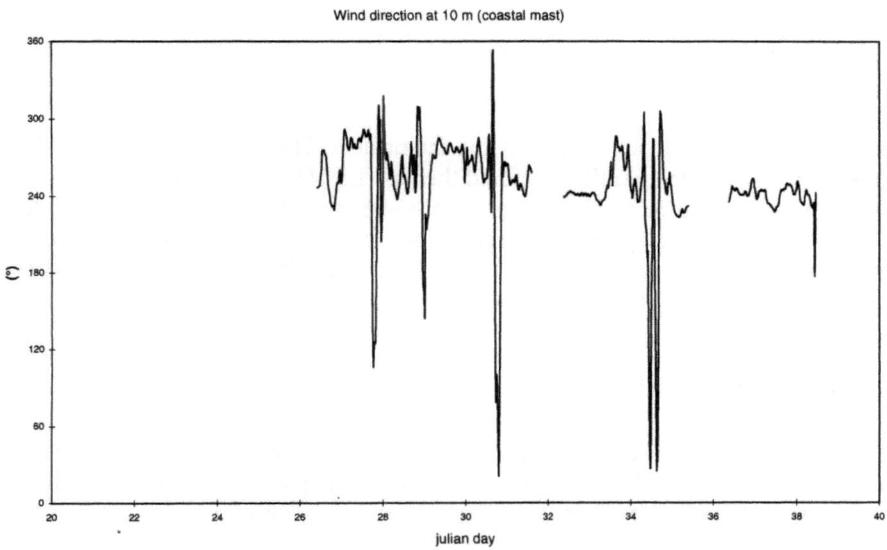

b

Fig. 4a,b. Wind direction measured at 10 m on the inner mast (**a**) and on the coastal mast (**b**)

3 Turbulent Sensible and Latent Heat Fluxes

The fast response ultrasonic and kripton instruments allow the evaluation of the turbulent sensible and latent heat fluxes via the direct method, i.e. the time integral of the cross products of fluctuations of absolute temperature (T') (or specific humidity q') with vertical wind component (w'). As far as the sensible heat flux is concerned, from the relationship between absolute temperature T and sonic temperature T_S [6, 7]:

$$T = T_S / (1+aq), \tag{1}$$

where a is a suitable dimensionless constant and q is the specific humidity (in kilograms per kilogram), the sonic temperature time series can be transformed in to the absolute temperature time series, from which the absolute temperature fluctuation T' can be obtained. As a matter of fact, in the present analysis, the cross products $\overline{w'T'}$ have been considered satisfactorily approximated by the terms $\overline{w'T'_s}$ [8], owing to the low levels of specific humidity during the experiment: $\underline{q} \equiv O(10^{-3} - 10^{-4})$ kg kg^{-1}. In fact, a simple calculation [6] shows that $\overline{w'T'_S}$ and $\overline{w'T'}$ are related by the approximate relationship:

$$\overline{w'T'_S} \cong \overline{w'T'} + 0.515\, T\, \overline{w'q'}. \tag{2}$$

By assuming a maximum value of 350 W m^{-2} for the latent heat flux ($\rho\, L_p\, \overline{w'q'}$), the difference between these two terms comes out as being in the order of 10%. Then, the latent (F) and sensible (H) heat fluxes have been expressed by the following relationships:

$$F = \rho L_p\, \overline{w'q'}; \tag{3a}$$
$$H = \rho c_p\, \overline{w'T'} \cong \rho c_p\, \overline{w'T'_{S}}, \tag{3b}$$

where r is the air density and L_p is the latent heat. The averaging time for the determination of the quantities defined through Eqs. (3a) and (3b) has been chosen as 30 min. In Fig. 5a and b the time trends of measured vertical turbulent fluxes of sensible heat at 5 and 10 m are represented, while Fig. 6 shows the time trend of the measured turbulent flux of latent heat at 10 m. Both fluxes refer to observations carried out on the inner mast.

4 Heat Energy Balance

In order to evaluate the enthalpy balance in lower layers of katabatic winds for the purpose of assessing the heat energy transfer at the air-ice surface boundary under non-stationary conditions, an atmospheric slab, 5 km long, oriented along the line linking the two masts and vertically limited by the horizontal surfaces at 10 and 5 m, has been selected. This allowed us to define a 'control volume' not connected to the ice surface, so that estimate of the flux in the ice was not necessary. In a previous paper, Heinemann [9] carried out a thorough investigation of the

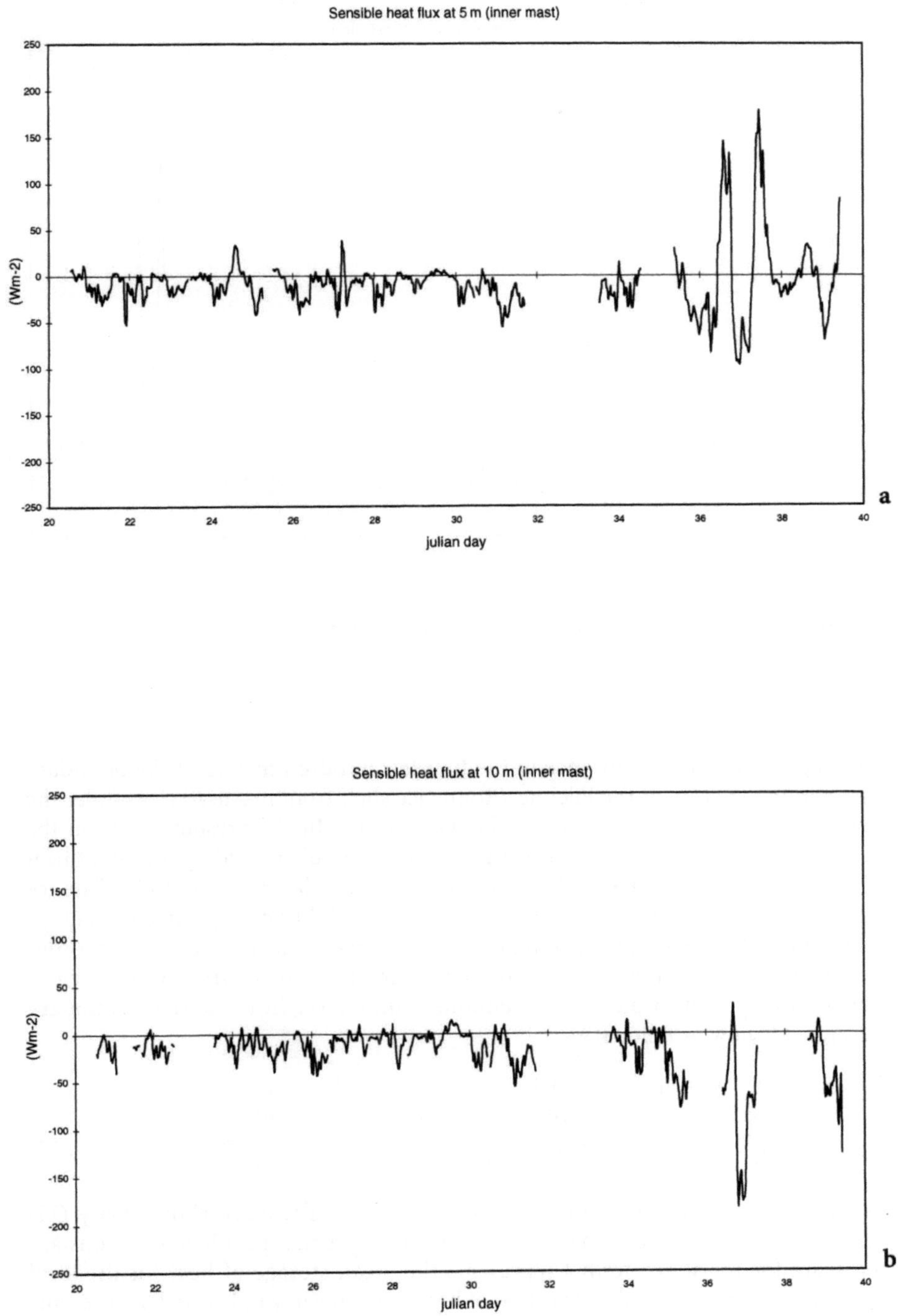

Fig. 5a,b. Sensible heat flux measured at 5 m (**a**) and at 10 m (**b**) on the inner mast

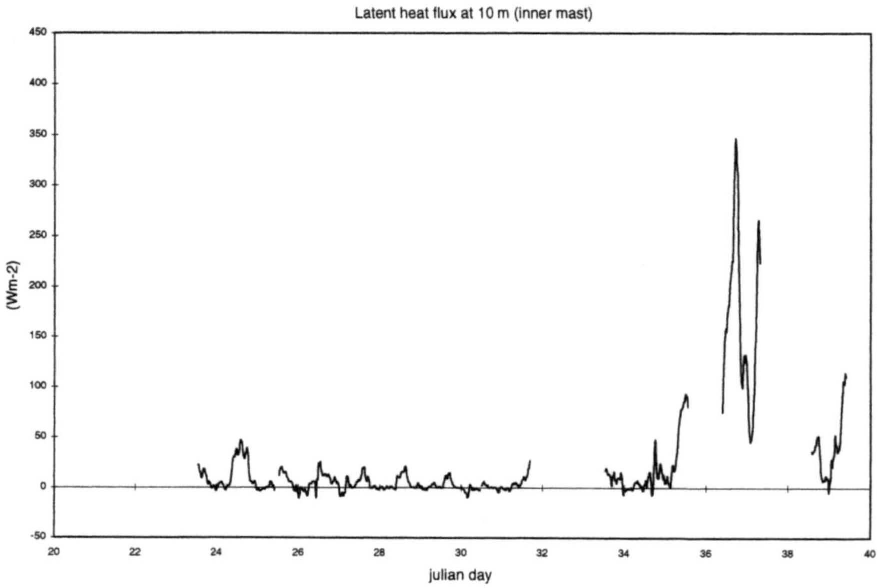

Fig. 6. Latent heat flux measured at 10 m on the inner mast

enthalpy budget in the absence in the first few hundred meters of the boundary layer, in the vicinity of the Filchner/Ronne ice shelf front in Antarctica, under the hypothesis of absence of phase exchange of water. In the present study, on the contrary, we are interested in investigating also the role played by the latent heat flux in the first few meters of the surface boundary layer on the Hells Gate ice sheet (even if some restraint on the present availability of experimental equipment did not allow us, for the moment, to come to definite conclusions). In order to also take the latent heat flux into account, the prognostic first law of thermodynamics, applied to the above-mentioned finite layer in the surface boundary layer, reads now as follows:

$$\underset{\text{I}}{\frac{\partial \overline{\vartheta}}{\partial t}} + \overline{u}_j \underset{\text{II}}{\frac{\partial \overline{\vartheta}}{\partial x_j}} = \kappa_\vartheta \underset{\text{III}}{\frac{\partial^2 \overline{\vartheta}}{\partial x_j^2}} + \underset{\text{IV}}{\frac{1}{\overline{\rho} c_p} \frac{\partial \overline{Q_j}}{\partial x_j}} - \underset{\text{V}}{\frac{L_p E}{\overline{\rho} c_p}} - \underset{\text{VI}}{\frac{\partial \overline{u'_j \vartheta'}}{\partial x_j}}, \tag{4}$$

where ϑ is the potential temperature, κ_ϑ is the molecular thermal diffusivity, $\overline{Q}$ is the net radiation, E is the phase change rate and c_p is the specific heat at constant pressure for moist air; term I represents the mean storage of heat (or the local term), term II describes the advection of heat by the mean wind, term III represents the mean molecular conduction of heat, term IV is the mean net body source associated with radiation divergence, term V is the body source term associated with latent heat release and term VI represents the divergence of turbulent heat flux.

All terms appearing in Eq. (4) can be computed from our measured physical quantities, except the vertical divergence of the turbulent latent heat flux, because, as mentioned above, the measurement of the latent heat flux at 5 m was not available. The vertical divergence of latent heat flux has then been computed as a residual of the other ones from the conservation Eq. (4). It is possible to assume [10] that: (1) the molecular conduction of heat $\kappa_\vartheta \frac{\partial^2 \overline{\vartheta}}{\partial x_j^2}$ is small compared with the turbulent conduction; and (2) the horizontal divergences of the turbulent $(\frac{\partial \overline{u'\vartheta'}}{\partial x}$ and $\frac{\partial \overline{v'\vartheta'}}{\partial y})$ and radiative $(\frac{\partial \overline{Q_1}}{\partial x}$ and $\frac{\partial \overline{Q_2}}{\partial y})$ fluxes are small compared with the vertical divergences. Finally, assuming a two-dimensional reference system with the wind vector projected into a component relative to the x axis aligned along the two masts and pointing offshore and its origin fixed at the inner one and the z axis upward, Eq. (4) becomes:

$$\rho c_p \frac{\partial \overline{\vartheta}}{\partial t} = - \frac{\partial H}{\partial z} + \frac{\partial \overline{Q}}{\partial z} - \rho c_p \overline{u} \frac{\partial \overline{\vartheta}}{\partial x} - \frac{\partial F}{\partial z}. \tag{5}$$

The local term $(\rho c_p \frac{\partial \overline{\vartheta}}{\partial t})$, the horizontal advection term $(\rho c_p \overline{u} \frac{\partial \overline{\vartheta}}{\partial x})$ and the vertical divergence terms $(\frac{\partial H}{\partial z}, \frac{\partial F}{\partial z})$ of turbulent fluxes [H and F, as defined in eqs. (2) and (3)] are approximated by the finite differences of the mean quantities measured (or computed) by the slow and fast response instruments, while the net radiative flux ($\overline{Q}$) is computed by the equation:

$$\overline{Q} = R_S (1 - \alpha_S) + R_{ld} - R_{lu}, \tag{6}$$

where R_S is the (global) short-wave radiation (from 0.3 to 1.1 μm), α_S is the albedo of the surface, R_{ld} is the downward long-wave atmospheric radiation and R_{lu} is the upward long-wave radiation from iced surface.

The global short-wave radiation R_S is measured by the Sitep solarimeter, while the other physical quantities in Eq. (6) are parametrized by some specific relationships [11-14] that, making use of the temperature and the humidity measured at 10 and 5 m by the other Sitep sensors, allow us to estimate the vertical divergence of the net radiative flux $\overline{Q}$. These relationships are: $R_{lu} = \varepsilon_s \sigma T_{sp}^4 + (1-\varepsilon_s) R_{ld}$ [11] and $R_{ld} = \varepsilon_{ac} \sigma T_a^4$ [12], where ε_s is the emissivity at the surface, σ is the constant of Bolzmann, T_{sp} is the skin temperature of ice surface, T_a is the air temperature and ε_{ac} is the atmospheric emissivity [$\varepsilon_{ac} = 0.67(e_a)^{0.08}$ [13], with $e_a = rh \cdot 6.112\exp[17.67(T_a - 273.15) / (T_a - 29.65)]$ [14], where e_a is the atmospheric vapour pressure and rh is the relative humidity].

The empirical equations we have used to estimate the vertical divergence of the radiative flux show two limitations; the first one is that they do not take the cloudiness into account; the second one is that they use the atmospheric humidity measured at about 5 km apart. Even if this is not important to the aim of the present work, because the vertical divergence of this flux is very small and the sky was

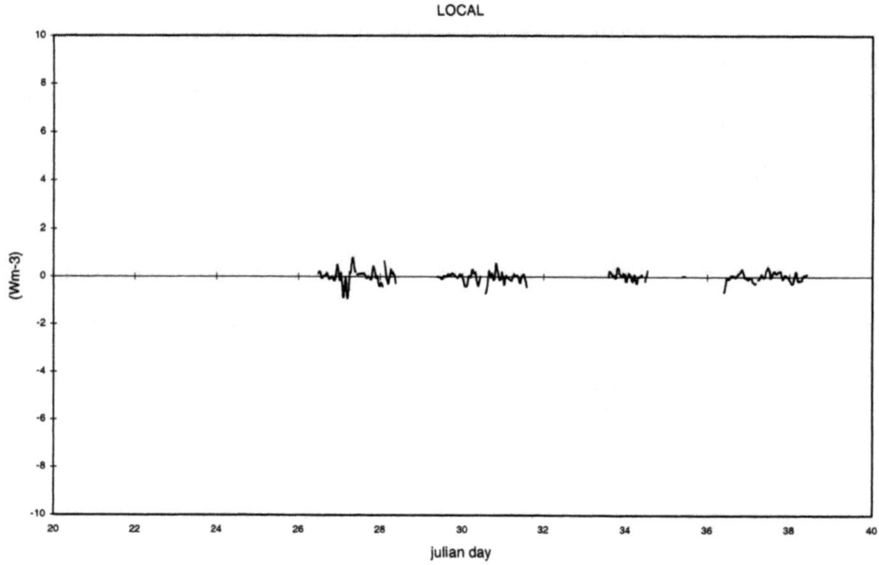

Fig. 7. LOCAL: local term

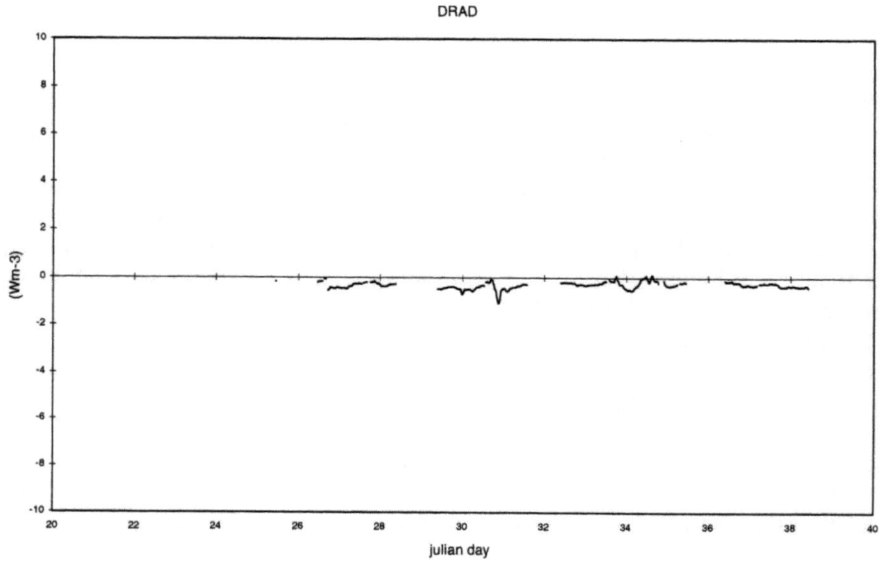

Fig. 8. DRAD vertical divergence of net radiation (inner mast)

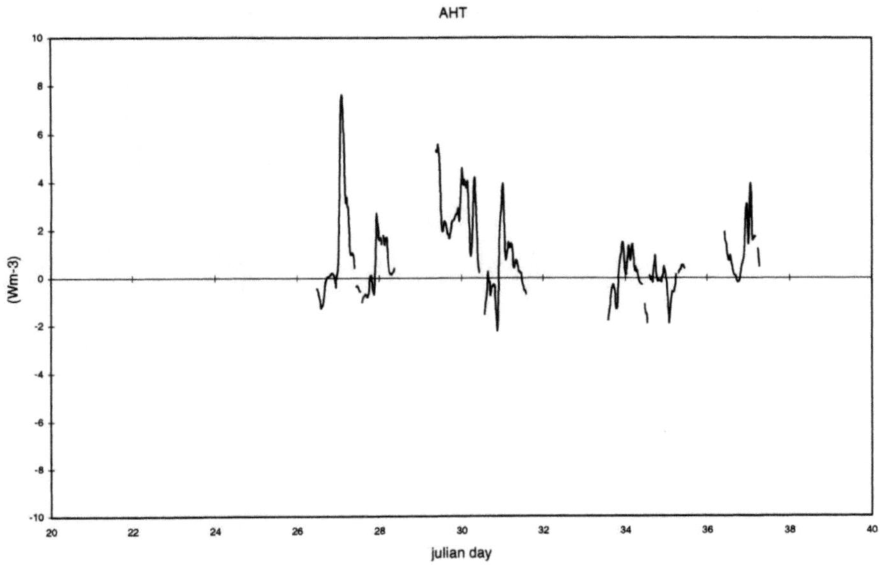

Fig. 9. AHT horizontal advection

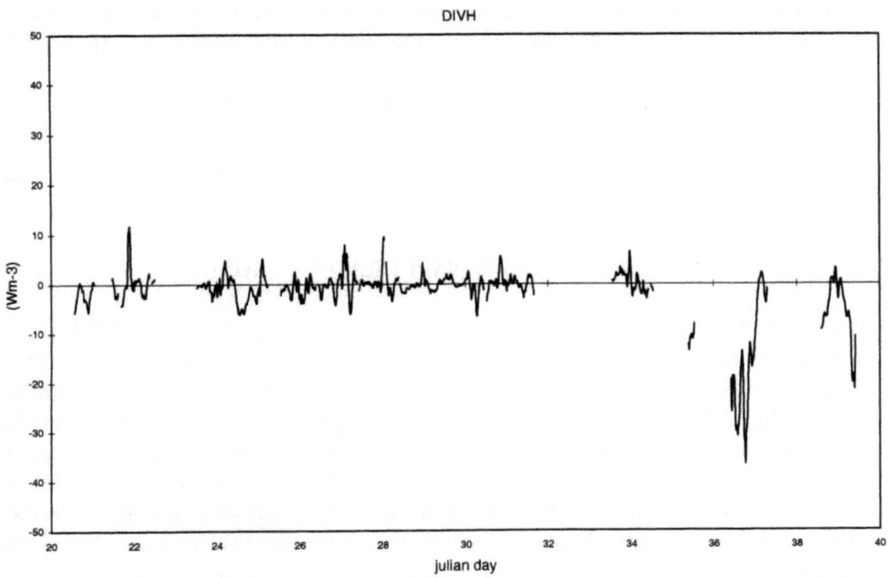

Fig. 10. DIVH vertical divergence of sensible heat flux (inner mast)

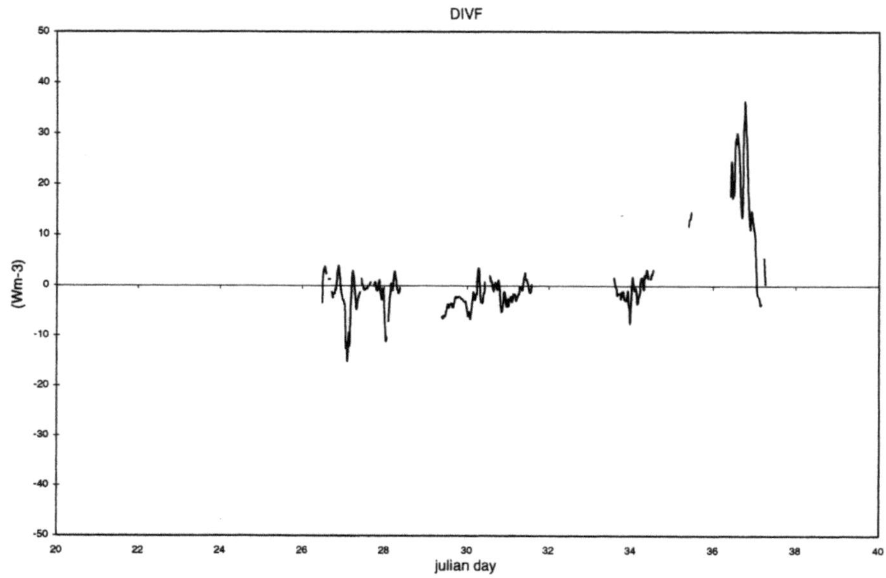

Fig. 11. DIVF vertical divergence of latent heat flux (inner mast)

cloudless during the observation period, nevertheless it must be recalled that for their validity, in general, it should be demonstrated that the horizontal difference of the radiative flux is smaller than its vertical difference over a vertical height of very few meters and under conditions of very low aerosol load and water vapour content.

With the above assumptions and approximations, Eq. (5) becomes:

$$\underset{\text{LOCAL}}{\underset{\downarrow}{\rho c_p \frac{\Delta \overline{\vartheta}}{\Delta t}}} = -\underset{\text{DIVF}}{\underset{\downarrow}{\frac{\Delta F}{\Delta z}}} - \underset{\text{DIVH}}{\underset{\downarrow}{\frac{\Delta H}{\Delta z}}} + \underset{\text{DRAD}}{\underset{\downarrow}{\frac{\Delta \overline{Q}}{\Delta z}}} - \underset{\text{AHT,}}{\underset{\downarrow}{\rho c_p u \frac{\Delta \overline{\vartheta}}{\Delta x}}} \tag{7}$$

where $\Delta F=F_{10}-F_5$, $\Delta H=H_{10}-H_5$, $\Delta\overline{Q}=\overline{Q}_{10}-\overline{Q}_5$, $\Delta\overline{\vartheta}=\overline{\vartheta}_{Coastal}-\overline{\vartheta}_{Inner}$, Δx=4754 m and Δz=5 m. Figures 7-11 show the time trend of the enthalpy balance components [LOCAL, DRAD, AHT, DIVH, DIVF respectively, the last one calculated from the balance Eq. (7)]. Starting from the measured components of the balance, it can be observed that, during the period of weak offshore wind, the principal contribution to the enthalpy budget is due to the turbulent and advection terms (DIVH and AHT). These terms, which are almost the same order and sometimes in balance, sometimes out of balance with each other, are one order of magnitude larger than the other two (LOCAL and DRAD). During this period of weak wind, the large values of AHT imply strong gradients of potential temperature J normal to the ice front, that could bring about strong forcing on the boundary layer along the fetch. In conclusion, according to Eq. (7), the magnitudes and signs of the divergence of turbulent heat flux and those of the horizontal advection term force the divergence of latent heat flux to take on a magnitude and a sign such as gives rise to only limited changes of the local temperature (LOCAL).

During the second period, characterized by the katabatic wind, the term of turbulent sensible heat flux becomes the dominant one and is almost one order of magnitude larger than the other terms. The DIVH term is large negative (up to about -36.5 W m^{-3}), showing a tendency to warm the layer. As the advection term is not able to offset it, so as to produce the limited local change of temperature observed during this period, Eq. (7) implies again large values of the divergence of the latent heat flux, with positive sign. Unfortunately, due to a breakdown of power supply, during the transitional period in which the katabatic wind took the place of weak winds and the average air temperature lowered from about 0 °C to about -6 °C degrees, the relevant data were lost, so that it has not been possible to analyze the relative contribution of each enthalpy budget component to the cooling of air. Our analysis had then to be limited to the 12-h episode of well-established katabatic wind, from 6 p.m. on day 36 to 6 a.m. on day 37. Figures 7-9 show that, during this episode, AHT was small positive (smaller than during the weak wind period), increasing from about 0 W m^{-3} to about 4 W m^{-3} in the first 7 and then decreasing again to about 1 W m^{-3} at the end of the period, LOCAL was small positive (about 0.2-0.4 W m^{-3}) in the first 4h and small negative (decreasing from about 0 W m^{-3} to about -0.3 W m^{-3}) in the last 8 h, while DRAD was almost zero all the time.

5 Conclusion

This chapter has described the experimental layout deployed during the tenth Antarctica Italian expedition to assess the heat energy budget within the non-stationary boundary layer of offshore winds blowing on the Hells Gate Ice Shelf, Terra Nova Bay, in the southern summer of 1994-1995. The temporal data series and the computed turbulent fluxes have been presented and an enthalpy budget including a possible role of the latent heat flux has been introduced. The results of this study are limited, at the present, by the availability of the measurement of the latent heat flux at only one level and by the remarkable difference of the sensible heat fluxes at 10 and 5 m.

The Gill sonic anemometers used in this experiment were calibrated by the manufacturer and calibrated again before and after the experiment by us, but, unfortunately, only in low wind conditions (<5 m s^{-1}).

Looking at the time series of the sensible heat fluxes provided by the two instruments, and at the time series of wind speed and latent heat flux, one can observe that their average difference is small when the wind speed is from 15 to 25 m s^{-1} (from day 37 to 37.30), moderate (~35 W m^{-2}) when the wind speed is from 10 to 20 m s^{-1} (from day 35.40 to 35.50) and high (~100 W m^{-2}) when the wind speed is from 20 to 30 m s^{-1} (from day 36.50 to 36.90).

This behaviour shows that the difference is not strictly a monotonic function of wind speed: in addition, when the difference is small, also the latent heat flux at 10 m is usually low, while when the difference is large, also the latent heat flux tends to increase. On the basis of the available information from this experiment, it seems that at least a qualitative suggestion of a role of latent heat can come from the results. Future experiments, with the measurement of all terms involved in the enthalpy balance, are needed to confirm or reject the above suggestion.

Acknowledgements. This research was supported by the Italian National Programme for Antarctic Research (PNRA).

References

1. Muench RD, CH Pease Polynya and leads: an overview of physical processes and environment. J Geophys Res 95(C6):9461-9479
2. Ficca G, Pangia M, Pierini S, Purini R, Sansone E (1996) A turbulent data analysis in the Antarctic boundary layer. Il Nuovo Cimento 19(C4):487-504
3. Hines KM, Bromwich DH, Parish TR (1995) A mesoscale modeling study of the atmospheric circulation of high southern latitudes. Mon Wea Rev 123:1146-1165
4. Bromwich DH (1991) Mesoscale cyclogenesis over the southwestern Ross Sea linked to strong katabatic winds. Mon W ae Rev 119:1736-1752
5. McMillen RT (1988) An eddy correlation technique with extended applicability to non-simple terrain. Boundary-Layer Met 43:231-245
6. Anfossi D, Cassardo C, Ferrero E, Longhetto A, Sacchetti D, Brusasca G, Colombo V, Marzorati A, Morselli MG, Rocchetti F, Tinarelli G (1993) On the analysis of ultrasonic anemometer data. Internal Report N; 282/93, Istituto di Cosmogeofisica, CNR, Torino
7. Cassardos C, Sacchetti D, Morselli MG, Anfossi D, Brusasca G, Longhetto A (1995) A study of the assessment of air temperature and sensible- and latent- heat fluxes from sonic-anemometer observation. Il Nuovo Cimento 18(C4):419-440
8. King JC (1990) Some measurements of turbulence over an Antarctic ice shelf. QJR Meteorol Soc 116:379-400
9. Heinemann G (1988) On the structure and energy budget of the boundary layer in the vicinity of the Filchner/Ronne Ice Shelf Front (Antarctica). Beitr Phys Atmos 244-258
10. Stull RB (1988) An introduction to boundary layer meteorology. Kluwer Academic Press, Dordrecht
11. Pielke RA (1984) Mesoscale meteorological modeling. Academic Press, London
12. Brutsaert W (1982) Evaporation into the atmosphere. D Reidel Publishing, Dordrecht
13. Stanley DO, Jurica GM (1972) Effective atmospheric emissivity under clear skies. J Appl Meteorol 11:349-355
14. Garratt JR (1992) The atmospheric boundary layer. Cambridge University Press, Cambrige

Meteorological Conditions During Snowfall at Terra Nova Bay (Antarctica)

A. Pellegrini[1], A.M. Della Vedova[2], P. Grigioni[1] and L. De Silvestri[1]

[1] ENEA AMB-GEM-CLIM, Via Anguillarese 301, [2] Via M. Bocassini 9, Roma, Italy

Abstract

Changes in the precipitation rate, resulting in a snow cover on sea ice, are not easily monitored by direct observations. Gauge determinations of snowfall are strongly contamineatd by wind transport of previously deposited snow. Glaciological accumulation values are a potentially important source of precipitation data, but the requirements for broad areal coverage, the lack of annual time resolution in many regions and the uncertain impact of wind reworking are problematic. In this paper a method to identify episodes of snowfall over North Victoria Land, and particularly over Terra Nova Bay, is described.

1 Introduction

Polar regions play a critical role in global change since major interactions between the atmosphere, ice, oceans and biota affect the entire global system through feedback, biogeochemical cycles, deep ocean circulation, atmospheric transport of energy and pollutants and changes in ice mass balance. Recent studies on the simulated climatic change due to the doubling of atmospheric CO_2 show the high altitudes to be one of the most sensitive regions for changes in variables such as temperature, precipitation, snow and sea-ice coverage [1]. In particular, the accumulation of snowfall on the Antarctic continent is one of the major factors determining the response of the ice sheet to climatic change. The Antarctic ice-sheet net accumulation and the total mass losses are each about 2000 km^3 $year^{-1}$, equivalent to about 6 mm $year^{-1}$ of sea level change [2]. Consequently, a significant change in the major components of the Antarctic mass balance could influence the current rate of sea-level rise up to 1.5 mm $year^{-1}$ [3]. Despite all available measurements of snow accumulation, ice velocities, basal melting and iceberg discharge, it is still not known for certain whether the ice sheet is growing or shrinking. The uncertainty in the estimate of the total mass balance is at least 20%, equivalent to a global sea-level change of about 1 mm $year^{-1}$ [4].

Understanding the response of present icecaps to climatic fluctuations depends on an effective means of modelling mass balance. Ablation can be reasonably well approximated from surface energy-balance calculations, but changes in snowfall, the primary mass-gain mechanism, are poorly understood. Besides, the snow can have a significant influence on seasonal variation of sea ice [5] and, consequently, on ocean-atmosphere energy exchange and temperature. Ledley [6] pointed out that the mechanisms causing this influence are: (1) snow acts as an insulator, keeping the sea ice relatively warm and thin during the winter and thus easier to melt away; (2) snow has a lower volumetric specific heat than the ice, causing it to cool and warm more rapidly than ice; and (3) snow generally has a higher albedo than sea ice, reflecting more solar radiation and thus producing thicker sea ice. As a result, the addition of a constant precipitation rate resulting in a snow cover on sea ice produces a cooling effect in the polar regions.

However, such changes are not easily monitored by direct observations. Gauge determinations of snowfall are strongly contaminated by wind transport of previously deposited snow. In addition, snowfall amounts at high polar elevations are often too small for confident measurement. Glaciological accumulation values are a potentially important source of precipitation data, but the requirements for broad areal coverage, the lack of annual time resolution in many regions and the uncertain impact of wind reworking are problematic [7].

In the following sections a method to identify episodes of snowfall over North Victoria Land, and particularly over Terra Nova Bay, is described. In Section 2 some characteristics of the Antarctic continent are presented. This is followed in Section 3 by a depiction of the space-time distribution of relative humidity in the studied area. Section 4 presents the method used to analyse episodes of precipitation. In Sections 5 and 6 some cases of mesoscale cyclogenesis and blowing/drifting snow are analysed. In Section 7 the agreement between the direct observations of precipitation over Terra Nova Bay area and the episodes of probably snowfall obtained by analysing the main meteorological parameters recorded at the different stations is explored. Section 8 summarises some conclusive considerations.

2 Study Area

Antarctica is essentially an ice-covered mountain surrounded by ocean with an average elevation of 2500 m and with very low moisture content. Zonal mean precipitable water at 80°S is only 1.5 kg m^{-2} in the annual average according to Peixoto and Oort [8]. The Antarctic Plateau is one of the two largest deserts of the world, the other one being the Sahara. Almost the entire South Polar Desert area comprises East Antarctic, while West Antarctica gets somewhat more precipitation. However, research on precipitation is hampered by the low rate of occurrence and the interference from wind-blown surface drift. Besides this, wherever a sufficient amount of snow is available, strong winds can drastically modify the surface, changing it from a smooth plane to a field of sastrugi, or compiling and moving snow dunes, increasing the surface roughness considerably [9].

Figure 1 shows the Victoria Land area, identifying points of interest and showing automatic weather station (AWS) sites. In this area, the Terra Nova Bay (TNB) Station (Italy) and McMurdo Station (USA) are located and both Antarctic Programmes deployed AWSs in the region: squares mark US ones, circles the Italian ones.

The meteorological measurements at these stations (including the manned sta-

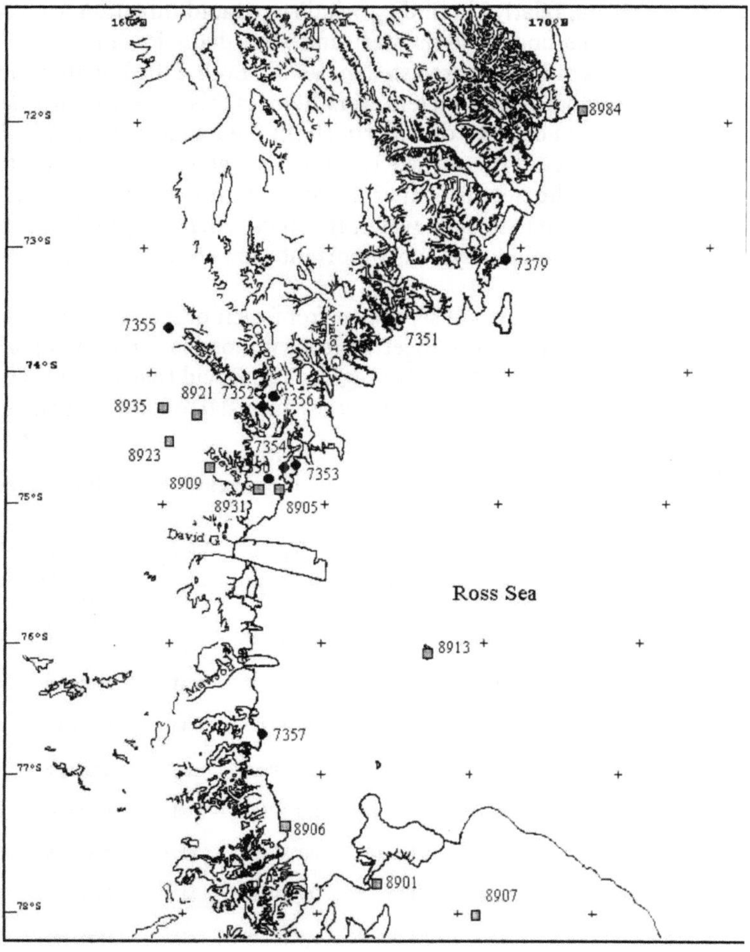

Fig. 1. Map of North Victoria Land and sites where automatic weather stations (AWS) are located. *Squares* mark US Programme AWSs; *circles* identify Italian ones

tion at TNB and McMurdo) do not support estimates of precipitation. It is possible to use other meteorological approaches to evaluate the precipitation rate. One technique is the atmospheric moisture balance where the precipitation is found as the residual from the budgeting of the fluxes of water vapour into and out of an atmospheric volume [10]. Another approach is to sum precipitation amounts calculated from estimates of synoptic-scale vertical motion [11]. Another different approach to inferring precipitation amounts and their variations over the polar ice sheets is to measure the amount of snow accumulated over several years, so meteorological and glaciological analyses have to be integrated [12].

The following analysis does not estimate precipitation amount, but is only intended to detect precipitation episodes. The main meteorological parameters such as temperature, relative humidity, wind speed and direction are used. The site names, locations and analysis start and stop dates for all the AWS units deployed in the area of interest are listed in Table 1. Even if the Italian stations give a record of meteorological parameters every hour, in this work 3-hourly data were used to give a uniform data set considering both the Italian and the US AWSs.

In order to understand the surface wind pattern at each individual station, it is essential to know the details of their location with respect to the ice cap, ice shelves, adjacent mountains, open water, etc. In fact, the topography of the environs of Antarctic stations plays an important part in determining prevailing surface winds [13].

Transantarctic mountains are separated from each other by glaciers descending from the East Antarctic Ice Sheet or from minor glacier basins [14]. Many large glaciers cross the Terra Nova Bay region. The David Glacier comes from the eastern slope of the Dome C and converges in a narrow outlet glacier which

Table 1. Site, location and dates of start and stop of operation for AWSs deployed in the area of interest

Argos ID number	Name	Location	Latitude (°S)	Longitude (°E)	Altitude (m)	Start date (yymmddhh)	End Date (yymmddhh)
8984	**Possession Island**	**Ocean Islands**	**71°54'00"**	**171°07'48"**	**30**	**92122900**	**94123123**
8906	**Marble Point**	**Ross Island Region**	**77°25'48"**	**163°45'00"**	**120**	**87010100**	**94123123**
8907	**Ferrell**	**Ross Island Region**	**78°01'12"**	**170°48'00"**	**45**	**87010100**	**94123123**
8901	**Willie Field**	**Ross Island Region**	**77°51'00"**	**167°04'48"**	**40**	**92012500**	**94123123**
8913	**Whitlock**	**Ocean Island**	**76°14'24"**	**168°42'00"**	**275**	**87010100**	**94123123**
8905	**Manuela**	**Inexpressible Island**	**74°52'48"**	**163°36'00"**	**80**	**87010100**	**94123123**
8909	**Shiristi**	**Reeves Glacier**	**74°42'00"**	**161°36'00"**	**1200**	**87122800**	**92093023**
8921	**Sushila**	**Reeves Névé**	**74°18'00"**	**161°18'00"**	**1430**	**88012000**	**91093023**
8923	**Sandra**	**Reeves Névé**	**74°29'24"**	**160°29'24"**	**1525**	**88011900**	**94123123**
8935	**Lynn**	**Reeves Névé**	**74°13'48"**	**160°22'12"**	**1772**	**88011900**	**94123123**
8931	**Pat**	**Nansen Ice Sheet**	**74°52'48"**	**163°00'00"**	**30**	**89010100**	**91010123**
7350	**Sofia**	**Nansen Ice Sheet**	**74°47'45"**	**163°18'46"**	**40**	**87020100**	**96063023**
7351	**Alessandra**	**Cape King**	**73°35'10"**	**166°37'16"**	**160**	**87020100**	**96063023**
7352	**Zoraida**	**Priestley Glacier**	**74°15'14"**	**163°08'41"**	**640**	**87020100**	**96063023**
7353	**Eneide**	**Terra Nova Bay**	**74°41'42"**	**164°05'36"**	**90**	**87020100**	**96063023**
7354	**Chiara**	**Enigma Lake**	**74°41'42"**	**164°07'23"**	**210**	**93010800**	**96063023**
7355	**Modesta**	**Priestley Névé**	**73°38'18"**	**160°38'35"**	**1900**	**89020100**	**96063023**
7356	**Lola**	**Tourmaline Plateau**	**74°10'59"**	**163°29'00"**	**1700**	**90010300**	**96063023**
7357	**Arelis**	**Cape Ross**	**76°43'56"**	**163°00'35"**	**150**	**90012200**	**96063023**
7379	**Silvia**	**Cape Philips**	**73°04'21"**	**169°36'55"**	**550**	**90120100**	**96063023**

extends from west to east into the sea with the Drygalski Ice Tongue. The Reeves Névé, which is fed by the East Antarctic Ice Sheet, and the Priestley Névé, which is nourished by the Talos Dome, flow into the Reeves and Priestley Glaciers, respectively. These two glaciers end in the Nansen Ice Sheet, an ice shelf which is pinned to Inexpressible Island and to Northern Foothills. Other main glaciers, such as the Aviator, the Campbell and the Mawson, and numerous tributary glaciers flow into the main outlet glaciers.

Ice accumulation in central East Antarctica has very low values, resulting in less than 2.5 cm/year^{-1} in depth. Due to orographic lifting of moist air masses flowing southward toward the edge of the ice sheet, and to decreasing distance from the moisture source, accumulation rates increase toward the coast. Another factor determining accumulation is the air temperature; moisture-carrying capacity of air diminishes strongly as its temperature decreases. So we can expect that when air temperatures increase snowfall will also increase, although a quantitative relationship has not been determined [15].

3 Humidity of Air Over North Victoria Land

At a given point, the snow surface can accumulate mass by precipitation, condensation, deposition and wind transport; on the other hand, snow can be removed by evaporation, sublimation and wind transport. Considering all these factors, up to the present the problem of finding some way to record the amount of snowfall has not been solved. Schwerdtfeger [9] reports that it has sometimes been observed that a gage was empty after an intensive snowfall and strong winds, whereas plenty of wind-driven snow had entered the collector on a day without any real precipitation. However, significant progress has recently been made in techniques to directly and indirectly determine Antarctic precipitation amounts.

In this chapter an approach is proposed to identify episodes of precipitation over North Victoria Land, in particular the Terra Nova Bay area.

Precipitation characteristics at the coast and in the high interior are very different. Coastal snowfall is episodic in association with synoptic-scale features such as cyclones and fronts [10]. In the interior, the portion of precipitation associated with clouds decreases, and the semicontinuous fall of ice crystals from cloud-free skies becomes progressively more important as elevation increases. In fact, in Antarctica the cloud-forming processes are weakly developed and the role of the air mass clouds in precipitation is small. Consequently, one can infer that the entire snow accumulation in Antarctica, estimated to be 2480 km^3 water year^{-1}, is due to this advection of water vapour from the oceanic regions. Thanks to this advection of water vapour, 61.7% of the world's reserves of freshwater are concentrated in Antarctica [16]. So information on humidity is indispensable for a study of the laws of moisture transport in the atmosphere, the processes of cloud and precipitation formation and heat balance [17].

Considering the monthly and annual averages of relative humidity at all available stations, some general features can be evidenced:

- The monthly trend of relative humidity, showing minimum values during the winter season and a maximum during summer, is similar to the monthly trend of temperature. Rusin [18] observed that weakening of the flow and the increase of evaporation from the snow surface in summer associated with the increase in air temperature are conducive to an increase in relative humidity, while the development of winter flow causes its decrease.

- The station that reports minimum monthly relative humidity values is 7352, where winter values are around 30%, and in summer mean monthly values of about 60% are recorded.
- Station 7379 results are subject to moister air masses than the others. In fact, at this site a mean monthly relative humidity value of about 70% is always recorded.
- Considering the annual averages, almost all stations have shown a descending trend of relative humidity in the past few years.

An analysis of the spatial distribution of the mean relative humidity in 1987-1995 reveals that throughout the year it is characterised by a distinctly expressed zonality and large gradients close to the coast. This conforms to the general predominance of zonality in the distribution of atmospheric processes and the temperature regime in the region in question.

4 General Analyses

Bromwich [10] showed that the intensity of cyclonic activity is the key factor governing the amount of precipitation and its variations. Besides this, precipitation generation in coastal regions is strongly influenced by the fact that moist maritime air masses, poleward moving, are deflected by the steep marginal ice slopes to blow parallel to the terrain contours. On Victoria Land, intrusions of moist air far into the continent are accompanied by southerly winds through a deep tropospheric layer.

Station 7352 is located on the Priestley Glacier and it is often subjected to katabatic winds, blowing from north-northwest to northwest (Fig. 2). Katabatic winds are due to the cooling of air on the slopes of the ice-plateau and to the consequent acceleration of this air down the slope under the action of gravity. Local orographic conditions and, particularly, the shape of the surface of the slope have a considerable influence on the strength of katabatic winds [19]. These winds originate in the interior of the continent and are prevalently dry; therefore when at station 7352 a high relative humidity is recorded, it can be argued either that air masses have a different origin or that blowing/drifting snow phenomena occur. From these considerations, the relative humidity recorded at station 7352 was analysed first.

Considering the whole time series of 3-hourly data from February 1987 to June 1996, we identified the cases in which station 7352 reported relative humidity (RH) values greater than or equal to 70%. Figure 3 shows the number of consecutive days with relative humidity greater than or equal to 70% at station 7352. The occurrence of 3-hourly consecutive data ranges from 2 to 52, i.e. from 0.25 to 6.5 consecutive days. An arbitrary selection of cases lasting a period of time greater than or equal to 2 days was done as being representative of well-established phenomena: these cases occurred one time during 1988, 1990 and 1994, two times in 1993, four times in 1989 and 1992 and five times during 1991. Considering differ-

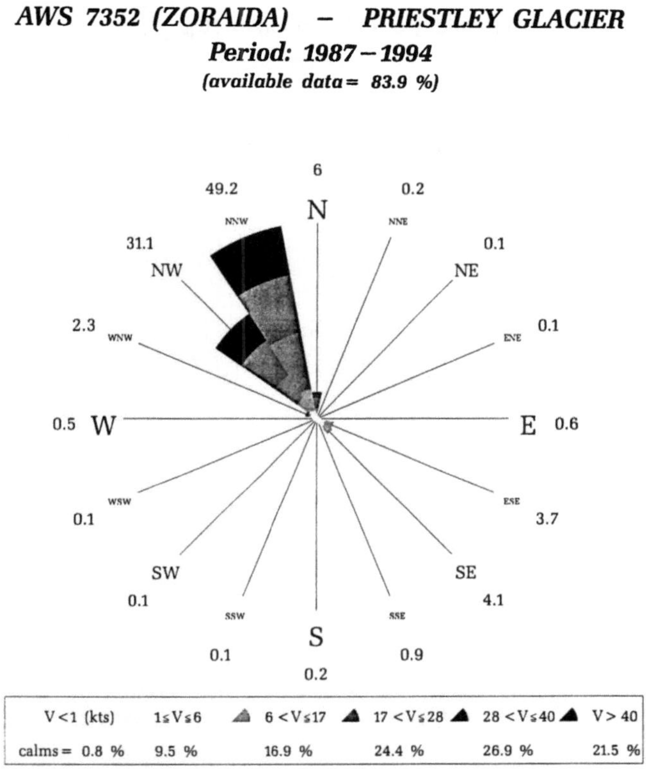

Fig. 2. Wind rose of station 7352

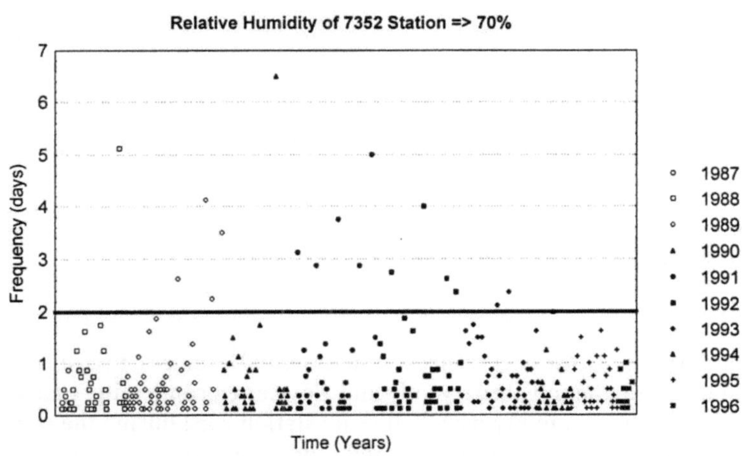

Fig. 3. Cases in which station 7352 recorded a relative humidity => 70% from 1987 to 1996

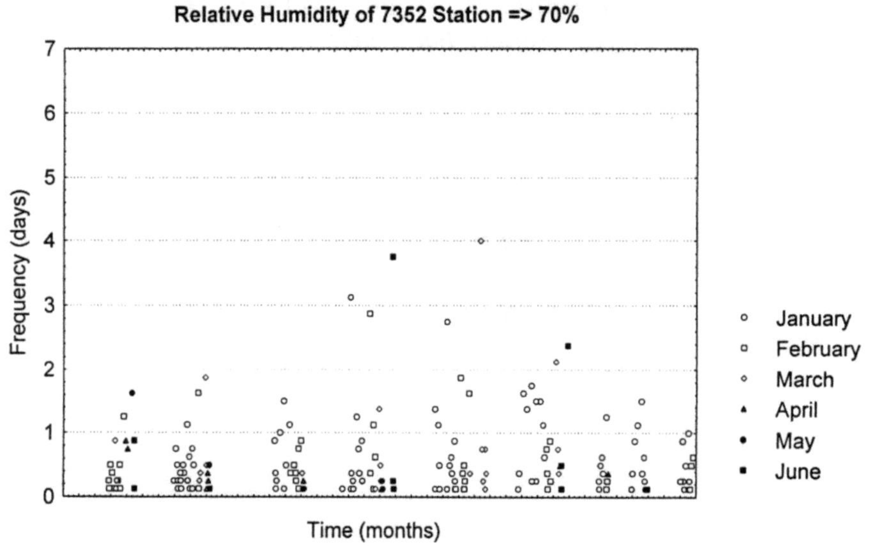

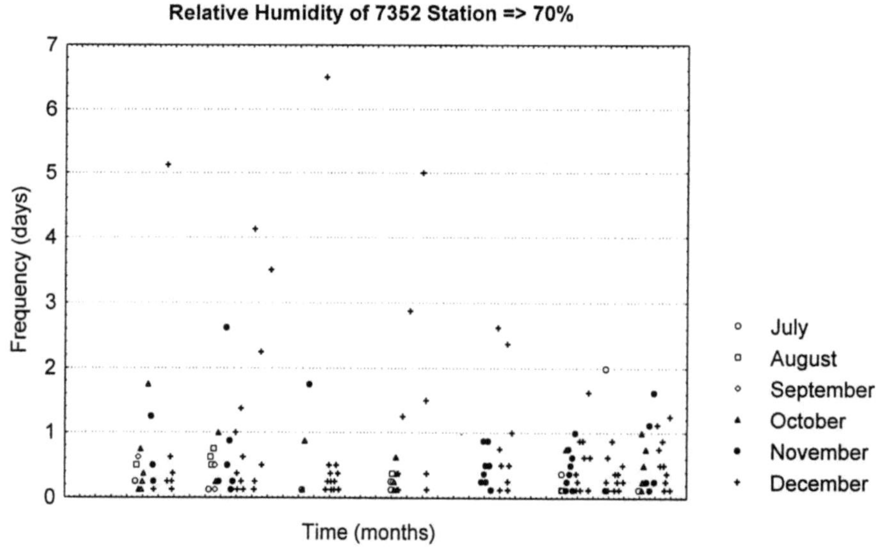

Fig. 4. Cases in which station 7352 recorded a relative humidity => 70% aggregated by month for the first (*above*) and second (*below*) half-year

ent months (Fig. 4), the occurrence is nine cases in December, two in January, June and March and one in February, July and November, confirming that air masses with high relative humidity usually flow on station 7352 during the summer season. To analyse in detail all these periods, air masses blowing from sea or from land were identified by considering wind direction.

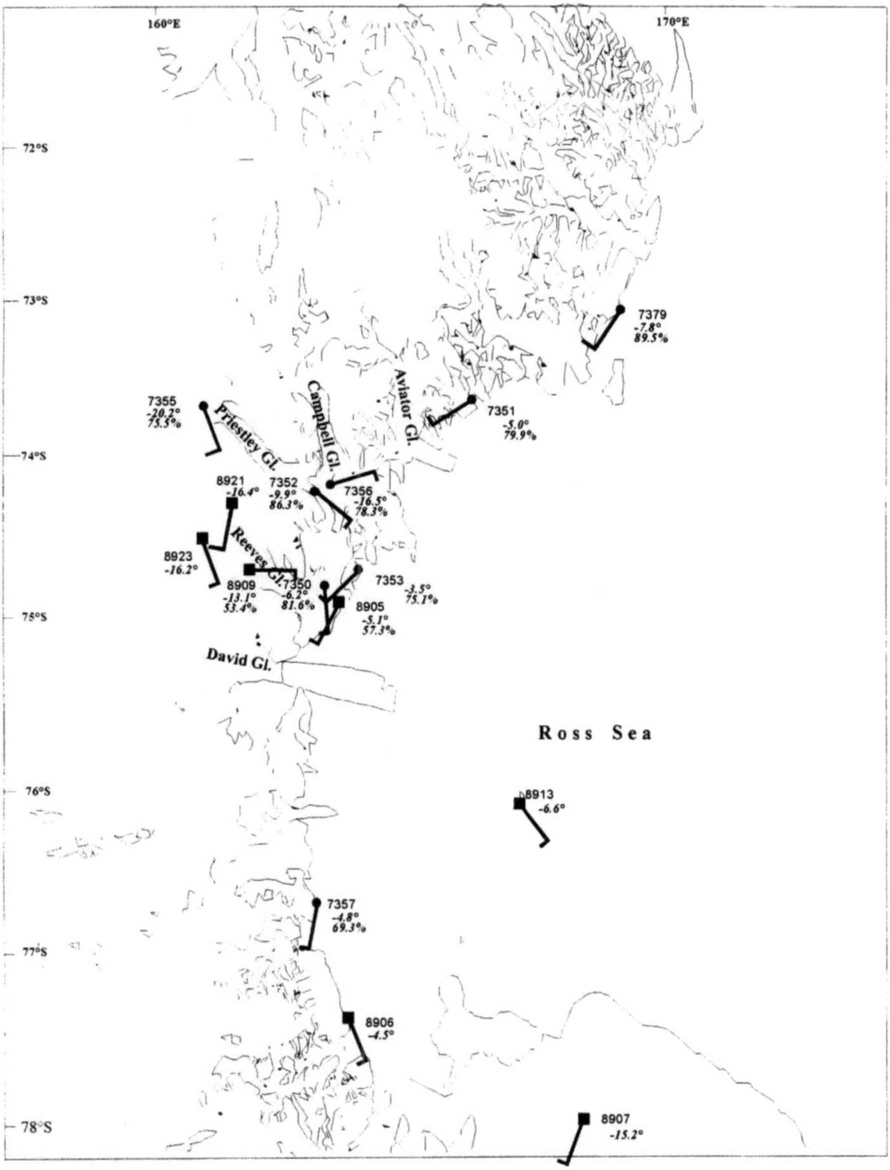

Fig. 5a. Ground level observations at available AWSs for a typical situation with marine flow, during which station 7352 recorded RH=> 70%

Again taking into consideration station 7352, the cases corresponding to flows blowing from the sea were found to have an RH ≧ 70% and wind direction DIR between 45° and 225°. This condition was verified for a maximum of 39 3-hourly consecutive observations. Winds were considered as blowing from the land when

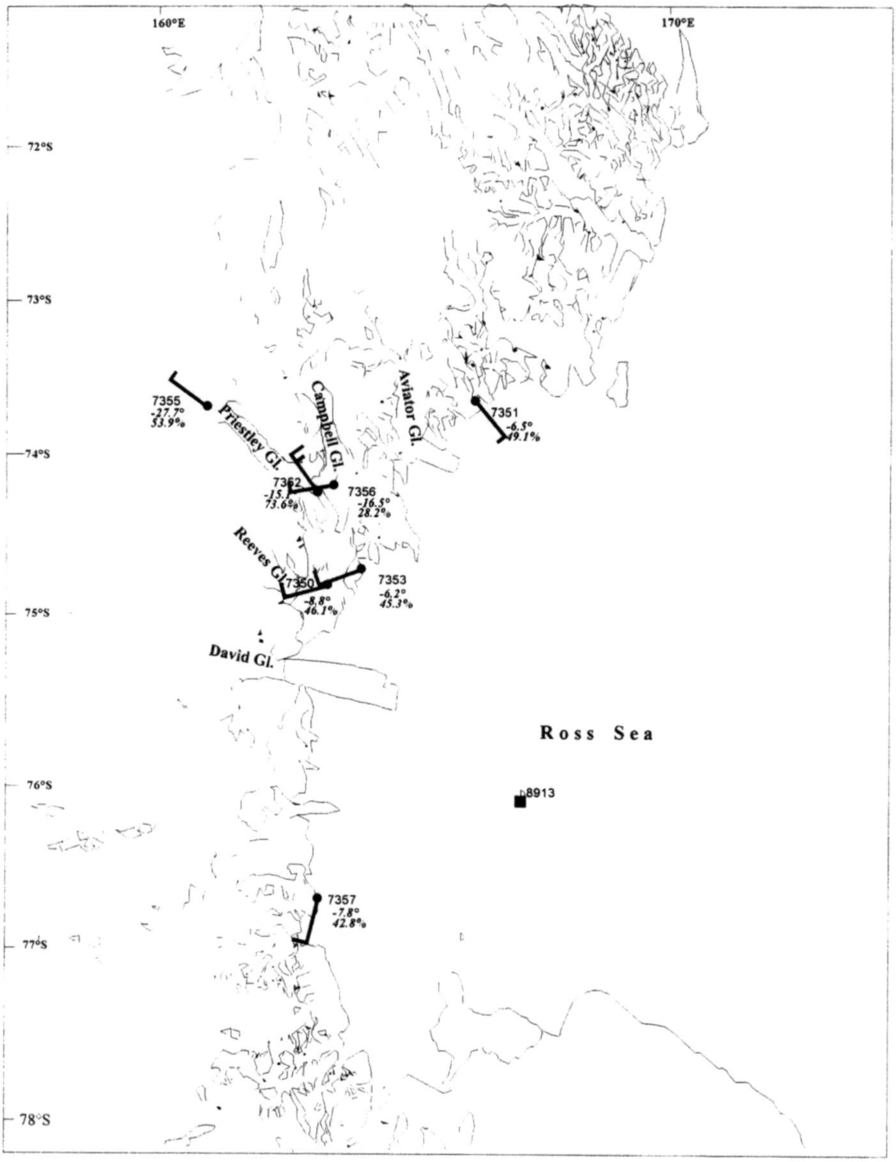

Fig. 5b. Ground level observations at available AWSs for a typical situation with land flow, during which station 7352 recorded RH=> 70%

$0° \leqq DIR \leqq 45°$ or $225° < DIR < 360°$: coupling this condition with an RH $\geqq$ 70%, from 2 to 24 3-hourly consecutive observations were found to occur.

Recalling that both conditions required relative humidity at station 7352 higher than 70%, we assumed that when wind direction indicates air blowing from the sea, precipitation is likely to occur, while in the case of air coming from the continent high humidity values can be due to drifting/blowing snow within the air flow. Two typical situations of marine and continental flow (Fig. 5a, b) were analysed: winds at surface level in Fig. 5a could lead to the hypothesis of cyclogenesis over the south-western Ross Sea, while the situation reported in Fig. 5b can be related to blowing or drifting snow, as indicated by the high wind speeds and by the low relative humidity recorded at stations other than 7352.

5 Cases of Mesoscale Cyclogenesis

When polar air masses move over relatively warm water, the fluxes of sensible and latent heat from the ocean surface can warm a deep layer of the atmosphere [20]. If the heating rate is not too great, baroclinic instability is the primary mechanism for formation and development of mesoscale cyclones [20, 21]. On the other hand, diabatic heating helps to reduce the static stability of the near-surface atmosphere, and mesoscale cyclones can grow faster and with a shorter wavelength [22].

The East Antarctic Plateau is located to the west of Ross Sea, and thus the prevailing westerly upper level flow [23] descends about 2000 m from the high Plateau toward the ocean. In a homogeneous incompressible fluid with constant density the potential vorticity is conserved. Carrasco and Bromwich [20] applied this concept to the Victoria Land/Ross Sea area, observing that a column of air over the Plateau expands vertically when it goes down onto the Ross Sea. This increase in depth implies a creation of negative (cyclonic) relative vorticity. Therefore the air column acquires cyclonic circulation over the western Ross Sea, offshore from Victoria Land.

The synoptic situation related to this phenomenon could be analysed by means of the Ertel potential vorticity maps [24]. Observing the maps corresponding to episodes of intrusion of moist air mass from the sea toward the continent presents a cell of negative isentropic potential vorticity over the region of North Victoria Land; instead when katabatic winds occur, an area with large values of potential vorticity develops offshore and over Victoria Land, and the contours tend to bunch together at the edge of the region. These air masses result in extremely dry conditions and conditions move at very high speed.

In addition to the broad-scale analysis, some case studies of mesoscale cyclogenesis over the Ross Sea presented in current literature have been considered, together with concurrent analysis of our data. Carrasco and Bromwich [20] examined four mesoscale cyclones that formed near Terra Nova Bay between 16 and 20 February 1988, describing their evolution thus:

- At 12:00 UTC, on 16 February a mesoscale vortex was located next to Inexpressible Island. Then the authors observed that the weak south-westerly winds recorded at AWS 7351 between 15:00 and 21:00 UTC on 16 February may signal the passage of the vortex along the Victoria Land coast.

- On 17 February the broad-scale cyclonic regime continues, affecting the southwestern Ross Sea. At 03:00 UTC another mesoscale cyclone developed over Drygalski Ice Tongue (see Fig. 7 from [20] and Fig. 10 from [25]). This cyclonic circulation moved warmer maritime air toward Terra Nova Bay as revealed by the 5 °C increase in air temperature at Inexpressible Island.
- By 06:00 UTC on 17 February a similar cyclone formed over the Plateau to the west of Terra Nova Bay. As Bromwich and Parish [25] underlined, the Plateau cyclone developed in a strong baroclinic zone between warm maritime air advected onto the Plateau by the coastal cyclone and cold katabatic air blowing southward along the mountains. The coastal cyclone propagated northward along the Victoria Land coast at 18.5 km h^{-1} and caused high wind speeds at Cape King for the 6-h period centred at 13:00 UTC on 17 February.
- By 00:00 UTC on 18 February the katabatic wind speed increased in the Inexpressible Island region and another cyclone had suddenly formed near Drygalski Ice Tongue.
- By 09:00 UTC on 18 February the katabatic speed at AWS 8905 had declined to 22 m s^{-1} and the mesocyclone had started to move slowly eastward. Then the katabatic wind speeds remained at around 20 m s^{-1} for the next 9 h, and by 18:00 UTC on 18 February the mesoscale cyclone had passed out of the station array [25]. This is confirmed by the fact that the last period relating to February 1988 ends at 12:00 UTC on 18 February.

Analysing the relative humidity at station 7352, the following consecutive observations with RH 70% can be individualised:

- From 06:00 on 16 Feb to 15:00 on 16 Feb
- At 21:00 on 16 Feb
- At 03:00 on 17 Feb
- From 09:00 on 17 Feb to 12:00 on 18 Feb

Another period, coincident with one selected with the proposed method, occurred from 17 to 23 January 1989 and was analysed in [26]. This paper analysed the possible interactions of the summer katabatic flow over Terra Nova Bay with the moving and developing Ross Sea cyclone.

Using pressure values, winds and the nephanalysis shows that a low with a developing frontal system was moving eastwards offshore from Adelie Land, between the 17 and 18 January 1989. Entering the Ross Sea on the 18th, this front deepened the semipermanent cyclone during the 19th. On the 22nd the pressure gradient was reinforced on the southern part of the Victoria Land coast. During the following 24-40 h the low moved northward into the middle of the Ross Sea, weakening [26].

Figures 5-8 in [26] show the mesoscale analyses, from which the following most important phenomena can be observed: (1) the upslope winds with precipitation along the Terra Nova Bay glaciers; (2) the moderate barrier winds in the McMurdo region and up to Franklin Island; and the katabatic flow restored over the Nansen Ice Sheet and temporarily at Franklin Island

The snowfall occurring on 17-18 January 1989 has been identified considering the RH 70% at station 7352; in fact, from all cases the following ones can be singled out:

- From 00:00 on 17 Jan to 06:00 on 17 Jan;
- From 12:00 on 17 Jan to 12:00 on 18 Jan;
- From 18:00 on 18 Jan to 21:00 on 18 Jan;
- From 03:00 on 19 Jan to 15:00 on 19 Jan;
- At 18:00 on 20 Jan.

The resulting situation is similar to that presented in Fig. 5a.

6 Cases of Blowing or Drifting Snow

It is important to separate the amount of snowfall and the flux rate of blowing snow. To date there are no data of annual precipitation at considered stations, because it is difficult to measure only the amount of snowfall under conditions of strong wind. Besides, also the changes of the snow surface level measured by the snow stakes method do not represent the precipitation, and even a visual observation of precipitation is often impossible in heavy drifting snow.

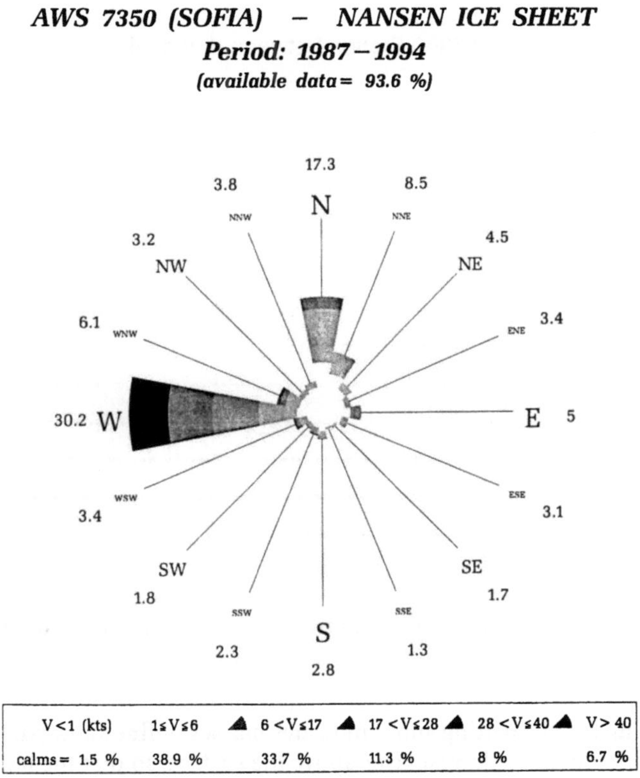

Fig. 6. Wind rose of Station 7350

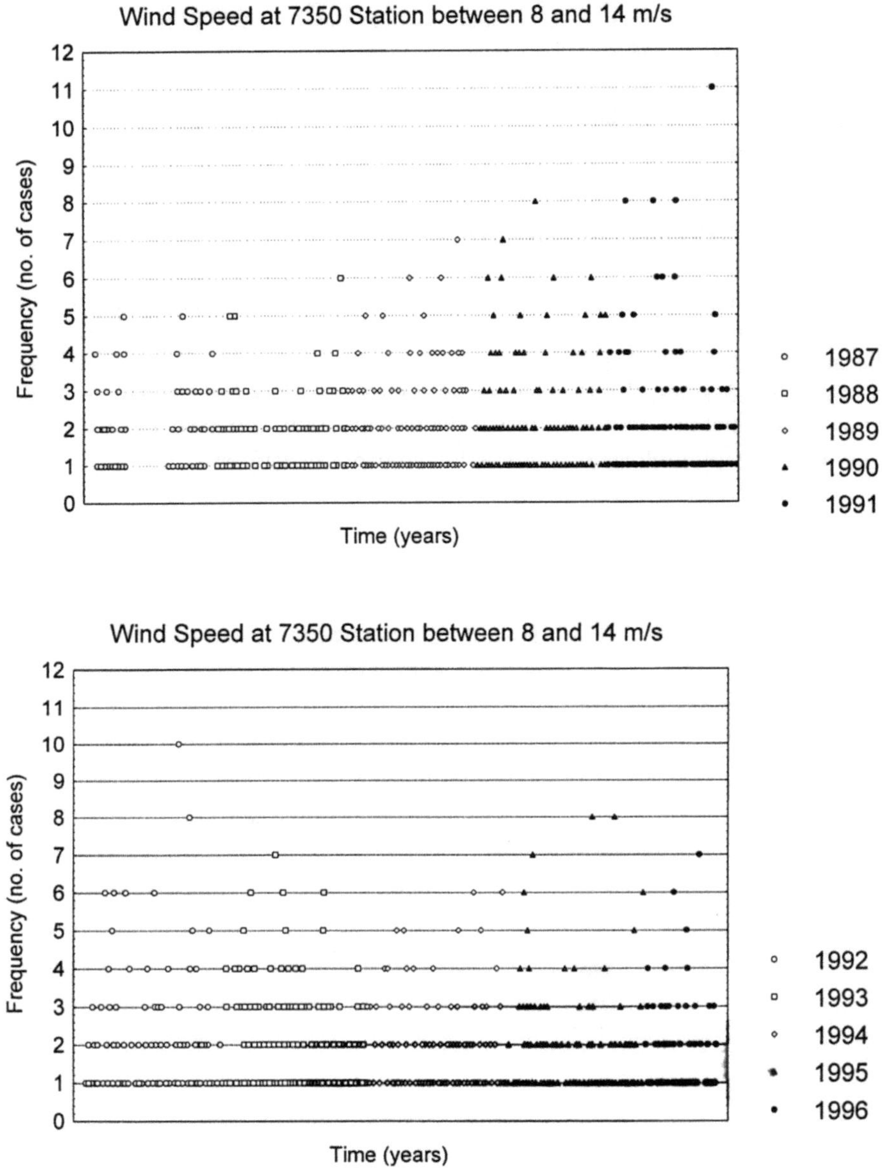

Fig. 7. Frequency of occurrence of wind speed between 8 and 14 m s^{-1} at AWS 7350

According to [9], 'drifting snow' indicates masses of fine snow particles carried by the wind only in the lowest layer above the surface, so that the horizontal visibility at the height of the observer's head is not impaired. When the wind velocity increases above a threshold, which depends on the nature of the snow surface

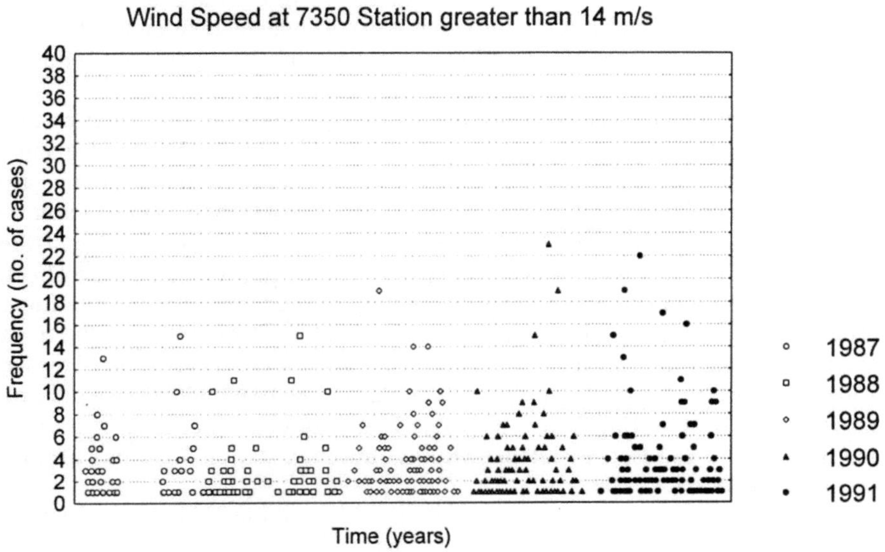

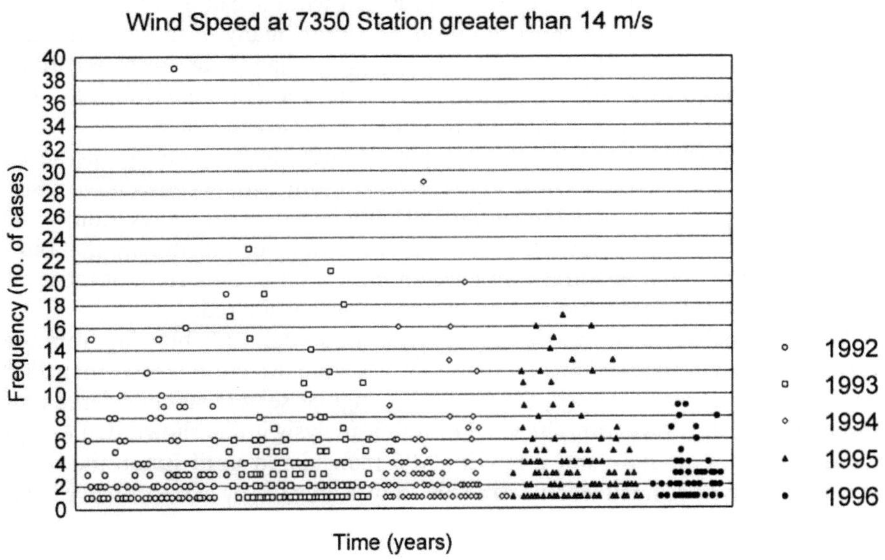

Fig. 8. Frequency of occurrence of wind speed greater than 14 m s^{-1} at AWS 7350

and perhaps other factors such as temperature, the snow drift begins to go into a state of suspension by turbulent diffusion. Threshold velocities have been quoted by various authors as generally falling within the range from 8 to 16 m s^{-1}. At larger velocities the layer of the atmosphere with drift-snow takes on the appearance of dense fog, and snow plumes can rise hundreds of metres even over level ground [27]. 'Blowing snow' indicates [9] a much more vehement type of wind-driven snow. Under extreme conditions the horizontal visibility can be strongly reduced, to less than 10 m.

For the area of Victoria Land, not having measurements of precipitation and least of all of drifting or blowing snow, the winds blowing on Nansen Ice Sheet recorded by station 7350 were considered. Speeds ranging from 8 to 14 m s^{-1} are assumed to be related to drifting snow and speeds greater than 14 m s^{-1} to blowing snow episodes. Station 7350 has been considered as representative of the whole area, since it is located at the central-northern part of Nansen Ice Sheet; therefore it is often subject to flows originating in the interior of the continent and blowing onto Nansen Ice Sheet from the west along Reeves Glacier and from the north along the Priestley Glacier (Fig. 6). This is a conservative hypothesis: if drifting or blowing snow episodes take place at station 7350, they would be more probably occurring at Inexpressible Island and on other sites of Nansen Ice Sheet more exposed to intense winds. However, station 7350 has a more complete set of data.

The number of consecutive observations in which station 7350 reported a wind speed between 8 and 14 m s^{-1} is shown in Fig. 7, while Fig. 8 reports consecutive episodes with wind speed greater than 14 m s^{-1}.

As can be seen from Fig. 7, episodes in which wind speed is high enough to cause drifting snow have a short duration, as a great many of these do not last more than 6 h, and only nine cases lasting more than 1 day were recorded. Typically, these episodes of drifting snow have a duration ranging from 1 to 11 3-hourly consecutive data (less than 3 h to 33 h).

The cases in which wind speed is compatible with blowing snow (wind speed greater than 14 m s^{-1}; Fig. 8) have an occurrence from 1 to 39 consecutive cases. The longest episode occurred in winter 1992 and lasted about 5 days, from 03:00 on 12 August to 21:00 16 August. It was characterised by dry air masses that blow with a speed of about 90 km h^{-1}. To have an indication of the origin of these air masses, also in this case the direction of winds was considered.

To individualise the air masses affect invest the station 7350 blowing from the Plateau and having a speed greater than 14 m s^{-1}, wind direction greater than 255° and less than 45° was analysed; the remaining angular sector was considered representative of the flows from the sea. From this analysis it was shown that except for one episode in 1987 and a few in 1996 (considered until June), the other episodes always occurred when Terra Nova Bay is affected by extraordinary katabatic winds.

7 Visual Observations of Precipitation

It has been pointed out [28] that prevailing conditions on the coasts of Victoria Land are alternatively determined by: East to south-easterly moist and relatively warm air flow of oceanic origin; and westerly dry and cold winds blowing down to the coasts from the Plateau. The weather associated with the former is typically cloudy with relatively important snowfall along the coasts and weak to moderate easterly or southerly winds, while the latter brings moderate to strong westerlies with little or no cloudiness. The development of logistic and research activities at the Antarctic bases requires reliable meteorological information, so real time data as well as weather forecasts up to several days in advance are decisive for outside working and living conditions.

The analyses presented in this work can be key inputs to evaluations of snowfall episodes even if they do not provide an indication of precipitation amount. To have an indication of reliability of the proposed methodology, daily meteorological information delivered by the Weather Office at Terra Nova Bay Station has been considered. This is available for summer only, as the Station is not manned during winter, and it includes also visual observations of weather and notes relative to precipitation and cloudiness.

The meteorological observations for the following summer periods were considered:

- 26 December 1988 - 16 February 1989
- 10 November 1989 - 15 February 1990
- 13 November 1990 - 16 February 1991
- 16 December 1991 - 24 January 1992

The days during which the snowfall has been explicitly indicated have been singled out and then these days and/or periods have been compared with those resulting from the described analysis. The results are shown in Fig. 9, where cumulative occurrences of verified and not-verified precipitation are plotted: as the duration of the phenomenon increases, the agreement results appears very good.

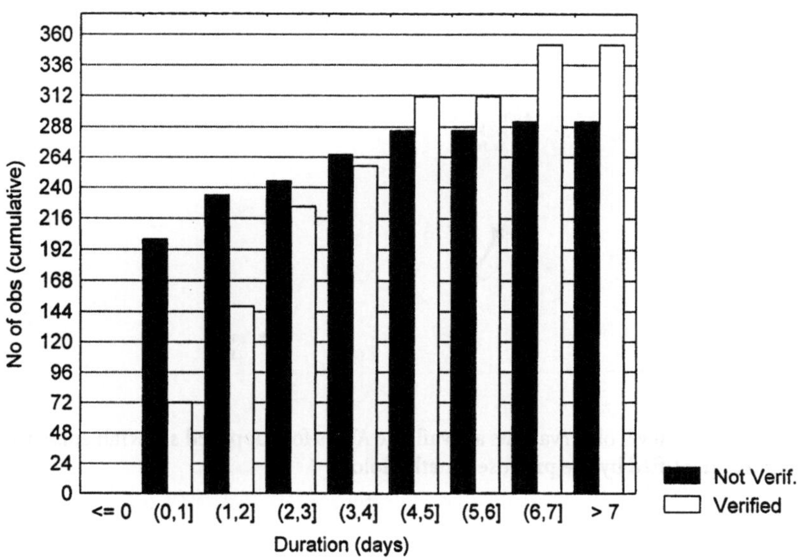

Fig. 9. Cumulative frequency of occurrence (3-hourly observations) of verified and not-verified snowfall episodes against duration of episodes

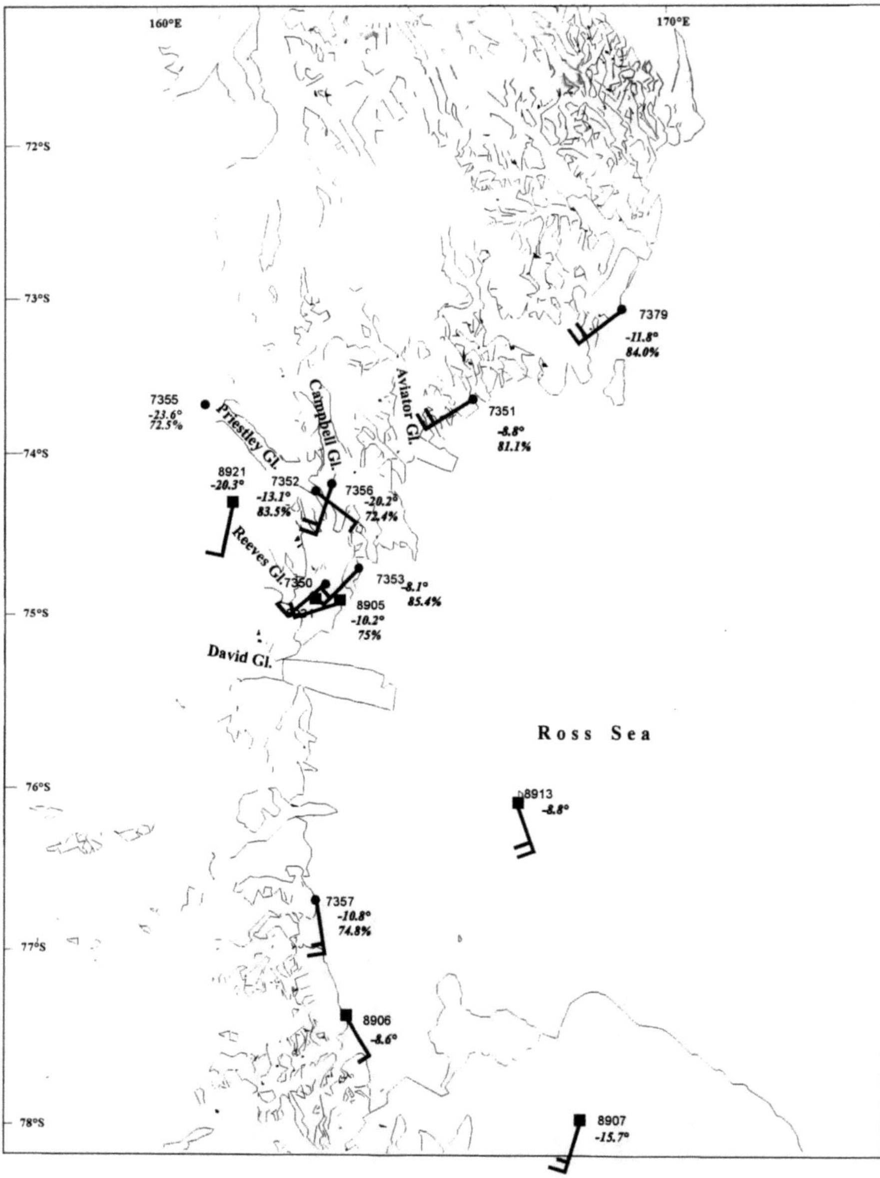

Fig. 10a. Ground level observations at available AWSs for supposed snowfall episodes during winter, identified by the proposed methodology

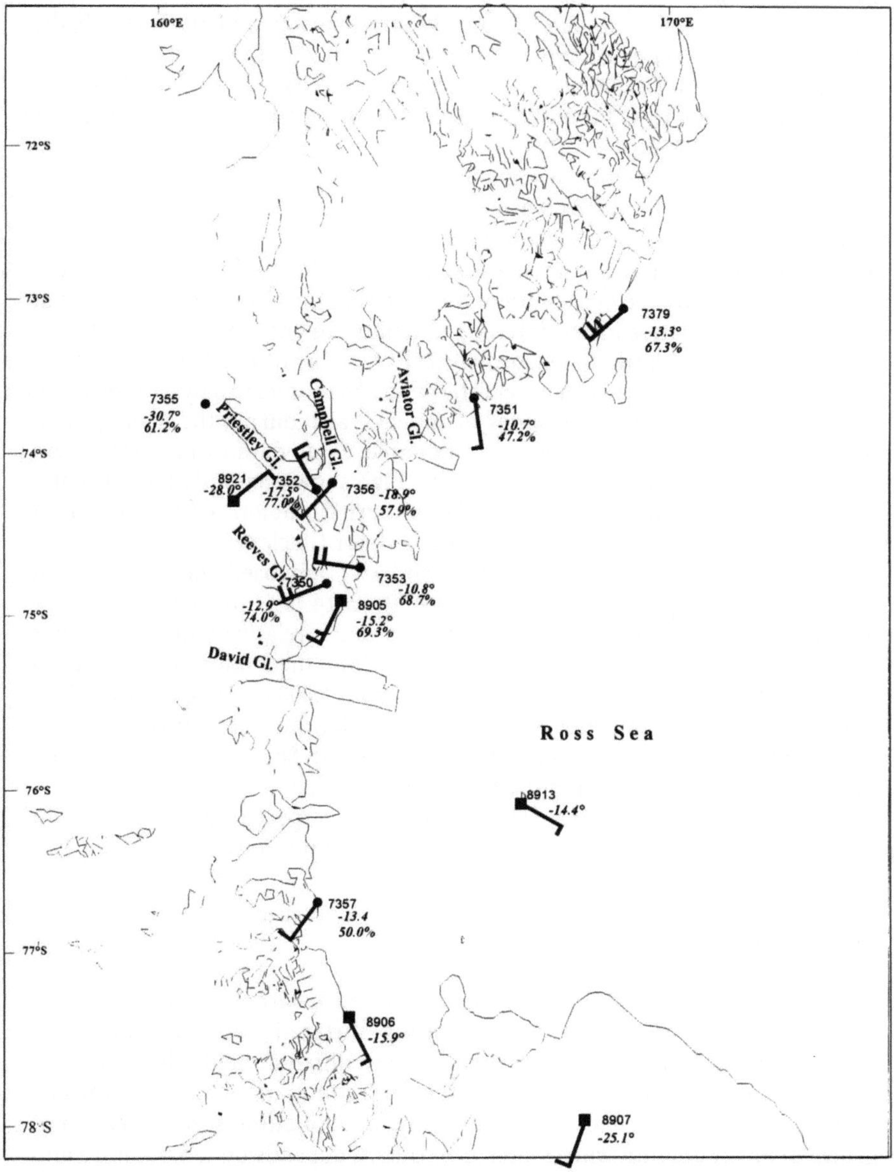

Fig. 10b. Ground level observations at available AWSs for supposed blowing snow episodes during winter, identified by the proposed methodology

In addition, Zibordi and Frezzotti [29] pointed out that during the 1992/1993 scientific expedition of the Italian National Programme for Antartic Research, the in-situ observations reported snowfalls (≈ 20 cm) on 23-24 December and (less than 5 cm) on 30-31 December 1992. These periods have been found also by analyses, from which it was found that at the following times an RH ≧ 70% was recorded at station 7352:

- from 09:00 on 22 December to 15:00 on 24 December
- from 09:00 on the 30th to 18:00 on the 30th
- from 00:00 on the 31th to 21:00 on the 31th

So, considering all these analyses, was shown to be the used methodology reliable for to establishing the possibility of precipitation in a given period, even if at present, estimates relative to precipitation amounts are not available, except for a few cases in the periods when meteorologists live at Terra Nova Bay.

The good agreement between observations and theoretical analysis allowed an attempt to identify supposed episodes of snowfall also during the winter, when direct observation is not possible. As an example, some days in June 1991 were considered. During winter 1991 a particularly, high snow accumulation was detected by the isotopic and chemical analyses carried out on a firn core from Drygalski Ice Tongue and on a snow pit from Aviator Glacier [30].

Applying our methodology, different episodes of supposed snowfall resulted during June 1991, in particular from 00:00 UTC on 14 June to 15:00 UTC on 17 June. During this time station 7352 always recorded a relative humidity greater than 70%; from 03:00 on 14 June to 12:00 on 16 June the station was subject to warm air masses flowing from the ocean, while from 15:00 on 16 June to 15:00 on 17 June the airstreams, blowing down from the Plateau, were dryer and wind speed was higher. According to the criteria described in Section 4, the latter period can be identified as a blowing/drifting snow episode.

Therefore, the two situations, (Fig. 10a,b) seem, respectively, to relate to an episode of cyclogenesis over the Ross Sea and to blowing snow.

8 Conclusions

In the Victoria Land area and more generally on the Antarctic continent routine measurements of precipitation amounts using surface-based collection techniques are a difficult task. Many investigators gave different methods of measuring or evaluating precipitation, and many these are summarised and described by Bromwich [10]. The dominant process affecting precipitation is the advection of moisture in an air column whose moisture content depends on temperature and altitude. Relative humidity controls evaporation and precipitation, which both increase with higher wind speeds.

Beginning with these considerations and analysing the more frequent meteorological conditions that can be found near the base of a steep mountain massif or, in particular, the prevailing surface winds on Victoria Land, the air masses with relative humidity greater than or equal to 70% that affect an automatic weather station on the Priestley Glacier were identified. By analysing the direction of these airstreams and comparing the periods resulting from the study with the synoptic situations described by different authors, it was found that marked east-southerly winds are associated with penetrations far into the continent of warm moist air masses.

To check the reliability of the proposed methodology, the meteorological observations provided by the Weather Office at the Italian Antarctic Station were considered, and the methodology was applied to a winter situation too. Understanding the precipitation distribution over Victoria Land and its temporal changes is especially important in order to improve the intense monitoring and research activity. It is desirable that, in future, quantitative analyses of snowfall over Terra Nova Bay area will be possible.

References

1. Mitchell JFB, Manbe S, Tokioka T and Maleshko V (1990) Equilibrium climate change. In: Houghton JT, Jenkins GJ, Enphraums JJ (eds) Climate change: the IPCC scientific assessment Cambridge University Press, Cambridge, pp 131-1720
2. Budd WF and Smith IN (1985) The state of balance of the Antarctic ice sheet: an updated assessment 1984. In Glaciers, ice sheets and sea level: effect of a CO_2 induced climatic change. Report of a workshop, Seattle, Washington, September 13-15, 1984, National Academy Press, Washington, DC, pp 172-177
3. Houghton JT, Jenkins GJ and Enphraums JJ (1990) Climate change: the IPCC scientific assessment. Cambridge (eds) University Press, Cambridge
4. Scientific Committee on Antarctic Research (SCAR) (1992) The role of Antarctica, in Global change. An introdution plan for a regional research programme. ICSU Press/SCAR,
5. Ledley TS (1985) Sea ice: multi-year cycles ands white ice. J geophys Res 90:5676-5686
6. Ledley TS (1991) The climatic response to meridional sea ice transport. J Clim 4
7. Bromwich DH (1991) Precipitation trends over Polar Ice Sheets from Antarctica. Proceedings of a Conference, June 11-15, 1990, University of Alaska, Fairbanks, p 190
8. Peixoto JP and Oort AH (1983) The atmospheric branch of the hydrological cycle and climate. In Street Perrott A, M Beran and R Ratchiffe eds. Variations in the global water budget, D Reidel Publishing Dordrecht, pp 5-66
9. Schwerdtfeger W, (1984) Weather and Climate of the Antarctic. Developments in Atmos Sci 15, Elsevier
10. Bromwich DH (1988) Snowfall in high southern latitudes. Rev Geophys 26(1):149-168
11. Bromwich DH, Carrasco JG, Liu Z and Yow Tzeng R (1993) Hemispheric atmospheric variation and oceanographic impacts associated with katabatic surges across the Ross Ice Shelf, Antarctica J Geophys Res 98(D7):13045-13062
12. Bromwich DH and Robasky FM (1993) Recent precipitation trends over the Polar Ice Sheets. Meteorol Atmos Phys 51:259-274
13. Mather KB (1969) The pattern of surface wind flow in Antarctica. Pure Appl Geophys 75:332-354
14. Orombelli G (1989) Terra Nova Bay: a geographic overview. Memorie della società geologica Italiana. Proceedings of meeting Geosciences in Victoria Land, Antarctica, Siena, 2-3 September 1987, pp 69-75
15. Van der Veen CJ and Oerlemans J (1985) The West Antarctic Ice Sheet: the need to understand its dynamics. from Dynamics of the West Antarctic Ice Sheet, pp 1-16
16. Voskresenskii AI (1976) On moisture content in the atmosphere over Antarctica (English translation). tr AANII 327:56-67
17. Zav'yalova IN (1987) Humidity of air in Antarctica. Climate of Antarctica. AA Balkama, Rotterdam, pp 92-101
18. Rusin NP (1961) Meteorological and radiation regime of Antarctica (English translation). Gidrometeoteoizdat, Leningrad, pp 447

19. Tauber GM (1960) Characteristics of Antarctic katabatic winds on Antarctic Meteorology, Proceedings of symposium Melbourne. Pergamon Press, Kidlington, pp 52-64
20. Carrasco JF and Bromwich DH (1993) Mesoscale cyclogenesis dynamics over the southwestern Ross Sea, Antarctica. JGR 98(D7):12973-12995
21. Carrasco JF and Bromwich DH (1996) Mesoscale cyclone activity near Terra Nova Bay and Byrd Glacier, Antarctica, during 1991. Global Atmos Ocean Syst 5:43-72
22. Craig G and Cho HR (1989) Baroclinic instability and CISK as the driving mechanisms for polar lows and comma clouds. In: Polar and Arctic lows, Twitchell PF, Rasmussen EA, Davison KL (eds) A Deepak, Hampton, Vancouver, pp 131-153
23. Taljaard JJ, Van Loon H, Crutcher HL and Jenne RL (1969) Climate of the upper air: southern hemisphere, vol 1. Temperature, dewpoints and heights at selected pressure levels. Rep NAVAIR, 50-1C-55 Nav. Weather Service Command, Washington, DC
24. Hoskins BJ, McIntyre ME and Robertson AW (1985) On the use of and significance of isentropic potential vorticity maps. Quat J R Meteor Soc 470:877-946
25. Bromwich DH and Parish TR (1988) Mesoscale cyclone interactions with the surface windfield near Terra Nova Bay, Antarctica. Antarct United States Review Issue, pp 1-5
26. Frustaci G, Bacci G and Pellegrini A (1992) Mesoanalysis and operational weather forecast for the western Ross Sea Conference Proceedings vol 34, Italian research on Antarctic atmosphere, Colacino M, Giovannelli G, Stefanutti L (Eds) SIF, Bologna, pp 49-64
27. Radok U(1977) Snow Drift. J Glaciol 19 81:123-139
28. Frustaci G and Pellegrini A (1995) Tropospheric and mesoscale circulation patterns related to barrier and katabatic winds at Terra Nova Bay Station, Antarctica. Paper presented at zist IUGG Conference, Boulder, Colorado July 1995
29. Zibordi G and Frezzotti M(1996) Orographic clouds in north Victoria Land from AVHRR images. Polar Record 32(183):317-324
30. Caprioli R, Della Vedova AM, Gragnani R, Grigioni P, Maggi V, Serra F, Stenni B and Torcini S (1998) Stratigraphic, isotopic and chemical profiles of a firn core from Drygalski Ice Tougue and of a snow pit from Aviator Glacier (northern Victoria Land, Antarctica). Progetto di Ricerca Glaciologia e Paleoclima, Casaccia, 20-21 Giugno 1994. Terra Antarctica, submitted

FSC
www.fsc.org
MIX
Papier aus verantwortungsvollen Quellen
Paper from responsible sources
FSC® C105338

If you have any concerns about our products,
you can contact us on
ProductSafety@springernature.com

In case Publisher is established outside the EU,
the EU authorized representative is:
Springer Nature Customer Service Center GmbH
Europaplatz 3, 69115 Heidelberg, Germany

Printed by Libri Plureos GmbH
in Hamburg, Germany